Lecture Notes in Computer Scie

T0237892

Commenced Publication in 1973
Founding and Former Series Editors:
Gerhard Goos, Juris Hartmanis, and Jan van Leeuwen

Editorial Board

Amihood Amir   Andrew Turpin
Alistair Moffat (Eds.)

String Processing and Information Retrieval

15th International Symposium, SPIRE 2008
Melbourne, Australia, November 10-12, 2008
Proceedings

 Springer

Volume Editors

Amihood Amir
Bar-Ilan University
Ramat-Gan, Israel
E-mail: amir@cs.biu.ac.il

Andrew Turpin
RMIT University
Melbourne, Australia
E-mail: aht@cs.rmit.edu.au

Alistair Moffat
The University of Melbourne
Carlton, Australia
E-mail: alistair@csse.unimelb.edu.au

Library of Congress Control Number: Applied for

CR Subject Classification (1998): H.3, H.2.8, I.2, E.1, E.5, F.2.2

LNCS Sublibrary: SL 1 – Theoretical Computer Science and General Issues

ISSN   0302-9743
ISBN-10   3-540-89096-3 Springer Berlin Heidelberg New York
ISBN-13   978-3-540-89096-6 Springer Berlin Heidelberg New York

Springer is a part of Springer Science+Business Media

springer.com

© Springer-Verlag Berlin Heidelberg 2008
Printed in Germany

Typesetting: Camera-ready by author, data conversion by Scientific Publishing Services, Chennai, India
Printed on acid-free paper   SPIN: 12557561   06/3180   5 4 3 2 1 0

Preface

This volume contains the papers presented at the 15th String Processing and Information Retrieval Symposium (SPIRE), held in Melbourne, Australia, during November 10–12, 2008.

The papers presented at the symposium were selected from 54 papers submitted in response to the Call For Papers. Each submission was reviewed by a minimum of two, and usually three, Program Committee members, who are experts drawn from around the globe. The committee accepted 25 papers (46%), with the successful authors also covering a broad range of continents. The paper "An Efficient Linear Space Algorithm for Consecutive Suffix Alignment Under Edit Distance" by Heikki Hyyrö was selected for the Best Paper Award, while Dina Sokol was awarded the Best Reviewer Award for excellent contributions to the reviewing process. The program also included two invited talks: David Hawking, chief scientist at the Internet and enterprise search company Funnelback Pty. Ltd. based in Australia; and Gad Landau, from the Department of Computer Science at Haifa University, Israel.

SPIRE has its origins in the South American Workshop on String Processing which was first held in 1993. Starting in 1998, the focus of the symposium was broadened to include the area of information retrieval due to the common emphasis on information processing. The first 14 meetings were held in Belo Horizonte, Brazil (1993); Valparaiso, Chile (1995); Recife, Brazil (1996); Valparaiso, Chile (1997); Santa Cruz, Bolivia (1998); Cancun, Mexico (1999); A Coruña, Spain (2000); Laguna San Rafael, Chile (2001); Lisbon, Portugal (2002); Manaus, Brazil (2003); Padova, Italy (2004); Buenos Aires, Argentina (2005); Glasgow, UK (2006); and Santiago, Chile (2007).

The annual SPIRE conference provides an opportunity for researchers to present original contributions on areas such as string processing (the searching, compression and mining of text, pattern matching, natural language processing, automata based string processing); information retrieval (indexing, ranking, filtering, cross-lingual IR systems, multimedia IR, digital libraries, collaborative retrieval, Web-related applications); interaction of biology and computation particularly related to string processing and retrieval; and information retrieval languages and applications (XML, SGML, information retrieval from semi-structured data, generation of structured data from text).

While many people have helped to make this conference possible, we particularly thank the members of the Program Committee and the additional reviewers who worked hard to ensure the timely review of all submitted manuscripts. We also thank William Webber, who compiled the proceedings, and Shane Culpepper, who maintained the website for the conference. We are grateful to

Yahoo! Research for providing funding in support of student attendees. Submissions were managed using the EasyChair conference system.

August 2008

Amihood Amir
Andrew Turpin
Alistair Moffat

Organization

Conference Chair

Alistair Moffat   University of Melbourne, Australia

Program Chairs

Amihood Amir   Bar-Ilan University, Israel
                     and Johns Hopkins University, USA
Andrew Turpin   RMIT University, Australia

Program Committee

Amihood Amir	Bar-Ilan University, Israel
Mike Atallah	Purdue University, USA
Gary Benson	Boston University, USA
Bodo Billerbeck	Microsoft Research, Cambridge
Carlos Castillo	Yahoo! Research, Spain
Charlie Clarke	University of Waterloo, Canada
Bruce Croft	University of Massachusetts, Amherst, USA
J. Shane Culpepper	University of Melbourne, Australia
Paolo Ferragina	University of Pisa, Italy
Edward Fox	Virginia Tech, USA
Jan Holub	Czech Technical University, Czech Republic
Shunsuke Inenaga	Kyushu University, Japan
Rao Kosaraju	Johns Hopkins University, USA
Avivit Levy	Shenkar College, Israel
Moshe Lewenstein	Bar Ilan University, Israel
Noa Lewenstein	Netanya College, Israel
Giovanni Manzini	University of East Piedmont, Italy
Massimo Melucci	University of Padua, Italy
Laurent Mouchard	University of Rouen, France
	and King's College, London, UK
Gonzalo Navarro	University of Chile, Chile
Igor Nor	University of Bristol, UK
Heejin Park	Hanyang University, Korea
Kunsoo Park	Seoul National University, Korea
Ron Y. Pinter	Technion, Israel
Ely Porat	Bar Ilan University, Israel
Simon Puglisi	RMIT University, Australia
Mathieu Raffinot	CNRS, France

Kunihiko Sadakane   Kyushu University, Japan
Falk Scholer   RMIT University, Australia
Ayumi Shinohara   Tohoku University, Japan
Fabrizio Silvestri   CNR, Italy
Steven Skiena   Stony Brook University, USA
Bill Smyth   McMaster University, Canada
                             and Curtin University, Australia
Dina Sokol   Brooklyn College, New York, USA
Wing-Kin Sung   NUS, Singapore
Andrew Turpin   RMIT University, Australia
Alexandra Uitdenbogerd   RMIT University, Australia
Esko Ukkonen   University of Helsinki, Finland
Anh Vo   University of Melbourne, Australia

External Reviewers

Miroslav Balik   Juha Kärkkäinen   Justin Tojeira
Hideo Bannai   Dong Kyue Kim   Raymond Wan
Peter Bruza   Joong Chae Na   Mingfang Wu
Gabriele Capannini   Kazuyuki Narisawa
Ben Carterette   Raffaele Perego
Michael Harris   Royi Ronen

Local Organization

Shane Culpepper   University of Melbourne, Australia
Alistair Moffat   University of Melbourne, Australia
Simon Puglisi   RMIT University, Australia
William Webber   University of Melbourne, Australia
Justin Zobel   NICTA and University of Melbourne,
                         Australia

Table of Contents

"Search Is a Solved Problem" and Other Annoying Fallacies

David Hawking

Chief Scientist
Funnelback Pty Ltd, Australia
David.Hawking@funnelback.com

Abstract. Since Google became a celebrity in the early noughties, many people with the power to control and direct research resources have taken the view that there is no more research to be done on the problem of information retrieval. In reality, there are so many variants of "the search problem" that not all have been catalogued, and few have been solved to the point where we can rely absolutely on the quality of results. Apparently no-one told the Web search companies that the problem was solved as, since that time, they have researched and developed a range of new search facilities and invested heavily in improving their basic products. Google, Yahoo! and Microsoft all maintain search research and development teams much larger than the biggest University computer science departments!

Through my involvement with the Funnelback internet and enterprise search company I have worked on many twists on the information retrieval problem which are not modelled in well-known test collections, and not encountered in basic Web search. In my talk I will try to outline some of the issues in trying to apply information retrieval and string processing theory into commercial practice.

A. Amir, A. Turpin, and A. Moffat (Eds.): SPIRE 2008, LNCS 5280, p. 1, 2008.
© Springer-Verlag Berlin Heidelberg 2008

Approximate Runs - Revisited

Gad M. Landau[1,2]

[1] Department of Computer Science,
University of Haifa, Haifa, Israel
landau@cs.haifa.ac.il
[2] Department of Computer and Information Science,
Polytechnic University, New York, USA
landau@poly.edu

Abstract. The problem of finding repeats within a string is an important computational problem with applications in data compression and in the field of molecular biology. Both exact and inexact repeats occur frequently in the genome, and certain repeats are known to be related to human diseases.

A multiple tandem repeat in a sequence S is a (periodic) substring r of S of the form $r = u^a u'$, where u (the period) is a prefix of r, u' is a prefix of u and $a \geq 2$. A run is a maximal (non-extendable) multiple tandem repeat. An *approximate* run is a run with errors (i.e. the repeated subsequences are similar but not identical).

Many measures have been proposed that capture the similarity among all periods. We may measure the number of errors between consecutive periods, between all periods, or between each period and a consensus string. Another possible measure is the number of positions in the periods that may differ.

In this talk I will survey a range of our results in this area. Various parts of this work are joint work with Maxime Crochemore, Gene Myers, Jeanette Schmidt and Dina Sokol.

A. Amir, A. Turpin, and A. Moffat (Eds.): SPIRE 2008, LNCS 5280, p. 2, 2008.
© Springer-Verlag Berlin Heidelberg 2008

Engineering Radix Sort for Strings*

Juha Kärkkäinen and Tommi Rantala

Department of Computer Science, University of Helsinki, Finland
{juha.karkkainen,tommi.rantala}@cs.helsinki.fi

Abstract. We describe new implementations of MSD radix sort for efficiently sorting large collections of strings. Our implementations are significantly faster than previous MSD radix sort implementations, and in fact faster than any other string sorting algorithm on several data sets. We also describe a new variant that achieves high space-efficiency at a small additional cost on runtime.

1 Introduction

Sorting is a fundamental problem in computer science that underlies a vast variety of computational tasks. When the sort keys are strings, it is possible to use any comparison based sorting algorithm but there are more efficient algorithms specialized for sorting strings. Among the best-known, simplest and fastest string sorting algorithms is the MSD (Most Significant Digit first) radix sort.

There are many possible ways of implementing the MSD radix sort. There exist extensive experimental studies on efficient implementation [2, 6] and recent new variants [7], but the possibilities have not been exhausted. We describe several new implementations, the best of which are significantly faster than any previous ones.

Radix sort and other string sorting algorithms tend to have irregular memory access patterns that are poorly suited for modern computer architectures with CPUs that are much faster than the main memory. Similar to several recent string sorting algorithms [7, 11, 12], our implementations reduce the number of slow memory accesses through better utilization of the cache memory. In addition, our algorithms reduce the *cost* of slow memory accesses by better utilization of the out-of-order execution capabilities of modern CPUs.

Some of our implementations are also very space-efficient. This is critical, for example, in several suffix array construction algorithms that rely on fast and space-efficient string sorting (see [9]).

Related Work. A seminal study on implementing MSD radix sort is by McIlroy, Bostic and McIlroy [6]. Andersson and Nilsson [2] describe more variants and provide another extensive experimental comparison. A recent, cache-efficient variant is by Ng and Kakehi [7]. Theoretical studies of radix sorting can be found in [1, 8].

---

* Supported in part by Academy of Finland grant 118653 (ALGODAN).

A. Amir, A. Turpin, and A. Moffat (Eds.): SPIRE 2008, LNCS 5280, pp. 3–14, 2008.

Two other fast string sorting algorithms are multikey quicksort [3] and burst-sort [11, 12]. Like radix sort, both distribute strings into buckets based on a single character. However, multikey quicksort uses character comparisons to distribute the strings into just three buckets (smaller, equal and larger), while burstsort organizes the buckets into a data structure called burst trie.

There is also an extensive literature on radix sorting *integers* (see [4, 10], e.g.), but these usually involve *LSD* radix sort, and the issues are quite different.

2   Problem and Experimental Setup

We consider the problem of sorting a set of strings $R = \{s_1, s_2, \ldots, s_n\}$ over the alphabet $\Sigma = \{0, 1, \ldots, \sigma - 1\}$ into the lexicographic order. Besides n and σ, an important parameter of the problem is D, the total length of the distinguishing prefixes of the strings. The *distinguishing prefix* of a string s_i is the shortest prefix of s_i that separates it from the other strings. Thus, D is the minimum number of characters that need to be inspected, and provides a lower bound for the problem complexity. The best theoretical variants of radix sorting have time complexity $\mathcal{O}(D + \sigma)$ [8].

The experiments use the standard representation of strings in the C programming language. Thus, $\sigma = 256$ and each string is terminated with 0, which does not appear elsewhere in the strings. The task is to sort an array containing pointers to the beginning of the strings. The actual strings are stored contiguously in one array and are not moved during the sorting. Besides the sorting time, we are interested in the amount of space needed in addition to the input.

The data sets used in the experiments are described in Table 1. The initial order of the strings is random.

Table 1. Description of the test data. The datasets URL, Genome, and Unique are from [12] while Random A and Random B we have generated ourselves.

Name	n	D	Description
URL	10^7	3.1×10^8	URL addresses with the protocol name stripped
Genome	3×10^7	3×10^8	strings of length 9 over the alphabet $\{a, c, g, t\}$ from real genomic data
Unique	3×10^7	2.8×10^8	unique words collected from English documents
Random A	3×10^7	4.6×10^8	strings of single character with the length chosen uniformly at random from $[0, 30)$
Random B	3×10^7	1.2×10^8	strings of length 30 with the characters chosen uniformly at random from $[32, 255)$

The experiments were carried out on a machine with an Intel Core 2 processor model E6400 running at 2.13 GHz. The sizes of the processor's L1 and L2 caches are 32 kilobytes and two megabytes, respectively. The caches have 8-way associativity, and they use a block size of 64 bytes. The data TLB (Translation Lookaside Buffer) has two levels: DTLB0 with 16 entries (supporting loads only) and DTLB1 with 256 entries. The page size is 4 kilobytes. The machine was equipped with two gigabytes of RAM.

We used the Linux operating system, kernel version 2.6.22. The compiler is GCC version 4.1.2 (Red Hat) with optimization flags -O3 and -march=core2. Debugging was disabled with -DNDEBUG. To measure the runtime of each algorithm, we calculate the CPU time using the getrusage function. Memory usage was measured using the memusage utility that is included in GNU libc[1]. Other measurements, including cache misses, are based on hardware counters, and were obtained using OProfile[2] version 0.9.3.

3   Basic Algorithm

The basic idea of MSD radix sort is simple. Start by distributing the strings into buckets based on the first character. Each bucket in turn is sorted recursively in the same way except that the second character is used in the second level of recursion, the third character in the third level of recursion and so on.

MSDRadixSort($R, depth$)
1   **if** $|R| < t$ **then** InsertionSort($R, depth$)
2   **for** $s \in R$ **do** $B[s[depth]] := B[s[depth]] \cup \{s\}$
3   **for** $c \in \Sigma \setminus \{0\}$ **do if** $|B[c]| > 0$ **then** MSDRadixSort($B[c], depth + 1$)

The first line switches to insertion sort for small buckets. We and others before us [2, 6] have found this to be an essential optimization in practice. It can be supported theoretically, too, as the analysis below shows.

Theorem 1. *The time complexity of MSDRadixSort is $\mathcal{O}((\sigma/t + t)D)$.*

Proof (sketch). The total time complexity is $\mathcal{O}(tD)$ for line 1, $\mathcal{O}(D)$ for line 2, and $\mathcal{O}((\sigma/t)D)$ for line 3.   □

Implementations are, of course, more compilicated and can differ significantly from each other but all efficient ones spend nearly all of their time in three kinds of activities that we can identify with the three lines of the pseudocode algorithm:

1. Insertion sort for small buckets.
2. Iterating through all strings.
3. Iterating through all buckets.

Based on the theoretical analysis, one would expect that either line 1 or line 3 dominates the run time depending on the value of the threshold t. Indeed, if t is too small, line 3 dominates, and if t is too large, line 1 dominates. However, if t is chosen somewhere close to the theoretically optimal value of $\sqrt{\sigma} = \sqrt{256} = 16$,

---

[1] http://www.gnu.org/software/libc/
[2] http://oprofile.sourceforge.net/

it is line 2 that dominates the run time in practice. We use $t = 32$ in the experiments; the precise value is not critical.

One explanation for the discrepancy between theory and practice is that the theoretical analysis represents the worst case, which for line 2 is usually close to the real behaviour, whereas lines 1 and 3 are often far from the worst case. Another significant reason is that line 2 produces most of the expensive cache and TLB misses. This can be seen in Table 2 that shows the experimental distribution of several performance measurements in the straightforward CE0 implementation described in Section 5. It is clearly line 2 where optimization efforts should focus.

Table 2. Distribution of performance measurements within CE0

Collection	Clock cycles			Instructions			TLB misses			L2 cache misses		
	Ln 1	Ln 2	Ln 3	Ln 1	Ln 2	Ln 3	Ln 1	Ln 2	Ln 3	Ln 1	Ln 2	Ln 3
URL	2%	95%	3%	8%	73%	19%	1%	99%	0%	1%	99%	0%
Genome	0%	99%	1%	0%	85%	15%	0%	100%	0%	0%	100%	0%
Unique	8%	88%	4%	32%	48%	20%	6%	94%	0%	4%	96%	0%
Random A	0%	100%	0%	0%	100%	0%	0%	100%	0%	0%	100%	0%
Random B	7%	92%	1%	50%	45%	5%	14%	86%	0%	3%	97%	0%

4   Variants

Despite the simplicity of the basic algorithm, there are many variants of MSD radix sort that differ significantly in implementation details. Here we outline several variants at a general level and describe them in more detail in the following sections.

One factor that complicates a radix sort implementation is that we do not know the sizes of the buckets in advance. There are two main ways of addressing this difficulty. We call these the C- and D-variants of MSD radix sort:

C (Counting): Perform the distribution twice, the first time only counting the bucket sizes without actually moving the strings.
D (Dynamic buckets): Implement the buckets using some dynamically expanding data structure.

The C-variants can be further categorized based on how the actual distribution is implemented. Variant CE (External array) distributes the strings into an external array, from where they are copied back to the input array in the end. A drawback is the extra space needed for the external array. Variant CI (In-place) performs the distribution by an in-place permutation. This variant is more complicated to implement and, unlike CE, does not produce a stable order.

Both kinds of C-variants are described by McIlroy et al. [6]. A significant drawback in all C-variants in comparison with D-variants is that the bucket of each key is computed twice. This can be an expensive operation since it is likely to cause cache and TLB misses. We describe new ways to reduce this cost in Section 5.

The D-variants can be further categorized by which data structure is used for implementing the dynamic buckets. McIlroy et al. [6] and Andersson and Nilsson [2] have described implementations using linked lists. The main drawback of D-variants is the time and space overhead due to dynamic data structures, which for lists is significant. In Section 6, we describe implementations based on more efficient data structures.

In both C- and D-variants, distributing the strings into buckets requires an access to one character in each string. This is typically the most expensive operation due to cache and TLB misses. However, whenever the algorithm accesses one character, it could cheaply (with low probability of additional misses) access the next few characters, too. In Section 7, we consider two ways of taking advantage of this possibility.

5   C-Variants

The C-variant of MSD radix sort has two phases. The first phase counts the bucket sizes and the second phase does the actual distribution. Here is a straightforward implementation of the CE-variant, which uses a temporary array in the second phase.

```
void CE0(unsigned char** strings, size_t n, size_t depth)
{
    if (n < 32) {
        insertion_sort(strings, n, depth);
        return;
    }
    size_t bucketsize[256] = {0};
    for (size_t i=0; i < n; ++i)                      /* Loop A */
        ++bucketsize[strings[i][depth]];
    unsigned char** sorted =
        (unsigned char**) malloc(n*sizeof(unsigned char*));
    static size_t bucketindex[256];
    bucketindex[0] = 0;
    for (size_t i=1; i < 256; ++i)
        bucketindex[i] = bucketindex[i-1]+bucketsize[i-1];
    for (size_t i=0; i < n; ++i)                      /* Loop B */
        sorted[bucketindex[strings[i][depth]]++] = strings[i];
    memcpy(strings, sorted, n*sizeof(unsigned char*));
    free(sorted);
    size_t bsum = bucketsize[0];
    for (size_t i=1; i < 256; ++i) {
        if (bucketsize[i] == 0) continue;
        CE0(strings+bsum, bucketsize[i], depth+1);
        bsum += bucketsize[i];
    }
}
```

As we saw in Table 2, most of the time is spend in loops iterating through all strings, i.e., the two for loops marked "Loop A" and "Loop B". The main culprit is the data access `strings[i][depth]` appearing in both loops. It accesses a single character in a string, and with no other accesses to the same string during the execution of the loop, it is quite likely to cause cache and TLB misses *both times*. Doing this slow access twice is the main drawback of the C-variants (both CE and CI) in comparison with the D-variants.

We introduce a new improvement based on the observation that both `for` loops do exactly the same sequence of slow character accesses. When we access each string for the first time in loop A, we also copy the character into a separate array called `oracle`.

```
for (size_t i=0; i < n; ++i)                    /* Loop A */
    ++bucketsize[oracle[i] = strings[i][depth]];
```

In loop B, we then replace `strings[i][depth]` with `oracle[i]`.

```
for (size_t i=0; i < n; ++i)                    /* Loop B */
    sorted[bucketindex[oracle[i]]++] = strings[i];
```

Because we access the `oracle` array sequentially, the new loop B generates much fewer cache and TLB misses (see Table 3), and the resulting implementation CE1 is much faster (see Fig. 1).

Loop A is still slow, though, but we can speed it up significantly using *loop fission*. We split loop A into two loops A1 and A2:

```
for (size_t i=0; i < n; ++i)                    /* Loop A1 */
    oracle[i] = strings[i][depth];
for (size_t i=0; i < n; ++i)                    /* Loop A2 */
    ++bucketsize[oracle[i]];
```

The implementation CE2 with the split loop is much faster than CE1 on most data sets (see Fig. 1). The explanation for this surprising behaviour is in the out-of-order execution capabilities of modern processors. While waiting for the slow memory load `strings[i][depth]` to finish, the processor can execute subsequent instructions as long as they do not depend on the result of the load. In particular, the next slow load `strings[i+1][depth]` is independent and can start before the previous one has finished. However, loop A in CE1 (and CE0) has an instruction between the two slow loads that interferes: a memory store whose address depends on the first load (updating the counter). Such a store might (as far as the processor knows) change the value of any memory location. Thus subsequent memory loads cannot proceed until the address of the store is known. After the split, loop A1 does not have such a blocking store and loop A2 does not have a slow load. The effect of the loop fission can be clearly seen in the number of loads blocked in Table 3.[3]

---

[3] TLB misses, too, are reduced as a result of the loop fission. The unsplit loop A generates, in fact, about two TLB misses per iteration, even though one would expect only one. We suspect that the load blocks somehow cause the TLB miss counter to be incremented twice.

Table 3. Hardware performance counter readings for the Unique data set ($n = 30 \times 10^6$, $D = 280 \times 10^6$). *Loads blocked* measures the number of instructions that load a value from memory but are stalled due to a preceding store instruction for which the store address is not yet known.

Algorithm	Clock cycles	Instructions	TLB misses	L1 misses	L2 misses	Loads blocked
CE0	36200×10^6	7600×10^6	800×10^6	530×10^6	400×10^6	520×10^6
Loop A	17900×10^6	1200×10^6	380×10^6	230×10^6	220×10^6	200×10^6
Loop B	13300×10^6	1800×10^6	360×10^6	210×10^6	160×10^6	220×10^6
CE1	23000×10^6	7700×10^6	450×10^6	330×10^6	265×10^6	395×10^6
Loop A	17700×10^6	1200×10^6	390×10^6	235×10^6	210×10^6	220×10^6
Loop B	1200×10^6	1800×10^6	1×10^6	20×10^6	5×10^6	100×10^6
CE2	11200×10^6	8500×10^6	206×10^6	330×10^6	190×10^6	180×10^6
Loop A1	5300×10^6	1200×10^6	150×10^6	230×10^6	150×10^6	0.1×10^6
Loop A2	700×10^6	1000×10^6	1×10^6	5×10^6	2×10^6	5×10^6
Loop B	1300×10^6	2100×10^6	2×10^6	20×10^6	5×10^6	120×10^6
CI	13700×10^6	11400×10^6	280×10^6	320×10^6	180×10^6	125×10^6

The additional space requirement of the CE2 algorithm is $4n$ bytes for the external pointer array and n bytes for the oracle array. The CI algorithm gets rid of the external pointer array by using the in-place distribution technique described in [6]. Otherwise, CI is identical to CE2.

An experimental comparison of the implementations in this section is shown in Fig. 1. Also included is the CI-type implementation by McIlroy, Bostic & McIlroy [6, Program C], which is commonly used as the reference implementation of MSD radix sort in the literature. Our best implementations are more than twice as fast on most data sets.

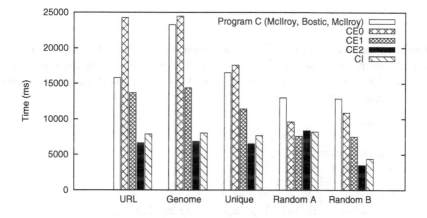

Fig. 1. Runtimes of C-variants

6   D-Variants

The D-variants of MSD radix sort use dynamic data structures to represent the buckets eliminating the need for a separate counting pass. The code looks like this:

```
template <typename Bucket>
void D(unsigned char** strings, size_t n, size_t depth, Bucket* buckets)
{
    if (n < 32) {
        insertion_sort(strings, n, depth);
        return;
    }
    for (size_t i=0; i < n; ++i)                          /* Loop C */
        buckets[strings[i][depth]].push_back(strings[i]);
    size_t bucketsize[256];
    for (size_t i=0; i < 256; ++i)
        bucketsize[i] = buckets[i].size();
    size_t pos = 0;
    for (size_t i=0; i < 256; ++i) {
        if (bucketsize[i] == 0) continue;
        std::copy(buckets[i].begin(), buckets[i].end(), strings+pos);
        pos += bucketsize[i];
    }
    for (size_t i=0; i < 256; ++i)
        buckets[i].clear();
    pos = bucketsize[0];
    for (size_t i=1; i < 256; ++i) {
        if (bucketsize[i] == 0) continue;
        D(strings+pos, bucketsize[i], depth+1, buckets);
        pos += bucketsize[i];
    }
}
```

The loop marked as "Loop C" is the expensive one here. It suffers from the load blocking phenomenon we saw in the previous section, and the cure is the same, too: loop fission. With a small trick, we can now manage with a much smaller oracle array:

```
size_t i=0;
for (; i < n-n%32; i+=32) {
    unsigned char oracle[32];
    for (size_t j=0; j < 32; ++j)
        oracle[j] = strings[i+j][depth];
    for (size_t j=0; j < 32; ++j)
        buckets[oracle[j]].push_back(strings[i+j]);
}
for (; i < n; ++i)
    buckets[strings[i][depth]].push_back(strings[i]);
```

The implementation takes the dynamic bucket data structure as a template argument. We have three variants using data structures from the standard C++ library, DL (`std::list`), DV (`std::vector`) and DD (`std::deque`), and one variant called DB using a custom data structure described below. As Fig. 2 shows, DV, DD and DB are quite competitive with CE2. The variant DL, on the other hand, turned out to be hopelessly slow and is excluded. Instead, the figure includes a list-based implementation by Andersson and Nilsson [2, MSD].

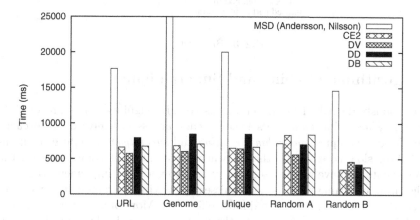

Fig. 2. Runtimes of D-variants

The slowness of DL is due to a large number of memory allocations and deallocations of list nodes. Andersson and Nilsson avoid this problem by passing the lists as an argument to the recursive calls instead of copying the lists back to the input array. Still, it is much slower than our implementations.

The variant DB is designed to reduce the space requirement by using a well-known data structure that we call the *block-list*. It is a linked list, where each list node holds a pointer to a fixed-size array (block). Using a block-list with a block size B, the buckets can be implemented using $\mathcal{O}(n/B + \sigma B)$ space in addition to the string pointers. With $B = \sqrt{n/\sigma}$, this is $\mathcal{O}(\sqrt{n\sigma})$. We use $B = 1024$.

Furthermore, our DB implementation gets rid of the need to store all the string pointers outside the input array by using the almost in-place distribution technique described in [5, Appendix B]. The idea is that as strings are distributed into the buckets, the space in the beginning of the input array becomes free, and can be used as storage for the blocks. The idea is illustrated in Fig. 3. The total additional space needed by the block-list is $\mathcal{O}(\sqrt{n\sigma})$ [5, Theorem B.1].

The DB variant has a significant overhead in processing empty or small buckets. To reduce this overhead, DB switches to CE2 when n drops below 2^{16}. This needs some further additional space. In total, though, the additional memory needed by the implementation remains less than 3 megabytes plus $3n/256$ bytes. A comparison of memory requirements is shown in Table 4 in Section 8.

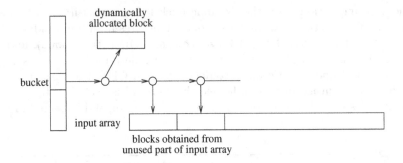

Fig. 3. Blocklist

7   Algorithmic Caching and Superalphabet

The previously described implementations spend a significant amount of their time simply accessing characters because each access is likely to cause cache misses. Two techniques, *algorithmic caching* and *superalphabet*, reduce the number of these slow accesses. Neither is a new technique, but combined with the improvements we have already seen, they lead to very fast implementations.

Algorithmic caching reduces cache misses by copying characters in advance to a place where they can be accessed more efficiently. More precisely, we represent each string not only by a pointer to the beginning but also by four characters stored next to the pointer. Initially, the characters are the first four characters of the string, but later they will be replaced by the next four characters, then by the next four and so on. The four characters in the cache are moved with the string pointer and can thus be accessed efficiently. Only every fourth level of recursion is slower due to refilling the cache.

The idea in the superalphabet technique is to treat pairs of characters as single characters in a larger alphabet. This effectively halves the number of characters and thus the number of character accesses. Increasing the alphabet size from 2^8 to 2^{16} can also reduce the speed of the algorithm due to the cost of iterating through all buckets (see Section 3). To avoid this, we switch from superalphabet to normal alphabet when the number of strings drops below 2^{16}.

Fig. 4 shows the performance of variants of CE2 described in Section 5 using algorithmic caching (CE2-A) and superalphabet (CE2-S). Using superalphabet improved performance in all cases. Also in the figure are an algorithmic caching implementation by Ng and Kakehi [7], which is a modification of the algorithm by McIlroy et al. in Fig. 1, and a superalphabet implementation by Andersson and Nilsson [2, Adaptive], which is a modification of the algorithm by the same authors in Fig. 2.

We also implemented superalphabet versions of CI, DV and DD, obtaining in each case a similar, usually small, improvement as with CE2. Some experimental results are shown in Section 8. We did not implement a superalphabet version of DB, since it cannot deal effectively with a large alphabet due to inefficiencies in handling small and empty buckets.

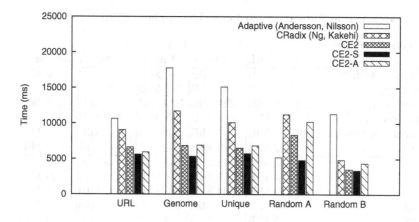

Fig. 4. Runtimes of A and S variants

8   Comparison to Other Algorithms

We finish with a comparison of some of our implementations with two other fast string sorting algorithms: multikey quicksort [3] and burstsort [11]. Like our algorithms, burstsort is designed to reduce cache misses and to run efficiently on modern hardware. The runtimes are shown in Fig. 5.

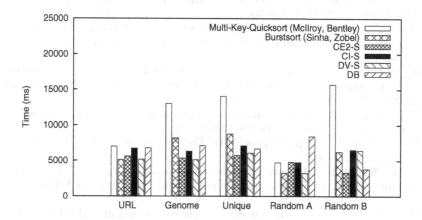

Fig. 5. Comparison to other algorithms

Finally, Table 4 shows the additional memory requirements of several implementations. Note, in particular, the implementation DB that combines a small space requirement with good runtime.

Table 4. Peak memory required in addition to input. MiB = 2^{20} bytes.

Implementation	Memory peak				
	URL	Genome	Unique	Random A	Random B
Program C [6]	0 MiB	0 MiB	0 MiB	0 MiB	0 MiB
MSD [2]	117 MiB	352 MiB	352 MiB	352 MiB	352 MiB
CRadix [7]	76 MiB	229 MiB	229 MiB	229 MiB	229 MiB
Adaptive [2]	117 MiB	352 MiB	352 MiB	352 MiB	352 MiB
Multi-Key-Quicksort [3]	0 MiB	0 MiB	0 MiB	0 MiB	0 MiB
Burstsort [11]	102 MiB	249 MiB	376 MiB	133 MiB	194 MiB
CE2	48 MiB	143 MiB	143 MiB	143 MiB	143 MiB
CI	10 MiB	29 MiB	29 MiB	29 MiB	29 MiB
DV	133 MiB	144 MiB	169 MiB	196 MiB	223 MiB
DD	39 MiB	117 MiB	117 MiB	117 MiB	117 MiB
DB	3 MiB	3 MiB	3 MiB	3 MiB	3 MiB
CE2-S	57 MiB	172 MiB	172 MiB	172 MiB	172 MiB
CI-S	19 MiB	57 MiB	57 MiB	57 MiB	57 MiB
DV-S	228 MiB	160 MiB	179 MiB	199 MiB	194 MiB

References

1. Andersson, A., Nilsson, S.: A new efficient radix sort. In: Proc. 35th IEEE Symposium on the Foundations of Computer Science, pp. 714–721. IEEE Computer Society, Los Alamitos (1994)
2. Andersson, A., Nilsson, S.: Implementing radixsort. ACM Journal of Experimental Algorithmics 3 (1998) Article No. 7
3. Bentley, J.L., Sedgewick, R.: Fast algorithms for sorting and searching strings. In: Proc. 8th ACM-SIAM Symposium on Discrete Algorithms, pp. 360–369. SIAM, Philadelphia (1997)
4. Franceschini, G., Muthukrishnan, S., Patrascu, M.: Radix sorting with no extra space. In: Arge, L., Hoffmann, M., Welzl, E. (eds.) ESA 2007. LNCS, vol. 4698, pp. 194–205. Springer, Heidelberg (2007)
5. Kärkkäinen, J., Sanders, P., Burkhardt, S.: Linear work suffix array construction. Journal of the ACM 53(6), 918–936 (2006)
6. McIlroy, P.M., Bostic, K., McIlroy, M.D.: Engineering radix sort. Computing Systems 6(1), 5–27 (1993)
7. Ng, W., Kakehi, K.: Cache efficient radix sort for string sorting. IEICE Transactions on Fundamentals of Electronics, Communications and Computer Sciences E90-A(2), 457–466 (2007)
8. Paige, R., Tarjan, R.E.: Three partition refinement algorithms. SIAM Journal on Computing 16(6), 973–989 (1987)
9. Puglisi, S.J., Smyth, W.F., Turpin, A.H.: A taxonomy of suffix array construction algorithms. ACM Computing Surveys 39(2) (2007)
10. Rahman, N., Raman, R.: Adapting radix sort to the memory hierarchy. ACM Journal of Experimental Algorithmics 6 (2001) Article No. 7
11. Sinha, R., Zobel, J.: Cache-conscious sorting of large sets of strings with dynamic tries. ACM Journal of Experimental Algorithmics 9 (2004) Article No. 1.5
12. Sinha, R., Zobel, J., Ring, D.: Cache-efficient string sorting using copying. ACM Journal of Experimental Algorithmics 11 (2006) Article No. 1.2

Faster Text Fingerprinting

Roman Kolpakov[1,*] and Mathieu Raffinot[2]

[1] Liapunov French-Russian Institute, Lomonosov Moscow State University,
Moscow, Russia
foroman@mail.ru

[2] CNRS, LIAFA, Univ. Paris Diderot - Paris 7, 75205 Paris Cedex 13, France
raffinot@liafa.jussieu.fr

Abstract. Let $s = s_1..s_n$ be a text (or sequence) on a finite alphabet
Σ. A fingerprint in s is the set of distinct characters contained in one
of its substrings. Fingerprinting a text consists in computing the set $\mathcal{F}$
of all fingerprints of all its substrings. A fingerprint, $f \in \mathcal{F}$, admits a
number of maximal locations $\langle i, j \rangle$ in S, that is the alphabet of $s_i..s_j$ is
f and s_{i-1}, s_{j+1}, if defined, are not in f. The set of maximal locations is
$\mathcal{L}$, $|\mathcal{L}| \leq n|\Sigma|$. Two maximal locations $\langle i, j \rangle$ and $\langle k, l \rangle$ such that $s_i..s_j =
s_k..s_l$ are named *copies* and the quotient of $\mathcal{L}$ according to the copy
relation is named $\mathcal{L}_C$. The faster algorithm to compute all fingerprints
in s runs in $O(n + |\mathcal{L}| \log |\Sigma|)$ time. We present an $O((n + |\mathcal{L}_C|) \log |\Sigma|)$
worst case time algorithm.

1 Introduction

We consider a finite ordered alphabet Σ and $s = s_1..s_n$ a sequence of n letters,
$s_i \in \Sigma$. The set of all sequences over Σ is denoted Σ^*. The rank of each letter
α in Σ is given by $f_\Sigma(\alpha)$ that ranges between 0 and $|\Sigma| - 1$. A sequence $v \in \Sigma^*$
is a factor or substring of s if $s = uvw$. The fingerprint $C(s)$ of a sequence s is
the set of distinct letters in s. By extension, $C_s(i, j)$ is the set of distinct letters
in $s_i..s_j$. A fingerprint is represented below by a binary table of F of size $|\Sigma|$. If
s contains the character α, $F[f_\Sigma(\alpha)] \leftarrow 1$, otherwise $F[f_\Sigma(\alpha)] \leftarrow 0$.

Definition 1. *Let $\mathcal{C}$ be a set of letters of Σ. A maximal location of $\mathcal{C}$ in $s =
s_1..s_n$ is an interval $[i, j]$, $1 \leq i \leq j \leq n$, such that*
(1) $C_s(i, j) = \mathcal{C}$; (2) if $i > 1, s_{i-1} \notin C_s(i, j)$; (3) if $j < n, s_{j+1} \notin C_s(i, j)$
This maximal location is denoted $\langle i, j \rangle$.

We denote by $\mathcal{F}$ the set of distinct fingerprints and by $\mathcal{L}$ the set of maximal
locations of all fingerprints of $\mathcal{F}$.

Definition 2. *Two maximal locations $\langle i, j \rangle$ and $\langle k, l \rangle$ of $s = s_1..s_n$ are copies
if $s_i..s_j = s_k..s_l$.*

---

* This work is supported by the Russian Foundation for Fundamental Research (Grant
05-01-00994) and the program of the President of the Russian Federation for supporting of young researchers (Grant MD-3635.2005.1).

A. Amir, A. Turpin, and A. Moffat (Eds.): SPIRE 2008, LNCS 5280, pp. 15–26, 2008.

The "copy" relation is obviously an equivalence relation. We denote $\mathcal{L}_C$ the set of equivalence classes. Let $q \in \mathcal{L}_C$ and $\langle i, j \rangle$ a maximal location in q, we denote $st_s(q)$ the string $s_i..s_j$. Table 1 shows an example of a copy relation.

In this paper, given a sequence s, we are interested in the following algorithmic problem:

+ Compute the set $\mathcal{F}$ of all fingerprints in s

This problem has many applications in information retrieval, computational biology and natural language processing [1]. The input alphabet Σ is considered to be the alphabet of the input sequence, thus $|\Sigma| \leq n$.

The problem has first be considered in [1] in which they presented a $O(n|\Sigma| \log n \log |\Sigma|)$ algorithm. This complexity has been improved to $\Theta(\min\{n|\Sigma| \log |\Sigma|, n^2\})$ time in [4]. The bound $\Theta(n|\Sigma| \log |\Sigma|)$ is that of the last algorithm in [4]. The $\Theta(n^2)$ bound is obtained using the first algorithm of [4], although this algorithm was first presented by Didier with $O(n^2 \log n)$ and $\Omega(n^2)$ time complexities in [3]. The $\log n$ gain between these two versions has been obtained using a lowest common ancestor algorithm (LCA). Surprisingly enough, these complexities were independent of the sizes of $\mathcal{F}$ and $\mathcal{L}$, although many sequence families have few fingerprints or few maximal locations. We thus proposed in [7] a new algorithm running in $O((n + |\mathcal{L}|) \log |\Sigma|)$ time. We improved it later to $O(n + |\mathcal{L}| \log |\Sigma|)$ time in [8]. As $|\mathcal{L}| \leq n|\Sigma|$, this algorithm is, at worst, as efficient as the last algorithm of [4], but much faster on many sequence families. However, in our will to deeply understand the problem, a question arises: what are the best parameters for this problem ? This paper is a new step toward the answer.

We present below an algorithm running in $O((n + |\mathcal{L}_C|) \log |\Sigma|)$ time, quite always faster than the previous algorithm because it depends of $\mathcal{L}_C$ instead of $\mathcal{L}$. Note that the number $|\mathcal{L}_C|$ can be significantly less than $|\mathcal{L}|$. As an example, we can consider the word w_k over the alphabet $\Sigma_k = \{a_1, a_2, \ldots, a_k\}$ which is defined in the following inductive way: $w_1 = a_1$ and $w_k = w_{k-1}(a_1 a_2 \ldots a_k)^k$ for $k > 1$. For this word we have $|w_k| = \frac{1}{6}k(k+1)(2k+1)$, $|\mathcal{L}| = \frac{1}{12}k(3k^3 + 2k^2 - 9k + 16) = \Theta(|w_k|^{4/3})$, and $|\mathcal{L}_C| = \frac{1}{6}k(k^2 + 5) = \Theta(|w_k|)$. Thus, in this case $|\mathcal{L}_C| = o(|\mathcal{L}|)$ as $k \to \infty$.

Table 1. Copy relation example for $s = a_1\, b_2\, a_3\, c_4\, e_5\, a_6\, b_7\, a_8\, c_9\, d_{10}$

Class q	Maximal locations	$st_s(q)$	Class q	Maximal locations	$st_s(q)$
I	$\emptyset$	ε	10	$a_1b_2a_3 \mid a_6b_7a_8$	aba
1	$a_1 \mid a_3 \mid a_6 \mid a_8$	a	11	$a_1b_2a_3c_4 \mid a_6b_7a_8c_9$	$abac$
2	$b_2 \mid b_7$	b	12	$a_3c_4e_5a_6$	$acea$
3	$c_4 \mid c_9$	c	13	$e_5a_6b_7a_8$	$eaba$
4	d_{10}	d	14	$a_6b_7a_8c_9d_{10}$	$abacd$
5	e_5	e	15	$a_1b_2a_3c_4e_5a_6$	$abacea$
6	$a_3c_4 \mid a_8c_9$	ac	16	$a_3c_4e_5a_6b_7a_8$	$aceaba$
7	c_9d_{10}	cd	17	$a_3c_4e_5a_6b_7a_8c_9$	$aceabac$
8	c_4e_5	ce	18	$a_3c_4e_5a_6b_7a_8c_9d_{10}$	$aceabacd$
9	e_5a_6	ea	19	$a_1b_2a_3c_4e_5a_6b_7a_8c_9d_{10}$	$abaceabcd$

Following the previous approaches, our algorithm improve a naming technique introduced in [6], adapted to the fingerprint problem in [1] and then successively improved in [4] and in [7]. The paper is organized as follows. In Section 2 we present a new structure called "participation tree" that is built from the suffix tree. This structure contains all the fingerprints that need to be coded using the new naming algorithm presented in Section 3.

We assume below without loss of generality that the input sequence does not contain two consecutive repeating characters. Such a sequence is named *simple*. The segments of repeating characters, say α, of any input sequence can be reduced to a unique occurrence of α. The two sequences have the same sets, $\mathcal{F}$, and the same sets, $\mathcal{L}$ and $\mathcal{L}_C$, up to small changes in the bounds. These changes can, however, be simply retrieved in $\Theta(1)$ per maximal location and the reducing algorithm is $\Theta(n)$. This technical trick really simplifies the algorithms we present by removing many straightforward technical cases.

2   Participation Tree

Let $s = s_1..s_n$ be a simple sequence of characters over Σ. In this first phase, for reasons that will appear clearly below, we add to the sequence a last character $s_{n+1} = \#$ that does not appear in the sequence. Thus $s = s_1..s_n\#_{n+1}$. Let i, j be a position in s, $1 \le i \le j \le n + 1$. We define $\mathrm{fo}_s(i, j)$ as the string formed by concatenating the first occurrences of each distinct character touched when reading s from position i (included) to position j (included). For instance, if $s = a_1b_2a_3c_4e_5a_6b_7a_8c_9d_{10}\#$, $\mathrm{fo}_s(3, 9) = aceb$ and $\mathrm{fo}_s(5, 10) = eabcd$.

Definition 3. *Let $s = s_1..s_ns_{n+1}$ with $s_{n+1} = \#$ and $1 \le i \le n$ a position in s. Let $j > i$ the minimum position such that $s_j = s_i$ if it exists, $j = n+2$ otherwise. We define $\mathrm{lfo}_s(i) = \mathrm{fo}_s(i, j - 1)$.*

For instance, if $s = a_1b_2c_3a_4d_5a_6b_7a_8c_9b_{10}e_{11}\#_{12}$, $\mathrm{lfo}_s(1) = abc$ and $\mathrm{lfo}_s(5) = dabce\#$.

The participation tree ressembles a tree of all $\mathrm{lfo}_s(i)$ in which we removed each last character (the need of this removal will appear clearly below). It contains the same paths labels. However, building this exact tree is too time consuming and the participation tree allows some redundancy in the path labels, the same path label might correspond to several paths from the root. Our tree is thus not always "deterministic" in the sense that a node can have several transitions by the same character. We define it and build it from the suffix tree by cutting and shrinking edges. We first succintly recall the properties of the suffix tree.

2.1   Suffix Tree

The suffix tree $\mathrm{ST}(s)$ is a compact representation of all suffixes of a given sequence $s = s_1 \ldots s_n$. It is basically a trie of all suffixes of s where all nodes with a single child are merged with their parents. Each transition of the tree is then coded as an interval $[i, j]$ corresponding to $s_i..s_j$. Its size is $O(n)$ and there exists many $O(n \log |\Sigma|)$ time contruction algorithm based on different paradigms.

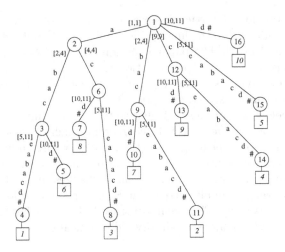

Fig. 1. Suffix tree of $s = a_1\ b_2\ a_3\ c_4\ e_5\ a_6\ b_7\ a_8\ c_9\ d_{10}\#_{11}$. Square boxes contain the initial position of the suffix. Each edge is labeled by a pair $[k, l]$ pointing to $s_k..s_l$ that we explicity write on the edge for clarity.

The three most important are chronically that of Weiner [11], McCreight[9] and Ukkonen [10]. An example of such a suffix tree is given in Figure 1.

We assume below that in the suffix tree each transition interval $[i, j]$ of ST(s) corresponds to the leftmost occurrence of the factor $s_i \ldots s_j$ in s. For instance, in Figure 1, the transition from 1 to 2 is the pointer $[1, 1] = s_1 = a$. This property is insured by Ukkonen [10] algorithm, but can also be insured on every suffix tree by a simple additional $O(n)$ step.

2.2 Participation Tree

Let $s = s_1..s_n s_{n+1}$ where $s_{n+1} = \#$. The participation tree $PT(s)$ is built from the suffix tree $ST(s)$ the following way. Imagine the suffix tree in an "expanded" version, that is each edge $[i, j]$ explicitly written by the corresponding factor $s_i..s_j$ (see Figure 1). Let us consider the sequence of characters on each path from the root and let α be the first character on this path. Let o be the second occurrence of α on this path if it exists. We perform the following steps:

1. we first reduce all characters on this path after o (included) to the empty string ε;
2. then, on the section from the root to the character before o we only keep the first occurrence of each appearing character, i.e. the others are reduced to ε;
3. we then replace the last character of each path from the root to a leaf by ε;
4. we replace all multi-characters edges by an equivalent serie of a single character and a node. An example of such a resulting tree is shown in Figure 2 (left);
5. as a last step, all ε edges (p, ε, q) are removed by merging p and q. The resulting tree is the participation tree. An example of this last tree is shown in Figure 2 (right).

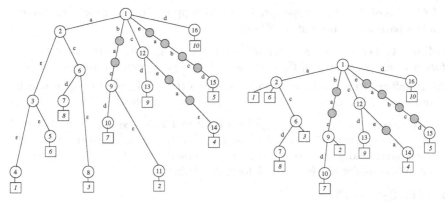

Fig. 2. From suffix tree to the participation tree (right picture) of $s = a_1b_2a_3c_4e_5a_6b_7a_8c_9d_{10}\#_{11}$. New nodes are in gray. The ε transitions are removed in the last step. Attached suffixes are shown in square boxes.

For each node q of $ST(s)$ and $PT(s)$ we denote $\mathrm{Suff}(q)$ the set of suffixes of s that appear as leaves of the subtree rooted in q. We consider below that the suffixes associated to a node in $ST(s)$ remains associated to the node in $PT(s)$, even after the merging. This is shown in Figure 2: the suffixes in the square boxes associated to nodes 4 and 5 in the left picture are associated to node 2 in the participation tree (right picture).

Lemma 1. *Let $s = s_1..s_n$. For all $i = 1..n$, each proper prefix of $lfo_s(i)$ labels a path from the root in $PT(s)$.*

Proof. The reduction of the path of suffix i in the suffix tree corresponds, when nodes are ignored, to $lfo_s(i)$ without its last character.   □

Note that a proper prefix of $lfo_s(i)$ might label several paths from the root in $PT(s)$.

Let $[i, j]$ be an interval on $s = s_1..s_n$ and let $\mathrm{Support}([i, j])$ be the minimum of the indices of the rightmost occurrences of $\alpha = s_p$, $i \le p \le j$, in the interval $[i, j]$. We define $O_s^{[i,j]}$ as $fo_s(\mathrm{Support}([i, j]), j)$. For instance, if $s = a_1b_2a_3c_4e_5a_6b_7a_8c_9d_{10}\#_{11}$, $\mathrm{Support}(\langle 1, 3\rangle) = 2$, $\mathrm{Support}([4, 10]) = 5$, $O_s^{\langle 1,3\rangle} = ba$ and $O_s^{[4,10]} = eabcd$.

Definition 4. *Let $s = s_1..s_n$ and $1 \le i \le j \le n$. We define $\mathrm{Extend}_s(i, j)$ as the maximal location reached when extending the interval $[i, j]$ to the left and to the right while the closest external characters s_{i-1} or s_{j+1} (if they exist) belong to $C_s(i, j)$.*

For instance, if $s = a_1\ b_2\ a_3\ c_4\ e_5\ a_6\ b_7\ a_8\ c_9\ d_{10}\#_{11}$, $\langle 1, 4\rangle = \mathrm{Extend}_s(2, 4)$ and $\langle 1, 9\rangle = \mathrm{Extend}_s(2, 7)$

Lemma 2. *Let $\langle i, j\rangle$ be a maximal location of $s = s_1..s_n$. There exists a permutation of all characters of $C_s(i, j)$ that labels a path from the root in $PT(s)$.*

Proof. $O_s^{\langle i,j \rangle}$ is obviously a proper prefix of $\text{lfo}_s(\text{Support}(\langle i,j \rangle))$, which, by lemma 1, labels a path from the root in $PT(s)$. □

Corollary 1. *Let* $s = s_1..s_n$. *For all* $i,j, 1 \leq i \leq j \leq n$, *there exists a permutation of all characters of* $C_s(i,j)$ *that labels a path from the root in* $PT(s)$.

Proof. It suffices to extend the segment $s_i..s_j$ to $\langle k,l \rangle = \text{Extend}_s(i,j)$ in which it is contained. Then $C_s(i,j) = C_s(k,l)$ and lemma 1 applies. □

Let $z = ((r,\alpha_1,p_1),\ldots,(p_{i-1},\alpha_i,p_i))$ be a path in $PT(s = s_1..s_n)$ from its root r. By notation extension, we denote $\text{Suff}(z) = \text{Suff}(p_i)$. Let $\text{SPref}(s)$ be the set of all such paths and $w(z) = \alpha_1\alpha_2..\alpha_i$. Let $\mathcal{P}(\mathcal{L})$ be the set of all sets of maximal locations. We consider the function Φ formally defined as:

$$\Phi : \begin{array}{l} \text{SPref}(s) \longrightarrow \mathcal{P}(\mathcal{L}) \\ \quad z \quad \longmapsto \{\langle k,l \rangle \in \mathcal{L} \mid O_s^{\langle k,l \rangle} = w(z) \text{ and } \text{Support}(\langle k,l \rangle) \in \text{Suff}(z)\} \end{array}$$

Lemma 3. *Let* $z = ((r,\alpha_1,p_1),\ldots,(p_{i-1},\alpha_i,p_i))$ *be a non empty path in* $SPref(s)$. *Then* $\Phi(z) \neq \emptyset$.

Proof. By construction of the participation tree, there exits $m \in \text{Suff}(z)$ such that $\alpha_1 \ldots \alpha_i$ is a proper prefix of $\text{lfo}(m)$. Let p be the first position of α_i in s following m. Then $\cup_{1 \leq f \leq i}\{\alpha_f\} = C_s(m,p)$. Let $\langle k,l \rangle = \text{Extend}_s(m,p)$.

We prove now that $\text{Support}(\langle k,l \rangle) = m$. As $\alpha_1 \ldots \alpha_i$ is a proper prefix of $\text{lfo}(m)$, there exist $\alpha = \text{lfo}(m)_{i+1}$ such that there is no occurrence of α in the interval $[m,p]$, and thus after the extension of $[m,p]$ to a maximal location $\langle k,l \rangle$, the indice l is strictly less than the indice of the first occurrence of α after m. As, by definition of $\text{lfo}(m)$, there is no occurrence of s_m before the indice of α after m in s, there is no other occurrence of s_m at the right of s_m in the interval $[m,l]$. Moreover, as all characters in $\alpha_1 \ldots \alpha_i$ and only them appears after m in $[m,l]$ in the order of $\alpha_1 \ldots \alpha_i$, and that the extension procedure insures that all characters in $[k,m]$ are characters of $\alpha_1 \ldots \alpha_i$, $\text{Support}(\langle k,l \rangle) = m$.

Finally, it is obvious that $O_s^{\langle k,l \rangle} = O_s^{[m,p]} = \alpha_1..\alpha_i = w(z)$, and thus $\langle k,l \rangle \in \Phi(z)$. □

Lemma 4. *Let* $z_1, z_2 \in SPref(s)$ *be two distinct non empty paths, then* $\Phi(z_1) \cap \Phi(z_2) = \emptyset$.

Proof. Assume *a contrario* that there exists $\langle k,l \rangle \in \Phi(z_1) \cap \Phi(z_2)$. Let $m = \text{Support}(\langle k,l \rangle)$, $m \in \text{Suff}(z_1)$ and $m \in \text{Suff}(z_2)$, thus one of the path is a prefix of the other. As $O_s^{[k,l]} = w(z_1) = w(z_2)$, the two paths must be equal, which contradicts the hypothesis. □

Lemma 5. *Let* $\langle i,j \rangle$ *and* $\langle k,l \rangle$ *be two distinct maximal locations of* $s = s_1..s_n$ *in the same equivalence class of* $\mathcal{L}_c$. *There exits* $z \in SPref(s)$ *such that* $\langle i,j \rangle \in \Phi(z)$ *and* $\langle k,l \rangle \in \Phi(z)$.

Proof. Let $m_1 = \text{Support}(\langle i,j \rangle)$ and $m_2 = \text{Support}(\langle k,l \rangle)$. As $s_i..s_j = s_k..s_l$, $u = s_{m_1}..s_j = s_{m_2}..s_l$ and m_1 and m_2 are thus in the subtree of the path h labeled by u in $ST(s)$. After reduction of this path in $PT(s)$, the resulting path z is such that $w(z) = O_s^{\langle i,j \rangle} = O_s^{\langle k,l \rangle}$ and $m_1, m_2 \in \text{Suff}(z)$. Thus $\langle i,j \rangle, \langle k,l \rangle \in \Phi(z)$. □

Theorem 1. *All maximal locations are in the image $\Phi(z)$ of a path z in $PT(s = s_1..s_n)$ and the size of $PT(s)$ is $O(|\mathcal{L}_C|)$.*

Proof. Lemma 2 directly implies that all maximal locations are in the image $\Phi(z)$ of a path z in $PT(s)$. As by lemma 4 the images $\Phi(z)$ are non overlapping, they form a partition of $\mathcal{L}$. Lemma 5 insures that $\mathcal{L}_C$ partition is a subpartition of the partition formed by the images of Φ. As by lemma 3 there is no empty image, the number of such images is smaller than or equal to $|\mathcal{L}_C|$.   □

Note that we considered the size of $PT(s = s_1..s_n)$ without the initial positions of suffixes (square boxes in Figure 2). With these positions, its size is $O(n+|\mathcal{L}_C|)$.

2.3   From Suffix Tree to Participation Tree

We extend the notion of $\text{fo}_s(i,j)$ keeping the positions of the characters in $s = s_1..s_n$. We define $\text{efo}_s(i)$ as the string formed by concatenating the first occurrences of each distinct character touched when reading s from position i (included) to position n (included) but indexed by the position of this character in the sequence. For instance, if $s = a_1b_2a_3c_4e_5a_6b_7a_8c_9d_{10}\#_{11}$, $\text{efo}_s(3) = a_3c_4e_5b_7d_{10}\#_{11}$ and $\text{efo}_s(5) = e_5a_6b_7c_0d_{10}\#_{11}$.

The idea of the algorithm is the following. For each transition (i,j) on the path of a longest suffix $v = s_f \ldots s_n$ we compute the "participation" of the edge to $\text{lfo}_s(f)$, that is, the number of new characters the edges brings in $\text{lfo}_s(f)$. For instance, in Figure 1 the participation of edge $(6,8) = [5,11]$ is e, since it is on the path of the longest suffix $s_3 \ldots s_n$ and $\text{lfo}_s(3) = ace$. The participation of edge $(12,14) = [5,11]$ is eab since $\text{lfo}_s(4) = ceab$. To compute the participation of interval $[i,j]$ on the path of a longest suffix $v = s_f \ldots s_n$, we use $\text{efo}_s(f)$ and also the next position of s_f after f in s, if it exists. Assume it is the case and let p be this position. Thus $s_p = s_f$. Let $\text{efo}_s(f) = s_f s_{l_1} s_{l_2} \ldots s_{l_z}$ and $l_h \leq p \leq l_{h+1}$. If $i \geq p$, the participation of $[i,j]$ is the empty word ε. Otherwise, $i < p$ and its participation is the string (potentially empty) $s_{l_a} \ldots s_{l_b}$ with

- $i \leq l_a$ and l_a is the smallest such indice;
- $l_b \leq min(j,p-1)$ and l_b is the greatest such indice.

For instance, on Figure 1, $\text{efo}_s(2) = b_2a_3c_4e_5d_{10}\#_{11}$ and $p = 7$ since 7 is the next position of b after position 2. Thus, participation of edge $(1,9) = [2,4] = b_2a_3c_4 = bac$, participation of $(9,11) = [5,11] = e_5 = e$ (since $p = 7$). For each suffix $[k,n]$, given $\text{efo}_s(k)$ and p, a bottom-up process from leaf k to the root of the suffix tree allows us to calculate the participation of each (not previously touched) edge on this path. We modify the suffix tree using successively $\text{efo}_s(k)$ for $k = 1..n$. A sketch of this algorithm is given in Figure 3.

At the end of this process, we first replace the last character of all paths from the root by ε. we finally remove all (p,ε,q) edges by merging p and q.

Theorem 2. *The participation tree of $s = s_1..s_n$ can be built in $O(n \log |\Sigma| + |\mathcal{L}_C|)$ time and $O(n + |\mathcal{L}_C|)$ space.*

BUILD_PART_TREE($ST(s = s_1..s_n s_{n+1}$ with $s_{n+1} = \#$))
  1. Compute $\text{efo}_s(1)$ and p_1
  2. **For** $i = 1..n$ **Do**
  3.   Current $\leftarrow$ Leaf(i) in $ST(s)$.
  4.   **While** Current not marked AND Current $\neq$ Root **Do**
  5.   Prec $\leftarrow$ Parent(Current) in $ST(s)$.
  6.   Compute the participation of edge (Parent, Current) in $\text{efo}_s(i)$
  7.   Mark Current
  8. $\text{efo}_s(i+1) \leftarrow$ Update $\text{efo}_s(i)$
  9. $p_{i+1} \leftarrow$ next position of s_{i+1} after $i+1$ in s
 10.   **End of while**
 11. **End of for**
 12. Replace each last character of all paths from the root by ε.
 13. Remove ε edges by node merging.

Fig. 3. Building the participation tree from the suffix tree

Proof. The algorithm is correct since it consists of directly compute the participation of each edge one after the other. We now study its complexity.

For each suffix $[k, n]$, given $\text{efo}_s(k)$ in an AVL tree and p, the bottom-up process from leaf k to the root of the suffix tree can be done in $O(\log |\Sigma|)$ time for each unmarked node. If the first $\text{efo}_s(1)$ is given as an AVL tree, initially built in $O(|\Sigma| \log |\Sigma|)$, $\text{efo}_s(2)$ can be obtained in $O(\log |\Sigma|)$, and so on for $k = 3..n$, assuming that for each k we know the next position of s_k in $s_{k+1} \ldots s_n$ if it exits. To know these positions, $|\Sigma|$ lists, one for each character α, of positions of α in s can be initialy computed in $O(|\Sigma| + n)$ time and consumed character after character. Thus, calculating the participation of each edge in the suffix tree can be done in $O(\log |\Sigma|)$ writing each time a unique part of the $PT(s)$ tree.

Replacing the last character of each path from the root by ε is $O(n)$. Merging each ε edges can also be performed in $O(n)$ since each such edge is either a previous edge of the suffix tree or was labeled by a single last character of a path from the root. The whole construction of $PT(s)$ is thus $O(n \log |\Sigma| + |\mathcal{L}_C|)$ time.

The size of the AVL tree is bounded by $|\Sigma|$, thus by n. The space required is the size of the suffix tree plus the size of the participation tree, thus $O(n + |\mathcal{L}_C|)$ space. □

3   Naming All Fingerprints

In this section we explain how to name all fingerpints from the participation tree. The naming technique itself is originally based on that of [6] that has been adapted for the fingerprint problem in [1]. The naming technique is used to give a unique name to each fingerprint of a substring of s. We first describe the naming technique and then we explain how to use it to name all fingerprints of s.

3.1 Naming Technique

We assume for simplicity, but without loss of generality, that $|\Sigma|$ is a power of two. We consider a stack of $\log|\Sigma| + 1$ arrays on top of each other. Each level is numbered from 1. The lowest, called the fingerprint table, contains $|\Sigma|$ names that might be only [0] or [1]. Each other array contains half the number of names that the array it is placed on. The highest array only contains a single name that will be the name of the whole array. Such a name is called a fingerprint name. Figure 4 shows a simple example with $|\Sigma| = 8$.

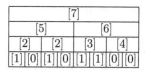

Fig. 4. Naming example

The names in the fingerprint table are only [0] or [1] and are given. Each cell, c, of an upper array represents two cells of the array it is placed on, and thus a pair of two names. The naming is done in the following way: for each level going from the lowest to the highest, if the cell represents a new pair of names, give this pair a new name and assign it to the cell. If the pair has already been named, place this name into the cell. In the example in Figure 4, the name [2] is associated to ([1], [0]) the first time this pair is encountered. The second time, this name is directly retrieved.

3.2 Naming a List of Fingerprint Changes

Assume that a specific set S of fingerprints can be represented as a list $L = (\alpha_1, \alpha_2, \ldots \alpha_p)$ of distinct characters such that $S = \{f_1, f_2, \ldots, f_p\}$ where $f_i = \cup_{1 \leq j \leq i}\{\alpha_j\}$. The core idea of the algorithm of [4] is to fill a fingerprint table bottom-up by building for each level an ordered list of new names that corresponds to the fingerprint changes induced at the previous level. A pseudo-code of this naming algorithm is given in Figure 5. We explain it below.

We number the level from 1, the lowest, to $\log|\Sigma| + 1$. The original list L is first tranformed into a list L_1 of changes on level 1 by replacing each character α_i by the pair $\{[1], f_\Sigma(\alpha_i)\}$. To initialize the process we add a list of $|\Sigma|$ pairs $\{[0], i\}$, $i = 1..|\Sigma|$ at the beginning of L_1.

This initial list is then used to compute all names of the cells in the second level. A table, FT, of $|\Sigma|$ names temporary records the pair of names to be coded. A list L'_1 of pairs of names is built as follows. The first $|\Sigma|$ elements of L_1 are read to initialize FT. The list L'_1 is initialized with $|\Sigma|/2$ pairs built by reading FT. Then, the remaining of the list L_1 is read and for each new element $\{[a], j\}$ *(1)* the table FT is changed in position j by $FT \leftarrow [a]$ and *(2)* the pair $\{(FT[2\lfloor j/2\rfloor], FT[2\lfloor j/2\rfloor + 1]), j/2\}$ if added to the end of L'_1. This means that in cell $j/2$ of the second level a name has to be given to the name pair $(FT[2\lfloor j/2\rfloor], FT[2\lfloor j/2\rfloor + 1])$.

NAME_LISTS($L = (\alpha_1, \alpha_2, \ldots \alpha_p)$ initial list of changes)
1. $L_1 \leftarrow (\{[0], 0\}, \ldots, \{[0], |\Sigma| - 1\})$
2. add $(\{[1], f_\Sigma(\alpha_1)\}, \ldots, \{[1], f_\Sigma(\alpha_p)\})$ to end of L_1
3. **For** $r = 1..\log|\Sigma|$ **Do**
4. $FT_r \leftarrow$ name table of size $|\Sigma|/2^{r-1}$
5. $E_{tp} \leftarrow$ first element of L_r
6.   **For** $l = 0..|\Sigma|/2^{r-1} - 1$ **Do**   /* initialization of table FT */
7. $\{[a], j\} \leftarrow E_{tp}$
8. $FT_r[j] \leftarrow [a]$
9. $E_{tp} \leftarrow$ next element in L_r
10.   **End of for**
11.   Let L'_r be an empty list
12.   **For** $l = 0..|\Sigma|/2^r - 1$ **Do**   /* initialization of L'_r list */
13.   add $\{(FT[2l], FT[2l + 1]), l\}$ to end of L'_r
14.   **End of for**
15. $E_{tp} \leftarrow$ first element of L_r
16.   **While** E_{tp} exists **Do**
17. $\{[a], j\} \leftarrow E_{tp}$
18. $FT_r[j] \leftarrow [a]$
19.   add $\{(FT_r[2\lfloor j/2 \rfloor], FT_r[2\lfloor j/2 \rfloor + 1]), j/2\}$ to end of L'_r
20. $E_{tp} \leftarrow$ next element in L_r
21.   **End of while**
22.   sort the pair of names in L'_r in lexicographical order
23.   give new names in each unique pair in L'_r
24.   build L_{r+1} by copying L'_r but replacing each pair by its new name
25. **End of for**

Fig. 5. Naming a list $L = (\alpha_1, \alpha_2, \ldots \alpha_p)$ of fingerprint changes

At this point L'_1 records the list of changes to be made in the cells at level 2 and the pairs of names that must receive a name. The pairs in this list are then sorted in lexicological order (through a radix sort) and a new name is assigned to each distinct pair of names (n_1, n_2). A new list L_2 is built from L'_1 (keeping the initial order of L'_1 and thus of L_1) by replacing each pair with its new name. For instance, if $\{([1], [0]), 1\}$ was in the list L'_1 and if the pair $([1], [0])$ received the new name $[2]$, then L_2 now contains $\{[2], 1\}$.

The list L_2 is the input at level 2 and the same process is repeated to obtain the names in the third level, and so on. The last list $L_{\log|\Sigma|+1}$ contains the names of all the fingerprints of $\mathcal{S}$.

Complexity. The initialization of L_1 is $\Theta(|L|)$ time. Then a linear sort of $\Theta(|L|)$ elements is performed for every level. As there are $\log|\Sigma| + 1$ levels, naming the list is $\Theta(|L| \log|\Sigma|)$ time.

3.3   Naming a Participation Tree

The naming approach of the previous section has been modified in [7] to name on the same set of names a table of lists of fingerprint changes. The main

DEPTH_FIRST_SEARCH(FT_k,$Current$)
1. **For** all α such that $\delta(Current, \alpha) \neq \Theta$ **Do**
2. $q \leftarrow \delta(Current, \alpha)$
3. $\{[a], j\} \leftarrow \Delta(Current, \alpha, q)$
4. $prec \leftarrow FT_k[j]$
5. $FT_k[j] \leftarrow [a]$
6. $\Delta(Current, \alpha, q) \leftarrow \{(FT_k[2\lfloor j/2 \rfloor], FT_k[2\lfloor j/2 \rfloor + 1]), j/2\}$
7.   DEPTH_FIRST_SEARCH(FT_k,q)
8. $FT_k[j] \leftarrow prec$
9. **End of for**

NAME_FINGERPRINT($PT(s)$)
10. $ninit_1 \leftarrow [0]$
11. **For** $k = 1 .. \log|\Sigma|$ **Do**
12. $FT_k \leftarrow$ name table of size $|\Sigma|/2^{k-1}$ all initialized to $ninit_k$
13.   DEPTH_FIRST_SEARCH(FT_k,Root($PT(s)$))
14. $Sl \leftarrow \Theta$ /* empty stack */
15.   **For** all edges $e = (p, \alpha, q)$ in $PT(s)$ **Do**
16. $\{(n_1, n_2), j\} \leftarrow \Delta(p, \alpha, q)$
17.   Add (n_1, n_2) to Sl.
18.   **End of for**
19.   add the couple $(ninit_k, ninit_k)$ to Sl
20.   sort Sl in lexicographical order
21.   give new names for each different couple in Sl
22.   replacing each pair in $\Delta(p, \alpha, q)$ by its new name
23. $ninit_{k+1} \leftarrow$ name of the pair $(ninit_k, ninit_k)$
24. **End of for**

Fig. 6. Naming all fingerprints in a participation tree $PT(s)$

modification is that the linear sorting is done for each level on all the pairs of all the lists of the table. We use a similar approach, but instead of a table of lists we consider the set of all paths from the root in the participation tree $PT(s)$. Each such path is considered as a list of fingerprint changes. The corollary 1 guaranty our approach. The NAME_FINGERPRINT algorithm names all fingerprints. Its pseudo-code is given in Figure 6.

As in the list naming of section 3.2, $\log|\Sigma|$ iterations are performed, one by fingerprint array level (loop 14-27), the lowest one excepted. With each edge (p, α, q) of $PT(s)$ is associated a value $\Delta(p, \alpha, q)$. At the end of iteration k, this value records the change corresponding to the edge in the fingerprint array of level $k + 1$. The value $\Delta(p, \alpha, q)$ is assumed to be initialized with $\{[1], f_\Sigma(\alpha)\}$ corresponding to the change induced by the edge at the lowest level 1.

In each iteration k, the recursive algorithm DEPTH_FIRST_SEARCH is called (line 16) on the participation tree to update all values $\Delta(p, \alpha, q)$ during a depth first search. The update operation on each such value is similar to the pair update in the naming of a simple list of fingerprint changes in section 3.2. Note that in

DEPTH_FIRST_SEARCH a special FT table is modified (line 6) before the recursive call but reinitialized to the previous value after the call (line 11). This permits to initialize the table FT only once before the first call to DEPTH_FIRST_SEARCH (line 15).

After the depth first search the values $\Delta(p, \alpha, q)$ are collected on all the edges (p, α, q) of the participation tree (lines 18-22) in a list Sl. This list is lexicographically sorted and a new name is given to each unique pair (line 25), similarly to the naming of a single list in section 3.2. The first pair of names of each $\Delta(p, \alpha, q)$ is then replaced by its new name.

To initialize the fingerprint array at the next level, the couple $(ninit_k, ninit_k)$ is added to the list of names (line 23) and its new name is retrieved after the sorting and the renaming (line 27).

At the end of the last iteration of the main loop (line 14-28), the last naming (line 25) returns the list of all the fingerprint names.

Theorem 3. *The* NAME_FINGERPRINT *algorithm applied on* $PT(s)$ *names all fingerprints of* s *in* $\Theta(|\mathcal{L}_C| \log |\Sigma|)$ *time.*

References

1. Amir, A., Apostolico, A., Landau, G.M., Satta, G.: Efficient text fingerprinting via parikh mapping. J. Discrete Algorithms 1(5-6), 409–421 (2003)
2. Bergeron, A., Chauve, C., de Montgolfier, F., Raffinot, M.: Computing common intervals of k permutations, with applications to modular decomposition of graphs. In: Brodal, G.S., Leonardi, S. (eds.) ESA 2005. LNCS, vol. 3669, pp. 779–790. Springer, Heidelberg (2005)
3. Didier, G.: Common intervals of two sequences. In: Benson, G., Page, R.D.M. (eds.) WABI 2003. LNCS (LNBI), vol. 2812, pp. 17–24. Springer, Heidelberg (2003)
4. Didier, G., Schmidt, T., Stoye, J., Tsur, D.: Character sets of strings (submitted, 2004)
5. Heber, S., Stoye, J.: Finding all common intervals of k permutations. In: Amir, A., Landau, G.M. (eds.) CPM 2001. LNCS, vol. 2089, pp. 207–218. Springer, Heidelberg (2001)
6. Karp, R.M., Miller, R.E., Rosenberg, A.L.: Rapid identification of repeated patterns in strings, trees and arrays. In: Proceedings of the 4th ACM Symposium on the Theory of Computing, Denver, CO, pp. 125–136. ACM Press, New York (1972)
7. Kolpakov, R., Raffinot, M.: New algorithms for text fingerprinting. In: Lewenstein, M., Valiente, G. (eds.) CPM 2006. LNCS, vol. 4009, pp. 342–353. Springer, Heidelberg (2006)
8. Kolpakov, R., Raffinot, M.: New algorithms for text fingerprinting. Journal of Discrete Algorithms (to appear, 2006)
9. McCreight, E.M.: A space-economical suffix tree construction algorithm. Journal of Algorithms 23(2), 262–272 (1976)
10. Ukkonen, E.: Constructing suffix trees on-line in linear time. In: van Leeuwen, J. (ed.) Proceedings of the 12th IFIP World Computer Congress, Madrid, Spain, pp. 484–492. North-Holland, Amsterdam (1992)
11. Weiner, P.: Linear pattern matching algorithm. In: Proceedings of the 14th Annual IEEE Symposium on Switching and Automata Theory, Washington, DC, pp. 1–11 (1973)

Context-Sensitive Grammar Transform: Compression and Pattern Matching

Shirou Maruyama[1], Yohei Tanaka[2], Hiroshi Sakamoto[3], and Masayuki Takeda[1]

[1] Department of Informatics, Kyushu University, Fukuoka 819-0395, Japan
{shiro.maruyama,takeda}@i.kyushu-u.ac.jp
[2] Graduate School of Computer Science and Systems Engineering,
Kyushu Institute of Technology, Iizuka 820-8502, Japan
t_youhei@donald.ai.kyutech.ac.jp
[3] Faculty of Computer Science and Systems Engineering,
Kyushu Institute of Technology, Iizuka 820-8502, Japan
hiroshi@ai.kyutech.ac.jp

Abstract. A framework of context-sensitive grammar transform is proposed. A greedy compression algorithm with the transform model is presented as well as a Knuth-Morris-Pratt (KMP)-type compressed pattern matching (CPM) algorithm. The compression performance is a match for gzip and Re-Pair. The search speed of our CPM algorithm is almost twice faster than the KMP type CPM algorithm on Byte-Pair-Encoding by Shibata et al. (2000), and in the case of short patterns, faster than the Boyer-Moore-Horspool algorithm with the stopper encoding by Rautio et al. (2002), which is regarded as one of the best combinations that allows a practically fast search.

1   Introduction

In this paper, we propose a framework of context-sensitive grammar (CSG) transform for fast compressed pattern matching. For this objective, we introduce a subclass of CSGs and construct an effective compression algorithm with a special case of the grammar transform model. We also implement Knuth-Morris-Pratt (KMP) pattern matching automaton on the compressed strings and show its performance by experiments. We thus refer to related work in both grammar-based compression and compressed pattern matching.

In the last several years, many researchers have tackled the minimum CFG problem as the optimum compression. The problem is defined as to find a smallest CFG which derives an input string only, which is NP-hard due to the relation with an algebraic problem called *the addition chain* [6]. Charikar et al. [3] proved the hardness of approximating the problem. They also showed an $O(\log n)$-approximation algorithm for a string of length n, which is currently the best ratio. Rytter [13] independently presented another $O(\log n)$-approximation algorithm using the suffix trees and the LZ-factorization technique. The same approximation ratio without suffix tree construction is achieved in [14], and the space efficiency is improved in [15] preserving a log-scale approximation ratio.

A. Amir, A. Turpin, and A. Moffat (Eds.): SPIRE 2008, LNCS 5280, pp. 27–38, 2008.

On the other hand, a large number of practical algorithms have been proposed. We specially refer to Re-Pair [7] since our compression algorithm is also based on the *recursive pairing*. The Byte-Pair-Encoding (BPE) is considered as simple implementation of Re-Pair with grammar symbols at most 256.

Such effective compression algorithms are closely related to the *compressed pattern matching* (CPM). Amir et al. [2] showed an algorithm of finding the first occurrence of a pattern on LZW compression in $O(n + m^2)$ time, where n and m are the lengths of text and pattern, respectively. Navarro and Raffinot [11] developed a more general technique, which abstracts both LZ77 and LZ78 and runs in $O(nm/w + m + occ)$ time, where w is the machine word length and occ is the number of pattern occurrences. Kida, et al. [5] proposed the *collage systems*: a formal system to represent a string by dictionary $\mathcal{D}$ and sequence $\mathcal{S}$ of variables, which unifies various dictionary-based compressions such as LZ family (LZ77, LZSS, LZ78, LZW), CFG transform based compressions, the Run-Length encoding, and so on. They also presented a general CPM algorithm on collage systems which runs in $O(h \cdot (d + s) + m^2 + occ)$ time, where d, h are the size and the maximum dependence of $\mathcal{D}$, respectively, and s is the length of $\mathcal{S}$, and the factor h disappears for the class of truncation-free collage systems that subsumes the CFG transform.

For practical speed-up of CPM, compressions with byte code are attractive since we can avoid any bitwise processing. BPE limits the number of the grammar symbols by 256 in order to represent in one byte each of them, and allows a fast search [16,17]. The compression ratio is, however, very poor. Matsumoto et al. [10] recently proposed to represent a large number of grammar symbols by byte-oriented Huffman code and improved both the compression and the search performances. However, the space requirement for the finite-state machine used grows linearly proportional to the number of grammar symbols.

Along this line of researches, our study is motivated by improving the present CPM performance in both theoretical and practical sense. Let us express our strategy for grammar-based compression by an intuitive example. If a text contains many occurrences of a digram AB, we can replace all of them by a single variable X which is associated with AB like $X \to AB$. The text is thus compressed to a shorter one according to the digram frequency. However the variables are incompatible among different digrams, i.e., we must produce k variables for k different digrams. Since this restriction is not avoidable in CFG transform, we relax the grammar class to context-sensitive grammars (CSGs) and introduce the *CSG transform*.

The introduced CSGs are *monotone grammars*[1] such that each of the production rules is of the form $aA \to \gamma$ or $A \to \gamma$ with a symbol a in the alphabet and a variable A. The production rules of the former form is called Σ-*sensitive*, so the grammar is also called Σ-sensitive. This grammar transform is related to the *context-dependent grammar* (CDG) by [18]. Indeed, a subclass of the Σ-sensitive grammars produced by our compression algorithm is included in the CDG transform.

Our contribution in this paper is as follows. We first analyze the expressiveness of Σ-sensitive grammars compared with CFGs by proving the upper/lower

---

[1] The length of string is not decreasing by any production rules.

bound of grammar size. We next give a compression algorithm by recursive pairing. In our method, digram AB is replaced by a variable X for every occurrence of trigram aAB with yielding a new production rule $aX \rightarrow aAB$. This strategy is potentially better than the standard recursive pairing since different digrams like $AA, AB, \ldots$ can be replaced by a same variable if they are appearing in different contexts. While the compression model is a special case of the Σ-sensitive grammars, we can show that, even when the number of grammar symbols is bounded by 256, the compression performance is a match for other practical methods such as Gzip and Re-Pair. Finally we develop a CPM algorithm on the compression model and show that it runs almost twice faster than the KMP type CPM algorithm on BPE by Shibata et al. [16], and that in the case of short patterns, it runs faster than the CPM technique of Rautio et al. [12], which is based on a variant of the Boyer-Moore-Horspool algorithm and the stopper encoding (SE), regarded as one of the best combinations of compression scheme and pattern matching technique that allow a fast search in practice.

2   Preliminaries

We assume a finite set Σ of alphabet symbols. The set of all strings over Σ is denoted by Σ^*, and $\Sigma^+ = \Sigma^* - \{\varepsilon\}$ for the empty string ε. The expression Σ^i denotes the set of all strings of length i. The length of a string $w \in \Sigma^*$ is denoted by $|w|$, and also for a set S, the notation $|S|$ refers to the size of S.

We recall the definition of context-free grammars (CFGs) and context-sensitive grammars (CSGs). A CFG is defined by $G = (V, \Sigma, P, S)$ with disjoint finite sets Σ and V, a finite set $P \subseteq V \times (V \cup \Sigma)^*$ of *production rules*, and the *start symbol* $S \in V$. Symbols in V are called *variables*. For any strings $\alpha, \beta \in (V \cup \Sigma)^*$, we write $\alpha X \beta \Rightarrow \alpha \gamma \beta$ if $(X \rightarrow \gamma) \in P$. Moreover, the reflexive, transitive closure of $\Rightarrow$ is denoted by $\overset{*}{\Rightarrow}$. The set of all strings defined by $S \overset{*}{\Rightarrow} w \in \Sigma^*$ is called the *language* of the grammar, and w is said to be *derived from* S.

A CSG is defined by $G = (V, \Sigma, P, S)$ such that any production rule is of $\alpha A \beta \rightarrow \alpha \gamma \beta$ for $A \in V$ and $\alpha, \beta, \gamma \in (V \cup \Sigma)^*$.

These grammars are almost[2] equivalent to the monotone grammars: each production rule $\alpha \rightarrow \beta$ satisfies $|\alpha| \leq |\beta|$. The semantics of the derivation by a CSG is analogously defined.

The *size* of a grammar, $|G|$, is the total length of all production rules. Without loss of generality, we can regard the size of a CFG G as $|V|$ in G, since there is an equivalent CFG G' in Chomsky normal form such that $|G'| \leq 2 \cdot |G|$.

3   CSG Transform and Grammar Size Analysis

In this section, we give the definition of a very restricted class of CSGs and show that the class is powerful enough to handle the *CSG transform* defined as follows.

---

[2] The difference is only that monotone grammars never derive ε.

We assume that any CFG G is restricted to be an *admissible grammar*: G derives exactly one string $w \in \Sigma^+$. This notion leads us to the CFG transform: G is an encoding of w and $S \overset{*}{\Rightarrow} w$ is the decoding to w. The notion of *CSG transform* is directly obtained by the same condition, i.e. any CSG derives exactly one string. We then assume that any CSG is also an admissible grammar.

3.1 Σ-Sensitive Grammars

For our grammar transform problem, we introduce a monotone CSGs each of which production rules is either of the forms:

$$aA \to \gamma \quad \text{and} \quad A \to \gamma,$$

where $a \in \Sigma$, $A \in V$, and $\gamma \in (V \cup \Sigma)^+$. Such a grammar is said to be Σ-*sensitive*. This notion is naturally extended to the Σ^n-sensitive grammars, where a production rule $uA \to \gamma$ for $u \in \Sigma^n$ is allowed. Moreover, the production $uA \to \gamma$ can be reduced to a short expression $XA \to \gamma$, where X is a variable in a CFG which derives u. The language class of Σ-sensitive grammars properly includes that of CFGs, e.g. the language $\{a^n b^n c^n \mid n \geq 1\}$ is derived by a Σ-sensitive grammar.

Next we mention a normal form of Σ-sensitive grammar. Any production rule $aA \to \gamma$ with $\gamma = A_1 \cdots A_k$ can be simulated by $B_i \to A_{i+1} B_{i+1}$ $(1 \leq i \leq k-3)$ and $B_{k-2} \to A_{k-1} A_k$. Thus, without loss of generality, we can assume that the length of right hand of any production rule is bounded by two.

3.2 Upper and Lower Bounds on Grammar Size

Let G_s be a minimum Σ-sensitive grammar for $w \in \Sigma^+$ and $|\Sigma| = k$, and let G_f be a minimum CFG equivalent to G_s.

Theorem 1. $\dfrac{|G_f|}{|G_s|} = O(k\ell)$, *where ℓ is the height of the derivation tree of G_s.*

Proof. Let T_s be the derivation tree of G_s and let n be a node of T_s whose right and left child are n_1 and n_2, respectively. Let T_f be the initial skeleton tree which is obtained by removing all labels from T_s. We can construct a derivation tree T_f of G_f from T_s by the following labeling function h.

1. $h(n) = a$ on T_f if n is a leaf on T_s whose label is $a \in \Sigma$.
2. $h(n) = A_{h(n_1),h(n_2)}$ on T_f if the production rule for n on T_s is $A \to \gamma$.
3. $h(n) = A_{a,h(n_1),h(n_2)}$ on T_f if the production rule for n on T_s is $aA \to \gamma$.

For the labeled trees T_s, T_f and any nodes n, m, we can easily show that $h(n) = h(m)$ on T_f iff $T_s(n) = T_s(m)$ on T_s, where $T(n)$ denotes the subtree of T rooted by n. Thus we obtain the derivation tree T_f of a CFG G_f equivalent to G_s. Finally we estimate the size of G_f. If the labels of T_s are different each other, then clearly, $|G_s| = |G_f|$. Let n, m be internal nodes on T_s such that $T_s(n), T_s(m)$ are maximal subtrees deriving a same string. Then the difference of labeling between $T_f(n)$ and $T_f(m)$ does not occur except in the leftmost path since two production rules $aA \to \alpha$ and $bA \to \beta$ are identical if $a = b \in \Sigma$. Thus, we obtain the upper bound $|G_f| \le k\ell|G_s|$. □

Theorem 2. $\dfrac{|G_f|}{|G_s|} = \Omega\left(\dfrac{k\log\frac{n}{k}}{k + \log\frac{n}{k}}\right)$, where n is the length of input string.

Proof. For $m \ge 1$ and $\Sigma = \{a_i, b_j \mid 0 \le i \le k, 0 \le j < k\}$, let us consider the following string.

$$w = (\overbrace{a_0 \cdots a_0}^{m} b_0 b_0) \cdot (\overbrace{a_1 \cdots a_1}^{m} b_1 b_1) \cdots (\overbrace{a_{k-1} \cdots a_{k-1}}^{m} b_{k-1} b_{k-1}) \cdot a_k a_k$$

For this w we can construct the Σ-sensitive grammar defined by

$$3k \text{ production rules} \begin{cases} a_i A \to a_i a_i \\ b_i A \to a_{i+1} a_{i+1} \ (0 \le i < k) \\ a_i B \to b_i b_i b_i \end{cases}$$

and $(\lceil \log m \rceil + \lceil \log k \rceil + 1)$ production rules for the derivations $X \overset{*}{\Rightarrow} A^m B$ and $S \overset{*}{\Rightarrow} a_0 X^k A$. On the other hand, it is clear that G_f must contain at least $k\lceil \log m \rceil$ production rules for deriving w. Thus, we obtain the lower bound by the relation $n = km + 2k + 2$. □

Finally we consider the CSG transform on constant alphabets. For this purpose, we begin with the notion of *LZ-factorization*. The factors $f_1, f_2, \ldots, f_k$ is called the LZ-factorization of a string w if $w = f_1 f_2 \cdots f_k$, $f_1 = w[1]$, and f_i is the longest prefix of $f_i \cdots f_k$ appearing in $f_1 \cdots f_{i-1}$. For example, if $w = ababaaba$, the LZ-factorization is a, b, ab, a, aba. By $\#LZ(w)$, we denote the number of factors of the LZ-factorization of w.

Theorem 3. *(Rytter [13]) For any string w and its admissible CFG G, it holds that $\#LZ(w) \le |G|$.*

Here we show a lower bound of the ratio CFG/CSG over a constant alphabet. For this proof, we mention the infinite *square-free string* over a three-letter alphabet.

A string is said to be *square-free* if it contains no squares α^2. For example, $abcacb$ is square-free but $ababc$ is not square-free. It is known (see e.g. [8]) that for a three-letter alphabet $\Sigma = \{a, b, c\}$, there exist infinite square-free strings, such as $abcbacbcabcbabcacbacabcacbcabacbabcabacbcacbacabcacb \cdots$. Using this string and its infinitely many prefixes, we prove the lower bound for our CSG transform for a constant alphabet.

Theorem 4. *Let $\Sigma = \{a, b, c\}$ and $w \in \Sigma^*$. Let G_s^* be a minimum Σ^n-sensitive grammar for w. Then, $\dfrac{|G_f|}{|G_s^*|} = \Omega\left(\dfrac{\sqrt[3]{n}\log n}{\sqrt[3]{n} + \log n}\right)$, where $n = |w|$.*

Proof. Let p_i be the i-th prefix of the infinite square-free string, that is, $p_0 = a, p_1 = ab, p_2 = abc, p_3 = abcb, \ldots$. For a sufficiently large m, we define the following string w.

$$w = \prod_{i=0}^{\frac{m}{2}} \left(p_{2i}^m p_{2i+1}^2 \right) \cdot p_{m+2}^2$$

$$= (\overbrace{p_0 \cdots p_0}^{m} p_1 p_1) \cdot (\overbrace{p_2 \cdots p_2}^{m} p_3 p_3) \cdots (\overbrace{p_m \cdots p_m}^{m} p_{m+1} p_{m+1}) \cdot p_{m+2} p_{m+2}$$

We first analyze the size of G_f. We consider the LZ-factorization of w. If p_i^m contains a period shorter than i, it must be of $(\alpha\beta)^k \alpha$ for some $k \geq 2$, which is not square-free. Thus, p_i^k is not appearing in p_j^m for any $i < j$ and $k \geq 2$. Hence, the LZ-factorization for p_i^m contains $\Omega(\log m)$ factors, that is, $\#LZ(w) = \Omega(m \log m) \leq |G_f|$.

On the other hand, we can construct a CFG deriving all $p_1, p_2 \ldots, p_{m+2}$ by m production rules defined by the variables $P_1, P_2 \ldots, P_{m+2}$. Then, this grammar encodes the string to the following string $\bar{w}$.

$$\bar{w} = (\overbrace{p_0 \cdots p_0}^{m} P_1 P_1) \cdot (\overbrace{P_2 \cdots P_2}^{m} P_3 P_3) \cdots (\overbrace{P_m \cdots P_m}^{m} P_{m+1} P_{m+1}) \cdot P_{m+2} P_{m+2}$$

By the similar technique in Theorem 2, we can construct a Σ^n-sensitive grammar which derives $\bar{w}$ within $O(\log m^2) = O(\log m)$ production rules. Thus, $|G_s^*| = O(m + \log m)$.

Therefore, since $|p_m| = m$ and $|w| = n = \Theta(m^3)$, we can obtain the lower bound $\frac{|G_f|}{|G_s^*|} = \Omega \left(\frac{m \log m}{m + \log m} \right) = \Omega \left(\frac{\sqrt[3]{n} \log n}{\sqrt[3]{n} + \log n} \right).$ □

From the analysis in this section, we can conclude that our CSG transform is powerful compared with the standard CFG transform.

4   Greedy Compression Algorithm

Re-Pair [7] is one greedy CFG transform algorithm based on the most-frequent-first strategy. It replaces every occurrence of a most frequent digram AB in the input string by a new variable symbol X and generates the production rule $X \rightarrow AB$. This process is repeated until no digram appear more than once.

We extend this algorithm to the CSG transform. Let $\Sigma = \{a_1, \ldots, a_k\}$ with $|\Sigma| = k$. The key idea is to select a digram $A_i B_i$ that occurs most frequently just after a_i for every $i = 1, \ldots, k$, and generate the following production rules.

$$a_1 X \rightarrow a_1 A_1 B_1, \quad a_2 X \rightarrow a_2 A_2 B_2, \quad \ldots, \quad a_k X \rightarrow a_k A_k B_k,$$

where X is a new variable symbol and A_i, B_i are either symbols in Σ or variable symbols introduced so far. Every occurrence of $A_i B_i$ preceded by a_i in the input string is replaced by one symbol X independently of i. The preceding symbol a_i of an occurrence of digram $A_i B_i$ is called its *context*. We remark that rewriting the input string yields occurrences of digrams preceded by a variable symbol (not a symbol in Σ), but their contexts remain unchanged and can be kept.

Although a straightforward extension of the algorithm of [7] requires $\Omega(|\Sigma||w|)$ time and space, our algorithm runs in $O(|w|)$ time and space. The key technique is a data structure which returns in constant time one of the most frequent digram for each context. Let $c \in \Sigma$, and let $L_c(f)$ be the list of active digrams with context c having frequency f. We use a specialized priority queue which stores $\mathcal{L}_c = L_c(f_1), \ldots, L_c(f_k)$, where $f_1, \ldots, f_k$ are the positive integers (priorities) in the increasing order such that $L_c(f_i)$ is not empty for all $i = 1, \ldots, k$. We maintain the priority queues $\mathcal{L}_c$ for all contexts c. An update of the priority queues takes constant time per a replacement operation. The total time and space complexity is thus $O(|w|)$.

5   Compressed Pattern Matching

One goal of the CPM problem (Goal 1) is a faster search in a compressed text, compared with decompression followed by an ordinary search [1]. A more ambitious goal (Goal 2) is a faster search in a compressed text in comparison with an ordinary search in the original text [9]. The aim of compression in the context of Goal 2 is not only to reduce disk storage requirement or data transmission cost but also to speed up string searching. In this section, we consider the CPM problem for restricted Σ-sensitive grammars and show a CPM algorithm based on [5]. We then discuss a Goal 2-oriented implementation of it.

Definition 1. COMPPATMATCH
Input: A pattern $\pi \in \Sigma^+$ and a Σ-sensitive grammar $G = (V, \Sigma, P, S)$ generating a string $w \in \Sigma^+$ such that every production rule in P takes either of the forms: $aA \to a\gamma$ and $A \to \gamma$ ($a \in \Sigma, A \in V, \gamma \in (V \cup \Sigma)^+$).
Output: All occurrences of π within w.

The production rule with lhs S is called the *start rule*, and the set of other rules in P is denoted by $P^\#$. Let $S \to b\mu$ be the start rule of G with $b \in \Sigma$ and $\mu \in (V \cup \Sigma)^*$. Similar to the collage system, we regard $P^\#$ as a dictionary and $b\mu$ as a variable sequence although the rules of $P^\#$ are not context-free. Denote by $\|P^\#\|$ the total length of the rhs's of rules in $P^\#$. Let $V^\# = V - \{S\}$.

For $a \in \Sigma$ and $X \in V$, let $\xi(a, X)$ denote the string u in Σ^+ such that $aX \overset{*}{\Rightarrow} au$. If no such a string u exists, $\xi(a, X)$ is undefined. For $X = c \in \Sigma$, let $\xi(a, X) = c$. Let $\lambda(a, X)$ be the rightmost symbol of $\xi(a, X)$.

Lemma 1. *The function* $\lambda : \Sigma \times (V^\# \cup \Sigma) \to \Sigma$ *can be constructed in* $O(\|P^\#\|)$ *time so that it responds in constant time.*

5.1   Application of Algorithm by Kida et al.

The input Σ-sensitive grammar $G = (V, \Sigma, P, S)$ is equivalent to the CFG $G' = (V_f \cup V_s \cup \{S\}, \Sigma, P_f \cup P_s \cup \{S \to \psi(b, \mu)\}, S)$, where

$$V_f = \{A \mid A \in V^\# \text{ and } A \to \gamma \in P \text{ for some } \gamma\},$$
$$V_s = \{A_a \mid A \in V, a \in \Sigma, \text{ and } aA \to a\gamma \in P \text{ for some } \gamma\},$$
$$P_f = \{A \to \gamma \mid A \in V_f \text{ and } A \to \gamma \in P\},$$
$$P_s = \{A_a \to \psi(a, \gamma) \mid A_a \in V_s \text{ and } aA \to a\gamma \in P\},$$

where $\psi(a, \gamma)$ denotes the string over $(V_f \cup V_s \cup \Sigma)$ obtained from γ by replacing every occurrence of $A \in V - V_f$ such that $\gamma = \alpha A \beta$ and $\alpha, \beta \in (V \cup \Sigma)^*$ by $A_c \in V_s$ such that $c = \lambda(a, Y)$ where Y is the rightmost symbol of $a\alpha$. Conversion of G into G' takes $O(\|P\|)$ time by using the function λ.

A naive solution to COMPPATMATCH would be to convert G into G' and apply the algorithm of Kida et al. [5]. The algorithm first preprocesses π and the rules in $P_f \cup P_s$ to build a finite-state machine M and then makes M run over the symbols of $\psi(b, \mu)$. The machine M consists of state-transition and output functions defined on the domain $Q \times (V_f \cup V_s \cup \Sigma)$, where Q is the set of states of the KMP automaton for π. It can be implemented in $O(|\pi|^2 + \|P_f \cup P_s\|) = O(|\pi|^2 + \|P^\#\|)$ time and space and runs in $O(|\mu| + occ)$ time over $\psi(b, \mu)$.

Theorem 5. COMPPATMATCH *can be solved in* $O(|\pi|^2 + \|P\| + occ)$ *time using* $O(|\pi|^2 + \|P^\#\|)$ *space.*

5.2 Practically-Fast Implementation

For practical speed-up, we want to implement the state-transition function as a two-dimensional array of size $|Q| \times |V_f \cup V_s \cup \Sigma|$ as in [16]. However, this is unrealistic for a large V since $|V_f \cup V_s|$ can be $|V^\#| \cdot |\Sigma|$. In what follows, we describe how to reduce the domain size.

Consider the set $Q = \{0, 1, \ldots, |\pi|\}$ of states of the KMP automaton for a pattern π, where j corresponds to the j-length prefix of π. The idea is to modify the KMP automaton by adding $|\Sigma|$ distinct states so that it memorizes the symbol read previously. Let $Q_\Sigma = \{q_a \mid a \in \Sigma\}$ be the set of these states. The state-transition function $\delta' : (Q \cup Q_\Sigma) \times \Sigma \rightarrow Q \cup Q_\Sigma$ of the modified KMP automaton is defined as follows.

$$\delta'(q, a) = \begin{cases} \delta(0, a), & \text{if } q = q_c \in Q_\Sigma \text{ for some } c; \\ q_a, & \text{if } q \in Q \wedge \delta(q, a) = 0; \\ \delta(q, a), & \text{otherwise}, \end{cases}$$

where δ is the state-transition function of the original KMP automaton. The function δ' is computed from the modified version of the goto and the failure functions. The modified goto function differs from the original one in that the arrows from the auxiliary state $\perp$ to q_a and the arrows from q_a to state 1 are added for all $a \in \Sigma$. The inductive computation of the modified failure function is performed in exactly the same way as the original one.

An example of the modified KMP automaton is shown in Fig. 1, together with the original one.

Based on the modified KMP automaton, we define functions *Jump* and *Output* on the domain $(Q \cup Q_\Sigma) \times (V^\# \cup \Sigma)$ by

$$Jump(q, X) = \delta'(q, \xi(a, X)),$$

$$Output(q, X) = \left\{ |\xi(a, X)| - |w| \;\middle|\; \begin{array}{l} w \text{ is a non-empty prefixes of } \xi(a, X) \\ \text{such that } \delta'(q, w) \text{ is the final state.} \end{array} \right\}$$

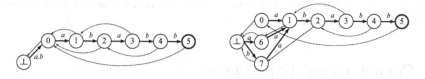

Fig. 1. KMP automaton for $\pi = ababb$ is displayed on the left, and the modified one is displayed on the right, where $\Sigma = \{a, b\}$ and the solid and the broken arrows represent the goto and the failure functions. We note that the values of the failure function for states 1 and 2 differ between the two automata.

$Jump(q, X)$

	a	b	A	B
0	1	7	—	—
1 (a)	1	2	2	7
2 (b)	3	7	—	4
3 (a)	1	4	2	5
4 (b)	3	5	—	4
5 (b)	1	7	—	2
6 (a)	1	7	2	7
7 (b)	1	7	—	2

$Output(q, X)$

	a	b	A	B
0	∅	∅	—	—
1 (a)	∅	∅	∅	∅
2 (b)	∅	∅	—	∅
3 (a)	∅	∅	{2}	{0}
4 (b)	∅	{0}	—	∅
5 (b)	∅	∅	—	∅
6 (a)	∅	∅	∅	∅
7 (b)	∅	∅	—	∅

Fig. 2. *Jump* and *Output* functions built from the modified KMP automaton of Fig. 1 for production rules $aA \to aBB, aB \to abb, bB \to bab$. Each parenthesized symbol following state s means the symbol read immediately before reaching s. Variable A represents $bbab$ in context a, while it represents nothing in context b. Also variable B means bb in context a, while it means ab in context b.

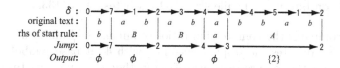

δ : 0→7→1→2→3→4→3→4→5→1→2

original text : b a b a b a b b a b

rhs of start rule: b | B | B | a | A |

Jump: 0→7——————→2————→4→3————————→2

Output: ϕ ϕ ϕ ϕ {2}

Fig. 3. Move of the machine of Fig. 2 over the rhs of $S \to bBBaA$.

where $q \in (Q - \{0\}) \cup Q_\Sigma$, $X \in V^\# \cup \Sigma$, and $a \in \Sigma$ is the context memorized by state q. For $q = 0$, $Jump(q, X)$ and $Output(q, X)$ are defined only for $X \in \Sigma$.

Fig. 2 shows the functions *Jump* and *Output* built from the modified KMP automaton of Fig. 1 for the production rules $aA \to aBB, aB \to abb, bB \to bab$. Fig. 3 shows the move of the machine over the rhs of $S \to bBBaA$.

We note that the domain $(Q \cup Q_\Sigma) \times V^\#$ is much smaller than the domain $Q \times (V_f \cup V_s \cup \Sigma)$. This is a big advantage in the sense that we can adopt the standard two-dimensional array implementation of *Jump*.

Theorem 6. *We can build in $O(|\pi| \cdot |V| + \|P^\#\|)$ time a two-dimensional table storing the values of Jump and a data structure for Output which responds in time linear in the answer size.*

If $|V^\# \cup \Sigma| \leq 256$, we can encode symbols in $V^\# \cup \Sigma$ in one byte. Compared to the CPM algorithm on BPE presented in [16], the number of production rules can be $|\Sigma|$ times larger whereas the table size of *Jump* is larger just by

$256 \cdot |\Sigma|$ table entries. Thus, both the compression and the search performances are expected to be improved drastically.

6   Computational Experiments

We implemented in C language the compression algorithm presented in Section 4 where the grammar symbols are bounded by 256 and encoded in one byte, and the CPM algorithm presented in Section 5.2. We evaluated their performances by a series of computational experiments. All the experiments were carried out on a SUN Ultra 20 M2 Workstation with a 2.2GHz Dual Core AMD Opteron 1214 and 2.0 GB RAM running Solaris 10. The text files we used are as follows.

Medline. A clinically-oriented subset of Medline, consisting of 348,566 references. The file size is 60.3 Mbytes. $|\Sigma| = 87$.

Genbank. The file consisting of accession numbers and nucleotide sequences taken from a data set in Genbank. The file size is 17.1 Mbytes. $|\Sigma| = 59$.

DBLP. A set of DBLP XML records. The file size is 130.7 Mbytes. $|\Sigma| = 96$.

Sources. The concatenation of all the .c, .h, .C, .java files of the linux-2.6.11.6 and gcc-4.0.0 distributions. The file size is 52.4 Mbytes. $|\Sigma| = 227$.

Pitches. A sequence of pitch values obtained from a myriad of MIDI files freely available on Internet. The file size is 52.4 Mbytes. $|\Sigma| = 133$.

Table 1 compares the compression ratios of our method and other compressors, where SE denotes the stopper encoding with 4-bit base symbols. Despite using byte codes, the compression ratio of our method is competitive to or slightly worse than the standard compressors for Medline, Genbank and DBLP. It is also much better than the other Goal 2-oriented compressors. On the other hand, the performance of our method is poor for Sources and Pitches. For Pitches, the performance of Re-Pair is also poor. Although Pitches is a mixture of pitch data with different nature, our method as well as Re-Pair depends on the substring statistics over the whole data and therefore shows poor performance. In fact, the performance of our method was improved by partitioning the file into fragments and then compressing them separately. The poor performance for Sources is mainly due to the large alphabet size ($|\Sigma| = 227$). We note that the number of production rules generated is upper-bounded by $|\Sigma|(256 - |\Sigma|)$, and the bounds for Medline, Genbank, DBLP, Sources and Pitches are, respectively, 14703, 11623, 15360, 6583 and 16359. Thus our compression scheme is suited when $|\Sigma|$ is closed to 128.

We compared the search time of our method with the KMP algorithm (KMP) and the BMH algorithm (BMH) over uncompressed text as well as existing Goal 2-oriented CPM methods: the KMP algorithm over BPE compressed text (KMP on BPE) [17] and the BMH algorithm over SE compressed text (BMH on SE) [12]. Fig. 4 displays the search times (including the preprocessing times) for Medline and DBLP. Our method runs faster than BMH on SE for short patterns.

Table 1. Compression ratio comparison (%)

	standard compressors			Goal 2-oriented compressors		
	gzip	bzip2	Re-Pair	SE [12]	BPE	ours
Medline	33.29	23.57	33.83	66.50	56.41	32.94
Genbank	21.98	22.17	31.32	51.74	31.37	28.22
DBLP	17.48	11.66	17.67	70.05	40.83	20.24
Sources	23.29	19.79	31.07	71.93	80.54	55.56
Pitches	30.27	35.73	58.23	74.77	78.34	63.36

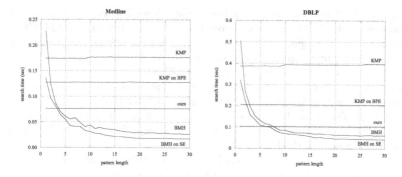

Fig. 4. Search-time comparison

7   Conclusion

We proposed a special case of CSGs called Σ-sensitive grammars for effective grammar transform and fast compressed pattern matching. While the Σ-sensitiveness is strong restriction, we show that this grammars is powerful enough to represent a compact formal model. Using a small subclass of this class, we obtained a sufficient compression ratio competitive with other practical models. Moreover we implemented the CPM algorithm on the compressed texts and confirmed its performance. In particular, compared to the BMH algorithm and the stopper encoding, regarded as one of the best combinations that allows a practically fast search, our method achieves much better compression and a faster search for short patterns.

References

1. Amir, A., Benson, G.: Efficient two-dimensional compressed matching. In: Proc. Data Compression Conference 1992 (DCC 1992), p. 279 (1992)
2. Amir, A., Benson, G., Farach, M.: Let sleeping files lie: Pattern matching in Z-compressed files. J. Comput. Syst. Sci. 52(2), 299–307 (1996)
3. Charikar, M., Lehman, E., Liu, D., Panigrahy, R., Prabhakaran, M., Sahai, A., Shelat, A.: The smallest grammar problem. IEEE Transactions on Information Theory 51(7), 2554–2576 (2005)

4. Crochemore, M., Hancart, C., Lecroq, T.: Algorithms on Strings. Cambridge University Press, Cambridge (2007)
5. Kida, T., Matsumoto, T., Shibata, Y., Takeda, M., Shinohara, A., Arikawa, S.: Collage systems: a unifying framework for compressed pattern matching. Theoret. Comput. Sci. 298(1), 253–272 (2003)
6. Knuth, D.E.: The Art of Computer Programming. Seminumerical Algorithms, vol. II. Addison-Wesley, Reading (1981)
7. Larsson, N.J., Moffat, A.: Off-line dictionary-based compression. Proceedings of the IEEE 88(11), 1722–1732 (2000)
8. Lothaire, M.: Combinatorics on Words. Cambridge University Press, Cambridge (1983)
9. Manber, U.: A text compression scheme that allows fast searching directly in the compressed file. ACM Transactions on Information Systems 15(2), 124–136 (1997)
10. Matsumoto, T., Hagio, K., Takeda, M.: A run-time efficient implementation of compressed pattern matching automata. In: Ibarra, O., Ravikumar, B. (eds.) CIAA 2008. LNCS, vol. 5148. Springer, Heidelberg (2008)
11. Navarro, G., Raffinot, M.: Practical and flexible pattern matching over Ziv-Lempel compressed text. J. Discrete Algorithms 2(3), 347–371 (2004)
12. Rautio, J., Tanninen, J., Tarhio, J.: String matching with stopper encoding and code splitting. In: Apostolico, A., Takeda, M. (eds.) CPM 2002. LNCS, vol. 2373, pp. 42–52. Springer, Heidelberg (2002)
13. Rytter, W.: Application of lempel-ziv factorization to the approximation of grammar-based compression. Theor. Comput. Sci. 302(1-3), 211–222 (2003)
14. Sakamoto, H.: A fully linear-time approximation algorithm for grammar-based compression. J. Discrete Algorithms 3(2-4), 416–430 (2005)
15. Sakamoto, H., Kida, T., Shimozono, S.: A space-saving linear-time algorithm for grammar-based compression. In: Apostolico, A., Melucci, M. (eds.) SPIRE 2004. LNCS, vol. 3246, pp. 218–229. Springer, Heidelberg (2004)
16. Shibata, Y., Kida, T., Fukamachi, S., Takeda, M., Shinohara, A., Shinohara, T., Arikawa, S.: Speeding up pattern matching by text compression. In: Bongiovanni, G., Petreschi, R., Gambosi, G. (eds.) CIAC 2000. LNCS, vol. 1767, pp. 306–315. Springer, Heidelberg (2000)
17. Shibata, Y., Matsumoto, T., Takeda, M., Shinohara, A., Arikawa, S.: A Boyer-Moore type algorithm for compressed pattern matching. In: Giancarlo, R., Sankoff, D. (eds.) CPM 2000. LNCS, vol. 1848, pp. 181–194. Springer, Heidelberg (2000)
18. Yang, E.-H., He, D.-K.: Efficient universal lossless data compression algorithms based on a greedy sequential grammar transform - part two: With context models. IEEE Transactions on Information Theory 49(11), 2874–2894 (2003)

Improved Variable-to-Fixed Length Codes

Shmuel T. Klein[1] and Dana Shapira[2]

[1] Dept. of Computer Science, Bar Ilan University, Ramat Gan 52900, Israel
[2] Dept. of Computer Science, Ashkelon Academic College, Ashkelon 78211, Israel

Abstract. Though many compression methods are based on the use of variable length codes, there has recently been a trend to search for alternatives in which the lengths of the codewords are more restricted, which can be useful for fast decoding and compressed searches. This paper explores the construction of variable-to-fixed length codes, which have been suggested long ago by Tunstall. Using a new heuristic based on suffix trees, the performance of Tunstall codes could be improved by more than 30%.

1 Introduction and Background

Huffman's classical algorithm [8] designs an optimal variable length code for a given distribution of frequencies of elements to be encoded. These elements can be simple characters, in which case this is a fixed-to-variable length code, but better compression can be obtained if the text at hand can be parsed into a sequence of variable length strings, the set of which is then encoded according to the probabilities of the occurrences of its elements, yielding a variable-to-variable length encoding.

If one considers, however, also other aspects of variable length codes, not just their compression ratio, there might be incentives to revert back to fixed length codes. Decoding, for instance, is more complicated with variable length, as the end of each codeword must be determined. Variable length codewords carry usually also a processing time penalty, especially for decoding and also for other desirable features, like the possibility to search directly within the compressed text, without the need to decompress first.

This lead to the development of several compromises. In a first step, the optimal binary Huffman codes were replaced by a 256-ary variant [12], in which the lengths of all the codewords are integral multiples of bytes. The loss in compression efficiency, which might be large for small alphabets, becomes tolerable and almost not significant as the alphabet size increases, in particular considering the trend set by the *Huffword* variant [11], of encoding entire words as basic elements rather than just individual characters. On the other hand, the byte-wise processing is much faster and easier to implement.

When searches in the compressed text should also be supported, Huffman codes suffer from a problem of synchronization: denoting by $\mathcal{E}$ the encoding function, the compressed form $\mathcal{E}(x)$ of an element x may appear in the compressed text $\mathcal{E}(T)$, without corresponding to an actual occurrence of x in the

A. Amir, A. Turpin, and A. Moffat (Eds.): SPIRE 2008, LNCS 5280, pp. 39–50, 2008.

text T, because the occurrence of $\mathcal{E}(x)$ is not necessarily aligned on codeword boundaries. To solve this problem, [12] propose to reserve the first bit of each byte as *tag*, which is used to identify the last byte of each codeword, thereby reducing the order of the Huffman tree from 256-ary to 128-ary. These *Tagged Huffman codes* have then been replaced by *End-Tagged Dense codes* (ETDC) in [3] and by (s, c)-*Dense codes* (SCDC) in [2]. An alternative code based on higher order Fibonacci numeration systems and yielding similar features is studied in [10]. The three last mentioned codes consist of fixed codewords which do not depend on the probabilities of the items to be encoded. Thus their construction is simpler than that of Huffman codes: all one has to do is to sort the items by non-increasing frequency and then assign the codewords accordingly, starting with the shortest ones.

This paper's objective is to push the idea of the compromise one step further by advocating again the use of fixed length codes. To still get reasonable compression, the elements to be encoded, rather than the codewords, will be of variable length, thus forming a variable-to-fixed length encoding. In a certain sense, this can be considered as the inverse of the basic problem solved by Huffman's algorithm. The original problem assumed that the set of elements to be encoded is given and sought for an optimal set of variable length codewords to encode those elements; the inverse problem considers the set of codewords as given, assuming furthermore that they are of fixed length, and looks for an optimal set of variable length substrings of the text which should be encoded by those codewords.

Dealing with the inverse problem can be justified by a shift in the point of view. The classical problem had as main objective to maximize the compression savings, or equivalently, using Huffman's formulation, to minimize the redundancy. The complementing approach considers as its main target a fast and easy decoding, for which a fixed length code is the best solution, but still tries to achieve good compression under these constraints. There are good reasons to view the coding problem as asymmetrical and to prefer the decoding side: in many applications, like for larger textual Information Retrieval systems, encoding is done only once while building the system, whereas decoding is repeatedly needed and directly affects the response time.

In the next section, we shall define the problem formally and also bring previous work, in particular on Tunstall codes [15], which are variable-to-fixed length codes. In Section 3 we suggest a new algorithm and bring experimental results in Section 4.

2   Variable to Fixed Length Encoding

Consider a text of length n characters $T = t_1 t_2 \cdots t_n$ to be encoded, where $t_i \in \Sigma$ and Σ is some general alphabet, for example ASCII. The text should be encoded by a fixed length code in which each codeword is of length k bits, k being the only parameter of the system. The objective is to devise a dictionary $\mathcal{D}$ of different substrings of the text, such that $|\mathcal{D}| \leq 2^k$ so that each of the elements of the dictionary can be assigned one of the k-bit codewords, and such that the text T can be parsed into a sequence of m dictionary elements, that is $T = c_1 c_2 \cdots c_m$, such that

$$m + \sum_{c_j \in \mathcal{D}} |c_j| \tag{1}$$

is minimized.

The size of the encoded text will be km, which explains why one wishes to minimize the number m of elements into which the text is parsed. The reason for adding the combined size of the elements in the dictionary $\sum_{c_j \in \mathcal{D}} |c_j|$ in (1) is to avoid a bias: usually, the size of the dictionary is negligible relative to the size of the text, so that m will be dominant in (1), but without the sum, one could define one of the dictionary elements to be the entire text itself, which would yield $m = 1$.

More specifically, we are looking for an increasing sequence of integers

$$1 \le i_1 < i_2 < \cdots < i_{m-1} < i_m = n,$$

which are the indices of the last characters in the parsed elements of the text, so that $c_1 = t_1 \cdots t_{i_1}$, and for $1 < j \le m$, $c_j = t_{i_{j-1}+1} \cdots t_{i_j}$. Denote by $\ell = |\{c_j, j = 1, \ldots, m\}|$ the number of $different$ strings c_j in the parsing, and by $d_1, \ldots, d_\ell$ the elements of $\mathcal{D}$ themselves. Each parsed substring c_j of the text, $1 \le j \le m$, is one of the elements d_i of the dictionary, $1 \le i \le \ell$, and the constraints are

$$\ell \le 2^k \qquad \text{and} \qquad m + \sum_{i=1}^{\ell} |d_i| \quad \text{is minimized.}$$

The number of possible partitions for fixed m is $\binom{n-1}{m-1}$, and if one sums over the possible values of m, we get $\sum_{m=1}^{n} \binom{n-1}{m-1} = 2^{n-1}$, so that an exhaustive search over the possible partitions is clearly not feasible for even moderately large texts. Choosing an optimal set of strings c_j might be intractable, since even if the strings are restricted to be the prefixes or suffixes of words in the text, the problem of finding the set is NP-complete [7], and other similar problems of devising a code have also been shown to be NP-complete in [5, 6, 9]. A natural approach is thus to suggest heuristical solutions and compare their efficiencies.

A well known variable-to-fixed length code has been suggested by Tunstall [15]. Assuming that the letters in the text appear independently from each other, the Tunstall code[1] is iteratively built as follows: if Σ denotes the alphabet, the dictionary $\mathcal{D}$ of elements to be encoded is initialized as $\mathcal{D} \longleftarrow \Sigma$. As long as the size of $\mathcal{D}$ does not exceed 2^k, the preset size of the desired fixed length k-bit code, one then repeatedly chooses an element $d \in \mathcal{D}$ with highest probability (where the probability of a string is defined as the product of the probabilities of its constituent characters), removes it from $\mathcal{D}$ and adds all its one letter extensions instead, that is, one performs

---

[1] Formally, a code is a set of codewords which encode some source elements, so in our case, the code is of fixed length and consists of the 2^k possible binary k-bit strings; what has to be built by Tunstall's algorithm is not the code, but rather the $dictionary$, the set of variable length strings to be encoded.

$$\mathcal{D} \longleftarrow \mathcal{D} - \{d\} \cup \left(\bigcup_{\sigma \in \Sigma} d\sigma \right).$$

The resulting set $\mathcal{D}$ is a prefix free set, where no element is the prefix of any other, implying *unique encodability*. This property may be convenient in practical applications, but is not really necessary: even if the parsing of the text into elements of $\mathcal{D}$ can be done in more than one way, there are several possible choices of parsing heuristics for actually breaking the text into pieces, for example a greedy approach, choosing at each step the longest possible match. On the other hand, removing the constraint of unique encodability enlarges the set of potential dictionaries, which might lead to better compression.

Tunstall's procedure has been extensively analyzed [1], and the assumption of independent character appearance has been extended to sources with memory [13, 14]. Our approach is a more practical one: instead of trying to model the text and choosing the dictionary based on the expected probabilities of the strings as induced by the model, we deal directly with the substrings that actually appear in the text and which can be processed by means of the text's *suffix tree*. This can be motivated by the fact that any theoretical model of the character generation process yields only an imperfect description of natural language text. For example, a Tunstall code assuming character independence may assign a codeword to the string eee according to its high associated probability, even though the string might possibly not appear at all in a real text; conversely, the probabilities of positively correlated strings like the or qu will probably be underestimated. The use of a suffix tree may restrict the strings to be chosen to such that actually appear, and possibly even frequently, in the text.

On the other hand, one might object that there is a considerable effort involved in the construction and the processing of a suffix tree. Though there are algorithms that are linear in the size of the given text, e.g. [16], the overhead relative to the simple Tunstall construction might be larger than could be justified. But for applications where encoding is done only once, as for large static IR systems, the additional time and space requirements might not be an issue. The details of the suggested heuristic are given in the following section.

3   A New Procedure for Constructing a Variable-to-Fixed Length Code

Given the text $T = t_1 \cdots t_n$, we start by forming the set $\mathcal{S}$ of all the suffixes $s_i = t_i t_{i+1} \cdots t_n \$$ of the string $T\$$, where $\$$ is a character not belonging to the original alphabet Σ. Each such string s_i is unique and may be used to identify the position i in the text T. The strings s_i are then stored in a *trie*, which is a labeled tree structure, as follows: every internal node of the trie has one or more children, and all edges are directed from a node to one of its children; the edges emanating from a node are labeled by different characters of Σ. Every node v of the trie is associated with a string $s(v)$, which is obtained by concatenating, in order, the labels on the edges forming the path from the root to node v. The *suffix tree* of $T\$$ is defined as the trie for which the set of strings associated to its leaves is the set $\mathcal{S}$ of the suffixes of $T\$$.

This basic definition may yield a structure of size $\Omega(n^2)$, which can be reduced to $O(n)$ by compaction, i.e., deleting, for every node v that has only a single outgoing edge to a node w, both the node v and this edge (v, w), appending the label of the deleted edge to the right of the label of the edge e which entered v, and directing e now to point to w. Schematically, a structure

$$x \xrightarrow{\;\alpha\;} v \xrightarrow{\;b\;} w \qquad \text{is transformed into} \qquad x \xrightarrow{\;\alpha b\;} w.$$

Thus in the compacted suffix tree, also called a *suffix trie*, edges may be labeled by strings, not just characters. Suffix trees have been used for a myriad of applications in string processing, including recently for data compression [4]. Our approach is different and will be described next.

Each node v can also be assigned a frequency $f(v)$. The frequency of a leaf node is 1, and that of an internal node is the sum of the frequencies of its children, so that all the frequencies can be evaluated in a post-order traversal of the suffix tree. As mentioned, both the construction and the assignment of labels and frequencies can be done in time linear in the size n of the text.

The problem of constructing a variable to fixed length code, once the size 2^k of the set of codewords is fixed, is to choose an appropriate subset of the nodes of the suffix tree and use the corresponding labels as elements of the dictionary. The choice of the subset will be guided by the following analogy: an element that appears with probability p is ideally encoded in $-\log_2 p$ bits. This is closely approached by an arithmetic encoder, and approximated in Huffman coding, because of the additional constraint that the length of a codeword is an integral number of bits. Looking at the relation between the probability and the corresponding codeword length, but reversing the roles of what is given (codewords of length k bits) and what we are looking for (elements to be encoded), we conclude in our case that all the strings to be chosen should have approximately probability 2^{-k}.

The frequencies $f(v)$ associated with the nodes in the suffix tree can be used to estimate the desired probabilities, but one has to be aware that several approximations are involved in the process. First, consider two strings x and y, such that a proper suffix of the first is a proper prefix of the second, in other words, there are non-empty strings x', y' and z, such that $x = x'z$ and $y = zy'$. The frequencies $f(x)$ and $f(y)$ give the number of occurrences of these strings in the text T, but not necessarily in any parsing of T into codewords. Since x and y may be overlapping, a part of the occurrences of y may not appear in the parsing. It does not even help to know the number of occurrences of the contracted superstring $x'zy'$, because the parsing is not necessarily forced to start the encoding of this string at its beginning: x' itself may have a proper prefix w, such that $x' = wx''$, which could be encoded as part of a preceding codeword, for example if $wx''zy'$ is preceded in the text by h and hw happens to be a dictionary item. This example can obviously be further extended.

Second, in order to translate the desired probabilities 2^{-k} into frequencies which can be compared to those stored in the suffix tree, one needs to multiply them with the total number of elements in the partition of T, which has been denoted in the introduction by m. However, this gives rise to a chicken and egg problem: one needs knowledge of m to evaluate the frequencies, with the help of which an appropriate subset of nodes can be selected; the corresponding strings then form the dictionary and induce a partition of the text T into m' occurrences of the dictionary terms. There is no guarantee that $m = m'$.

We still shall base our heuristic on the frequencies $f(v)$, but do not claim that these values reflect the actual number of occurrences. They can, nevertheless, serve as some rough estimate if one assumes that the overlaps mentioned above, which will bias the counts, are spread evenly over all the processed strings. To describe the heuristic, we need some definitions.

Definition 1. Given a compacted suffix tree, we define a *cut* C of the tree as an imaginary line crossing all the paths from the root to each of the leaves at exactly one edge.

Definition 2. The *lower border* of a cut, $LB(C)$, is the set of nodes of the suffix tree just below the cut, where we refer to the convention of drawing (suffix) trees with the root on the top and with edges leading from a parent to a child node pointing downwards.

Definition 3. The *upper border* of a cut, $UB(C)$, is the set of nodes of the suffix tree which are parent nodes of at least one node of the lower border.

Figure 1 is a schematic representation of a small suffix tree visualizing these definitions. The cut is the broken line traversing the full breadth of the tree. Nodes above the cut are drawn as circles and those below the cut as squares. The nodes of the borders are filled with color (black squares for the lower border

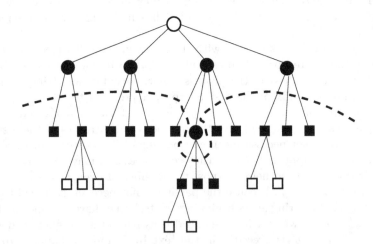

Fig. 1. Schematic representation of a cut in a suffix tree

and black circles for the upper one), and those not belonging to any of the borders are only outlined.

The following properties will be useful below.

Theorem 1. Given a compacted suffix tree with n leaves, and any cut C of the tree, we have

$$\sum_{v \in LB(C)} f(v) = n,$$

that is, the sum of the frequencies of the nodes of the lower border of all possible cuts is constant, and equal to the size of the underlying text n.

Proof: By induction on $|UB(C)|$, the number of nodes in the upper border of C. If the upper border includes only one node, it must be the root. The children of the root are labeled by the single characters[2], and the sum of their frequencies is clearly n.

Assume the truth of the claim for every cut C such that $|UB(C)| = i - 1$, and consider a cut C_0 for which there are i nodes in its upper border. Let v be one of these i nodes, choosing v as one of the nodes of maximal depth in $UB(C_0)$, and denote by $w_1, \ldots, w_r$ the children of v, and by p its parent node. Because of the maximal depth, we have that $w_j \in LB(C_0)$ for all j, and $f(v) = \sum_{j=1}^{r} f(w_j)$. If the cut C_0 is moved upwards so as to cross the path from the root to w_j not on the edges (v, w_j), but on the edge (p, v), yielding a new cut C_1, this removes the nodes w_j from the lower border, but adds their parent v to it. We thus have that

$$\sum_{x \in LB(C_1)} f(x) = \sum_{x \in LB(C_0)} f(x) - \sum_{j=1}^{r} f(w_j) + f(v) = \sum_{x \in LB(C_0)} f(x),$$

so that the sum of frequencies on the lower bound of the cut remained constant. But we cannot yet apply the inductive hypothesis, because the size of the upper bound of C_1 is not necessarily $i - 1$. Indeed, v has been removed from $UB(C_0)$, but its parent node p could have been added. That will happen if all the sibling nodes of v were also in the upper border of C_0.

We therefore repeat the above procedure of pushing the cut upwards until a cut C_h is reached for which $|UB(C_h)| = i - 1$. This must ultimately happen, because each iteration reduces the number of nodes above the cut, which is a superset of the upper border, by 1. Each step leaves the sum of the frequencies constant, so we conclude that

$$\sum_{x \in LB(C_0)} f(x) = \sum_{x \in LB(C_h)} f(x) = n,$$

the last equality following by the inductive assumption.   ∎

---

[2] Unless there is a character which is always followed by the same one, like q and u, but in that case, the string qu should be considered as if it were a single element of Σ.

Theorem 2. Given a compacted suffix tree with n leaves, and any cut C of the tree, the strings associated with the nodes of the lower border $LB(C)$ form a dictionary $\mathcal{D}$ which ensures unique encodability.

Proof: For the fact that at most one encoding is possible, it suffices to show that the strings form a prefix set. Assume on the contrary that there are two strings v and w in $\mathcal{D}$ such that v is a prefix of w, and denote the corresponding nodes in the tree by n_v and n_w, respectively. Then n_v is an ancestor node of n_w in the tree, and since both nodes belong to $LB(C)$, the cut C crosses the path from the root to n_w twice: once at the edge entering n_w and once at the edge entering n_v, in contradiction with the definition of a cut. Thus no string of $\mathcal{D}$ is the prefix of any other.

At least one encoding is always possible due to the *completeness* of the set, in the sense that a cut has been defined as a line crossing every path in tree. As a constructive proof, suppose that the prefix of length $i - 1$ of T has already been uniquely encoded, we show that exactly one codeword can be parsed starting at the beginning of the remaining suffix $t_i t_{i+1} \cdots$. Consider a pointer pointing to the root of the suffix tree, and use the characters $t_i t_{i+1} \cdots$ to be processed as guides to traverse the tree. For each character x read, follow the edge emanating from the current node and labeled by x. Such an edge must exist, because the tree reflects all the substrings that appear in the text. This procedure of stepping deeper into the tree at each iteration must ultimately cross the cut C, and the string associated with the first node encountered after crossing the cut, is the next element in the parsing.   ∎

Note that the strings associated with the *upper* border of a cut do not always form a prefix set, as can be seen in the example in Figure 1.

From Theorem 2 we learn that it might be a good idea to define the dictionary as the lower border of some cut, so we should look for cuts C for which $|LB(C)| = 2^k$. Our heuristic, which we call DynC for Dynamic Cut, extends the Tunstall procedure, but working on the suffix tree with actual frequencies instead of an artificial tree with estimated probabilities. DynC will traverse the tree left to right and construct the lower border of the desired cut according to the local frequencies. One of the problems mentioned earlier was that one cannot estimate the frequencies without knowing their total sum. But because of Theorem 1, we know that if we restrict ourselves to choose the elements of the lower border of a cut, the sum of all frequencies will remain constant. We can therefore look for nodes v in the tree for which $f(v)/n \simeq 2^{-k}$, that is $f(v) \simeq n2^{-k}$.

Ideally, there should be 2^k such elements, but in practice, there is a great variability in the frequencies. We therefore suggest to build the dictionary incrementally in a left to right scan of the tree, adapting the target value of the desired frequency for the current element dynamically, according to the cumulative frequencies of the elements that are already in the dictionary. More formally:

DynC: Left-to-right construction of $\mathcal{D}$

$\mathcal{D} \longleftarrow \emptyset$
target $\longleftarrow n2^{-k}$
cumul $\longleftarrow 0$
scan the suffix tree in DFS order while $|\mathcal{D}| < 2^k$
    $v \longleftarrow$ next node in scan order for which $f(v) \leq$ target
    $\mathcal{D} \longleftarrow \mathcal{D} \cup \{v\}$
    cumul $\longleftarrow$ cumul $+ f(v)$
    target $\longleftarrow \dfrac{n - \text{cumul}}{2^k - |\mathcal{D}|}$
end while

The updated target value is obtained by dividing the expected sum of the frequencies of the remaining elements to be added by their number. This allows the procedure to set the target higher than initially, in case some elements have been chosen with very low occurrence frequency.

The strict compliance with the constraints imposed by deciding that $\mathcal{D}$ should be a complete prefix set turned out to be too restrictive. In many cases, a node with quite high frequency could have several children with very low occurrence rate, and including the strings associated with these children nodes in $\mathcal{D}$ would eventually clog the dictionary with many strings that are practically useless for compression. To avoid the bias caused by the low values, a lower bound B has been imposed on $f(v)$ for the string $s(v)$ to be considered as a candidate to be included in $\mathcal{D}$. As a result, the dictionary was not complete any more, so to ensure that the text can be parsed in at least one way, $\mathcal{D}$ was initialized with the set of single characters. This, in turn, implied the loss of the prefix property, so the parsing with the help of the suffix tree needed to be supplemented with some heuristic, for example a greedy one, trying at each step to parse the longest possible dictionary element.

Decoding of the fixed length code is of course extremely simple. All one has to do is to store the strings of $\mathcal{D}$ consecutively is a string S, and refer to each element by its offset and length in the string S. These (*off,len*) pairs are stored in the dictionary table DT, which is accessed by 2-byte indices (in the case $k = 16$) forming the compressed text. Formally:

Decoding of DynC encoded text

while ($i \longleftarrow$ read next 2 bytes) succeeds
    (*off,len*) $\longleftarrow DT[i]$
    output $S[off \cdots off+len-1]$

For $k = 12$, in order to keep byte alignment, one could process the compressed text by blocks of 3 bytes, each of which decodes to two dictionary elements.

4   Experimental Results

To empirically test the suggested heuristic, we chose the following input files of different sizes and languages: the Bible (King James version) in English, and the French version of the European Union's JOC corpus, a collection of pairs of questions and answers on various topics used in the ARCADE evaluation project [17]. To get also different alphabet sizes, the Bible text was stripped of all punctuation signs, whereas the French text has not been altered.

Table 1. Comparison of compression performance between Tunstall and DynC

File	Bits	Tunstall		DynC	
		expected	actual	B	compression
English 2.96 MB 52 chars	12	0.617	0.614	50	0.616
				200	0.477
				300	0.591
	16	0.589	0.587	2	0.477
				7	0.391
				10	0.399
French 7.26 MB 131 chars	12	0.744	0.751	300	0.549
				500	0.496
				700	0.561
	16	0.691	0.689	20	0.404
				25	0.401
				30	0.414

Table 1 compares the compression efficiency of the suggested heuristic with that of the Tunstall codes, for both $k = 12$, corresponding to a small dictionary of 4096 entries, and $k = 16$, for a larger dictionary with 65536 elements. The first column lists also relevant statistics, the size of the files and the size of the alphabets, and the second column gives the parameter k. All compression figures are given as the ratio of the compressed file to the full size before compression. The column headed *expected* gives the expected compression of the Tunstall code: let $A = \sum_{v \in \mathcal{L}} p_v \ell_v$ be the average length of a Tunstall dictionary string, where $\mathcal{L}$ is the set of leaves of the Tunstall tree, p_v is the probability corresponding to leaf v, and ℓ_v is its depth in the tree. Then n/A is the expected number of codewords used for the encoding of the text, so the expected size of the compressed text is $kn/8A$, and the compression ratio is $k/8A$. As noted above, the Tunstall dictionary contains many strings that are not really used. The column headed *actual* is the results of actually parsing the text with the given dictionary, giving quite similar results.

The results for DynC are given for several bounds. Interestingly, on all our examples, compression first improves with increasing bound, reaches some optimum, and then drops again. This can be explained by the fact that increasing the bound leads to longer strings in the dictionary, but increasing it too much will imply the loss of too many useful shorter strings. Table 1 brings 3 examples of different bounds B for each file and each k. One can see that DynC reduces the file by additional 20–40% relative to Tunstall.

Table 2. Comparison of compression performance of DynC with other methods

	Tunstall	Huffman	DynC	SCDC	Fib3
English	100	89.9	66.4	43.1	39.2
French	100	83.9	58.2	34.7	31.2

A comparison of DynC with other methods can be found in Table 2, which arranges the methods by order of decreasing performance. The sizes are given as a percentage of the size of the file compressed by Tunstall, corresponding to 100%. Regular Huffman coding, based on the individual characters, is only 10–12% better than Tunstall. The values for DynC correspond to the parameters B and k that gave the best results in Table 1. Much better compression can be obtained if one does not insist on fixed length codes, as for SCDC [2], where the values are given for the best (s, c) pair, or Fib3 [10].

We conclude that if one has good reasons to trade compression efficiency for the simplicity of fixed length codes, the suggested heuristic may be a worthwhile alternative to the classical Tunstall codes.

References

1. Abrahams, J.: Code and parse trees for lossless source encoding. Comm. in Information and Systems 1(2), 113–146 (2001)
2. Brisaboa, N.R., Fariña, A., Navarro, G., Esteller, M.F. (s,c)-dense coding: an optimized compression code for natural language text databases. In: Nascimento, M.A., de Moura, E.S., Oliveira, A.L. (eds.) SPIRE 2003. LNCS, vol. 2857, pp. 122–136. Springer, Heidelberg (2003)
3. Brisaboa, N.R., Iglesias, E.L., Navarro, G., Paramá, J.R.: An efficient compression code for text databases. In: Sebastiani, F. (ed.) ECIR 2003. LNCS, vol. 2633, pp. 468–481. Springer, Heidelberg (2003)
4. Crochemore, M., Ilie, L., Smyth, W.F.: A Simple Algorithm for Computing the Lempel Ziv Factorization. In: Proc. Data Compression Conference DCC 2008, Snowbird, Utah, pp. 482–488 (2008)
5. Fraenkel, A.S., Klein, S.T.: Complexity Aspects of Guessing Prefix Codes. Algorithmica 12, 409–419 (1994)
6. Chrobak, M., Kolman, P., Sgall, J.: The greedy algorithm for the minimum common string partition problem. ACM Transactions on Algorithms 1(2), 350–366 (2005)

7. Fraenkel, A.S., Mor, M., Perl, Y.: Is text compression by prefixes and suffixes practical? Acta Informatica 20, 371–389 (1983)
8. Huffman, D.: A method for the construction of minimum redundancy codes. Proc. of the IRE 40, 1098–1101 (1952)
9. Klein, S.T.: Improving static compression schemes by alphabet extension. In: Giancarlo, R., Sankoff, D. (eds.) CPM 2000. LNCS, vol. 1848, pp. 210–221. Springer, Heidelberg (2000)
10. Klein, S.T., Kopel Ben-Nissan, M.: Using Fibonacci compression codes as alternatives to dense codes. In: Proc. Data Compression Conference DCC 2008, Snowbird, Utah, pp. 472–481 (2008)
11. Moffat, A.: Word-based text compression. Software – Practice & Experience 19, 185–198 (1989)
12. de Moura, E.S., Navarro, G., Ziviani, N., Baeza-Yates, R.: Fast and flexible word searching on compressed text. ACM Trans. on Information Systems 18, 113–139 (2000)
13. Savari, S.A., Gallager, R.G.: Generalized Tunstall codes for sources with memory. IEEE Trans. Info. Theory IT–43, 658–668 (1997)
14. Tjalkens, T.J., Willems, F.M.J.: Variable to fixed length codes for Markov sources. IEEE Trans. Info. Theory IT–33, 246–257 (1987)
15. Tunstall, B.P.: Synthesis of noiseless compression codes, Ph.D dissertation, Georgia Institute of Technology, Atlanta, GA (1967)
16. Ukkonen, E.: On-line construction of suffix trees. Algorithmica 14(3), 249–260 (1995)
17. Véronis, J., Langlais, P.: Evaluation of parallel text alignment systems: The arcade project. In: Véronis, J. (ed.) Parallel Text Processing, pp. 369–388. Kluwer Academic Publishers, Dordrecht (2000)

Term Impacts as Normalized Term Frequencies for BM25 Similarity Scoring

Vo Ngoc Anh[1], Raymond Wan[2], and Alistair Moffat[1]

[1] Department of Computer Science and Software Engineering
The University of Melbourne, Victoria 3010, Australia
[2] Bioinformatics Center, Kyoto University,
Gokasho, Uji, Kyoto 611-0011, Japan

Abstract. The BM25 similarity computation has been shown to provide effective document retrieval. In operational terms, the formulae which form the basis for BM25 employ both term frequency and document length normalization. This paper considers an alternative form of normalization using document-centric impacts, and shows that the new normalization simplifies BM25 and reduces the number of tuning parameters. Motivation is provided by a preliminary analysis of a document collection that shows that impacts are more likely to identify documents whose lengths resemble those of the relevant judgments. Experiments on TREC data demonstrate that impact-based BM25 is as good as or better than the original term frequency-based BM25 in terms of retrieval effectiveness.

1 Introduction

Given a natural language query q and a text collection D of N documents, the principal task of information retrieval systems is to rank the documents in decreasing order of their similarity to the query. Several models for document ranking have been investigated over the years, including vector space and probabilistic models. These models have been compared both experimentally in the context of the Text REtrieval Conference (see http://trec.nist.gov/ for results using these models) and theoretically (for example, see Fuhr [2001] for a probabilistic interpretation of these models). While the themes underlying these various models differ, they are quite alike in their implementations: in operational terms, they all end up with a formula to calculate the *similarity score* $S(d, q)$ between $d \in D$ and q. The majority of them also employ a *bag of terms* approach, which takes into account only the frequencies of terms in documents and queries, and ignores the order in which terms appear.

In the bag of terms approach, the raw values employed in the similarity computation include the *term frequencies* $f_{d,t}$ and $f_{q,t}$ of term t in a document d and the query q, respectively; and the *document frequency* f_t, which is the number of documents in D that contain t. Another value that is frequently used is the *document length* $|d|$, which is conventionally calculated from the values $f_{d,t}$ and f_t across the set of terms that appear in d. The formulation of $S(d, q)$ for both the vector space and probabilistic models is then a combination of the underlying statistics. Moreover, each of the raw values typically appears in a way that can be viewed as being a *normalized* version of the initial raw value. This notion is discussed further in the next section.

A. Amir, A. Turpin, and A. Moffat (Eds.): SPIRE 2008, LNCS 5280, pp. 51–62, 2008.
© Springer-Verlag Berlin Heidelberg 2008

Within the framework of vector space ranking, Anh and Moffat [2005] explored the idea of using the ranks of the term frequencies $f_{d,t}$ in each document d, instead of the $f_{d,t}$ values themselves. This led to the development of document- (and query-) centric *impacts*, which simplified the similarity computation, and enhanced both the efficiency and effectiveness of vector space ranking.

This paper explores the use of document impacts in similarity computations in which there are explicit and varied normalizations of the various components, such as the BM25 method [Robertson et al., 1994], which has been shown to be successful in achieving good retrieval effectiveness in a range of experimental environments [Spärck Jones et al., 2000]. Like the vector space model, BM25 employs both term frequency and document length normalization.

Section 2 provides necessary background. Section 3 employs a document collection and a set of queries and their relevance judgments to illustrate empirically how term frequencies differ from impacts, and speculates that applying impacts to BM25 might be effective. Section 4 puts that idea into practice, and describes further variations. The experiments reported in Section 5 then quantify those benefits.

2   Term Frequencies, Document Lengths and Normalization

The similarity scores of both the vector space and BM25 models can be operationally expressed as

$$S(d, q) = \sum_{t \in d \cap q} \text{QTF}(q, t) \cdot \text{TF}(d, t, D) \cdot \text{IDF}(t, D) \tag{1}$$

where the first factor is essentially defined through $f_{q,t}$; the second from $f_{d,t}$ and possibly f_t plus collection-wide statistics; and the third through f_t, and further use of collection-wide statistics. The appearance of d in $\text{TF}(d, t, D)$ is normally expressed in terms of document length $|d|$.

In the vector space model, various formulations of $S(d, q)$ are possible [Salton and Buckley, 1988]. In the experiments described by Zobel and Moffat [1998] one of the better performing formulations defines the components of Formula (1) via

$$\text{IDF}(t, D) = \log \left(1 + f_t^{\text{avg}} / f_t\right)$$
$$\text{QTF}(q, t) = 1 + \log f_{q,t}$$
$$|d| = \sqrt{\sum_{t \in d} \left((1 + \log f_{d,t}) \cdot \text{IDF}(t, D)\right)^2} \tag{2}$$
$$\text{TF}(d, t, D) = (1 + \log f_{d,t}) / \left((1 - s) + s \cdot |d| / |d|^{\text{avg}}\right),$$

where f_t^{avg} and $|d|^{\text{avg}}$ are the average values of f_t and $|d|$ over the collection, and s is a constant with a typical value of 0.7. This particular form is referred to in this work as the *standard vector space model*, or Std-VSM.

The Std-VSM formulation employs several normalization schemes. The first is the shift from using raw term frequencies to their logarithm in $\text{QTF}(q, t)$ and $\text{TF}(d, t, D)$ [Buckley et al., 1993]. Note that this shift reduces the relative gap between $\text{TF}(d, t, D)$ values as $f_{d,t}$ increases. The second normalization scheme is the scaling of f_t in

$\text{IDF}(t, D)$ using the logarithm and f_t^{avg}. Finally, in $\text{TF}(d, t, D)$, the pivoted document length normalization (as described by Singhal et al. [1996a,b]) is used, where the document lengths $|d|$ are adjusted using the slope constant s.

In probabilistic models, weighting methods normally rely on the estimation of various probabilities [Robertson and Spärck Jones, 1976]. And, as in the case of the vector space model, there are a number of schemes to choose from. In this work, we employ a version of BM25, referred to as Std-BM25 (that is, *standard* BM25), which defines the components of Formula (1) as

$$
\begin{aligned}
\text{QTF}(q, t) &= f_{q,t}/(k_3 + f_{q,t}) \\
|d| &= \sum_{t \in d} f_{d,t} \\
\text{TF}(d, t, D) &= f_{d,t}/\left(k_1 \cdot ((1 - b) + b \cdot |d|/|d|^{\text{avg}}) + f_{d,t}\right) \\
\text{IDF}(t, D) &= \log_e\left((N - f_t + 0.5)/(f_t + 0.5)\right)
\end{aligned}
\tag{3}
$$

where k_1, k_3 and b are parameters with the default values of 1.2, 1000, and 0.75, respectively [Robertson et al., 1998]. It can be seen that several normalization schemes are also present in this formulation.

In the *impact-based vector space model*, or Imp-VSM, the similarity score is calculated as

$$
S(d, q) = \sum_{t \in d \cap q} \omega_{q,t} \cdot \omega_{d,t} .
\tag{4}
$$

where $\omega_{q,t}$ (the impact of t in query q) and $\omega_{d,t}$ (the impact of t in document d) are integer numbers, valued between 1 and k_ω, typically a small integer like 8 [Anh and Moffat, 2005]. Compared with Formula (1), this approach normalizes each of $\text{QTF}(q, t) \cdot \text{IDF}(t, D)$ and $\text{TF}(d, t, D)$ to small integers. Impact normalization also has two key distinguishing features: first, like the impacts themselves, the set of resultant score values has a small cardinality; and second, the normalization process employs ranks rather than exact values. In the case of document impacts, the set of all term frequencies within a document is normalized with respect to each other, so that, within any single document, terms having the same frequencies are assigned the same impact (but it does not follow that terms sharing the same impact have the same frequency). Only a small number of terms are assigned high impacts, and the great majority of the terms in each document are assigned low impacts. A different process is applied to query impacts, employing the products $\text{QTF}(q, t) \cdot \text{IDF}(t, D)$ ($t \in q \cap d$) in a quantitative way (see Anh and Moffat [2005] for details).

3 Term Frequencies Versus Impacts

To explore the differences between term frequencies and rank-based impacts in representing documents, and the consequential effects on retrieval effectiveness, we made use of the WSJ collection. A component of the TREC data, WSJ consists of 173,252 articles from *The Wall Street Journal* published in 1987–1992; and is accompanied by a set of 50 topics and a corresponding set of pool-based human-generated relevance judgments. For the purpose of our experimentation, 50 queries were formed by taking the TREC <title> fields for topics 051-100. The corresponding relevance judgments (covering 6,228 relevant documents in total) then allowed retrieval effectiveness to be evaluated.

As a first step, we compared the correlation between each document's term frequency vector and the corresponding impact vector, taking only the non-zero entries. The correlation is based on document length $|d|$, defined as the total number of word tokens in d, that is, $|d| = \sum_{t \in d} f_{d,t}$. This choice is justified by our expectation that changes to the document length would affect the distributions of frequencies and impacts differently. The correlation between the vectors is calcu-

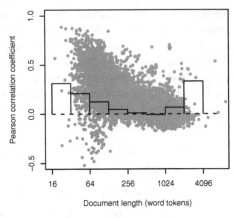

Fig. 1. Pearson correlation between two vector representations of documents (term frequencies and term impacts)

lated using Pearson coefficient, where $r = -1$, 0, and 1 indicate negative, no, and positive correlation, respectively. A randomly selected sample of 10,000 of these correlation values are plotted in Figure 1 as a function of the length of the document they relate to, together with the average (over all of the documents in the collection that fall within that binary-power range of document lengths) of the Pearson coefficients. Most of the documents in the collection contain fewer than 4,000 tokens, and only a few are longer than 6,000 tokens. Because of this, we restrict the analysis in the rest of this section to documents with a maximum length of 4,000.

It is unsurprising that, in general, impacts are positively correlated with term frequencies. However, the correlation is on average not especially strong, and there are many uncorrelated document vectors (with r close to zero); and a sizeable minority of documents that demonstrate negative correlations. The reason for the varied behavior is, of course, that impacts take into account the relative importance of each term in the document, rather than just the raw occurrence frequency. The lesson to be learned from Figure 1 is that substituting frequencies by impacts might result in a non-linear change in document vectors, and to (potentially) changes in the document ranking outcome.

The next step was to compare the pattern of documents retrieved by the vector space and BM25 models with the pattern exhibited by the relevant documents. We implemented the three retrieval schemes described earlier (Std-VSM, Std-BM25, and Imp-VSM) and compared their performance with systems built by others. Specifically, we compared the results of all three implementations with the results described by Anh and Moffat [2005] for the collections TREC12, TREC45-CR and wt10g (see Section 5 for a description of the data sets). In addition, the results from our Std-BM25 implementation were also cross-checked with the baselines established by Tao and Zhai [2007] for the collections (used in their work) AP, FR and TREC8 (the latter is also TREC45-CR, but with a smaller query set). All of the cross-checks showed our implementations to be competitive with their counterparts, and that the differences in retrieval effectiveness were marginal.

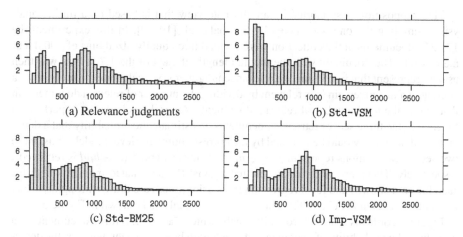

Fig. 2. Distribution of document lengths in tokens for (a) the documents that were relevant to the 50 queries and (b), (c), (d), for the retrieved results using three retrieval models. The total number of documents is 6,228 in (a) and 46,885 in the other graphs. The retrieval effectiveness as MAP for parts (b), (c) and (d) are respectively 0.2175, 0.2218, and 0.2308.

Each of the retrieval models was then applied to the 50 queries with a cut-off depth of 1,000 results per query, following the usual TREC approach. In the end, there were 46,885 items in each result set. The histograms in Figure 2 plot the length distributions (using the same definition of $|d|$ as in Figure 1) of the relevant documents, and of the three sets of retrieved documents. The vertical axis of the histograms is expressed as percentages of the corresponding set size, rather than raw frequency counts. Figure 2(a) shows that the majority of relevant documents are less than 1,000 tokens in length, with peaks at 200, 600, and 800 tokens. The Std-VSM and Std-BM25 show a noticeable bias towards short documents of around 200 tokens (Figure 2(b) and Figure 2(c)). On the other hand, in Figure 2(d), the Imp-VSM approach appears to select more documents that are around 800 tokens in length. Out of the three length distributions of retrieved documents, it is Imp-VSM that most closely resembles the relevant judgments.

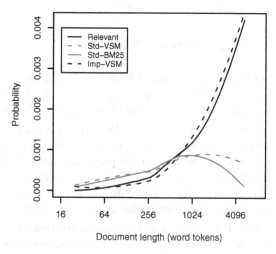

Fig. 3. Smooth plots using Loess to represent the probability of a document being relevant, or retrieved, as a function of document length

This supposition is supported by an alternate view that is based on a further analysis resembling the one undertaken by Singhal et al. [1996a]. In this experiment the 173,252 documents in the collection were divided into equally-sized bins of 100 documents each. The document with the median length of each of the 1,733 bins was then used to represent that bin. Using the relevance judgments, we calculate the probability a document d in a given bin b_i is relevant by dividing the number of relevant documents in that bin by the total number of relevant documents. A similar calculation is performed for each of the three sets of retrieved documents, to estimate the probability that a document of that length would be returned by the corresponding retrieval model. Thus, these two sets of calculations represent $P(d \in b_i | d$ is relevant$)$ and $P(d \in b_i | d$ is retrieved$)$, respectively. These probabilities are plotted against the designated median length for each bin (not shown) and local regression lines using Loess are drawn through each set of points [Cleveland et al., 1992]. The result of this analysis is shown in Figure 3.

Figure 3 confirms that the probability with which Imp-VSM returns documents is a good fit to the probability of documents of that length being relevant, whereas the deviation arising from the use of Std-VSM and Std-BM25 is pronounced for long documents. On the other hand, the number of long documents in the collection is small (a fact illustrated in Figure 1), meaning that the overall effect of the length distribution mismatch on retrieval effectiveness might not be significant.

Combining the tendencies presented in Figure 2 and Figure 3 with the fact that Imp-VSM is the "impact version" of Std-VSM, it is natural to speculate that combining impacts with the Std-BM25 approach in a reasonable way might give rise to useful improvements in effectiveness.

4   Impact-Based BM25

We have explored several BM25 variants that make use of impacts rather than term frequencies, while keeping QTF(q, t) and IDF(t, D) for the most part unchanged. The variants are summarized in Table 1 and described below.

Naive: In this variant we blindly replace the bag of term frequencies by the corresponding bag of term impacts. That is, we suppose that each document d is rewritten so that term t now has the frequency $\omega_{d,t}$ instead of $f_{d,t}$. The document length is redefined as $|d_\omega| = \sum_{t \in d} \omega_{d,t}$. This version is Naive because various normalizations have already been integrated into the impacts, and the distributions of impacts and document lengths might be quite different from those of frequency and frequency-based lengths.

Unit: This variant is based on the supposition that in the impact-based document space, all documents have the same length. With rank-based impact assignments, each document is associated with a small number of high impact terms and a much larger number of low impact terms. Moreover, the highest impact is just $k_\omega = 8$, smaller than the highest term frequency in most documents. As a result, in the impact space, document lengths are much closer to each other than in the frequency space. The Unit variant takes these observations a step further by supposing that the transformation process aims to have the ideal state when all documents have the same length, yielding a formula that no longer depends on the parameter b.

Table 1. Using ranked impacts to normalize the $\mathtt{TF}(d, t, D)$ component of BM25. The $\mathtt{Unit}$ and $\mathtt{QueryImp}$ approaches differ because the latter uses $\omega_{q,t}$ as the normalized value for $\mathtt{IDF}(t, D) \cdot \mathtt{QTF}(q, t)$, whereas the other methods retain the original BM25 computation.

Method	$\mathrm{TF}(d, t, D)$
$\mathtt{Naive}$	$\omega_{d,t} \ / \ (k_1 \cdot ((1 - b) + b \cdot |d_\omega| / |d_\omega|^{\mathrm{avg}}) + \omega_{d,t})$
$\mathtt{Unit}$	$\omega_{d,t} \ / \ (k_1 + \omega_{d,t})$
$\mathtt{Avg}$	$\omega_{d,t} \ / \ (k_1 \cdot |d_\omega|^{\mathrm{avg}} + \omega_{d,t})$
$\mathtt{UnitLog}$	$\log(1 + \omega_{d,t}) \ / \ (k_1 + \log(1 + \omega_{d,t}))$
$\mathtt{AvgLog}$	$\log(1 + \omega_{d,t}) \ / \ (k_1 \cdot |d_\omega^*|^{\mathrm{avg}} + \log(1 + \omega_{d,t}))$
$\mathtt{QueryImp}$	$\omega_{d,t} \ / \ (k_1 + \omega_{d,t})$

$\mathtt{Avg}$: This approach is also based on the supposition that all documents have the same length. However, we pull back a bit, and reason that even though there is no difference in document length, it is still appropriate to include a collection-dependent value as a normalizing constant.

$\mathtt{UnitLog}$: This variant is a modification of $\mathtt{Unit}$ in which the document impacts are adjusted by the taking of logarithms, further lowering the role of large impact values.

$\mathtt{AvgLog}$: This variant is similar to $\mathtt{Avg}$ in the way that $\mathtt{UnitLog}$ is similar to $\mathtt{Unit}$.

$\mathtt{QueryImp}$: This variant uses the same $\mathtt{TF}(d, t, D)$ as $\mathtt{Unit}$, but differs (from all of the previous four variants) in that it adopts the impact retrieval approach of Anh and Moffat [2005], by combining the $\mathtt{QTF}(q, t)$ and $\mathtt{IDF}(t, D)$ components together, and mapping that value to an integer.

Note that in all of these BM25-based methods the document and query impacts are both integers, but the computations shown in Table 1 mean that the similarity scores assigned to documents are not integers, and the fast query processing methods designed for $\mathtt{Imp-VSM}$ cannot be directly applied.

5 Experiments

This section describes the experiments conducted to check the effectiveness of the described impact-based BM25 models by comparing against the original BM25 and the impact-based VSM approaches. We employ the TREC document collections and queries that were used for ad-hoc retrieval. Since the performance of a retrieval method might vary with different data sets, we try to provide a good picture by using all data collections and query sets provided by TREC to date for this purpose. The data sets are named in the TREC conventional way, with the maximal number of judged topics provided by TREC. Thus, we have the data sets TREC12, TREC23, TREC24, TREC45, TREC45-CR, wt10g, and .GOV2. The topics for these data sets are 051-200, 201-250, 251-300, 301-350, 301-450 plus 601-700 minus 672 (topic 672 has no relevant judgement), 451-550, and 701-850, respectively. For details, see http://trec.nist.gov/.

All of the retrieval models are implemented within the same programming environment. In processing the data, a "light" stemmer (only stripping regular word endings like "s", "ies", and "ed") was applied, and no words were stopped. Queries are normally short and were formed by taking the title fields of the TREC topics. However, in the case of queries 201-250 only long description fields for these topics were supplied by TREC, and so these were employed as the queries. Three effectiveness metrics were monitored: mean average precision (MAP), mean precision at 10 documents retrieved (P@10), and mean reciprocal rank (RR).

BM25 Parameter Selection

The first experiments aimed to set values for the parameters needed – the ones used in Std-BM25 may be inappropriate for the Imp-BM25 variants. Since we focus only on the term frequency and document length factors, we fix the parameter k_3 to the same value that was used in the original BM25. Similarly, the parameter b only appears in Naive, and its usual value of 0.5 is retained. On the other hand, the relationship between k_1 and effectiveness needs to be explored.

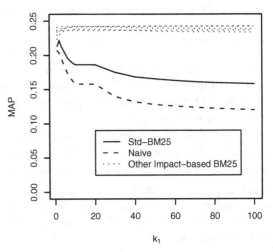

Fig. 4. Effectiveness, measured as MAP, of different BM25 models as a function of k_1

The WSJ collection and queries 051-100 were used for this training exercise. Retrieval effectiveness as a function of k_1 is plotted in Figure 4, and shows that the performance of Std-BM25 and the Naive method varies markedly with the value of k_1. On the other hand, k_1 has almost no effect on the suite of other impact-based versions, giving them an absolute advantage. Performance patterns in one data set do not necessarily indicate that the same trends will be observed in other data sets; nevertheless, stability of the lines at the top of Figure 4 give certain level of confidence that k_1 can be dropped away from the new similarity mechanisms. For definiteness, we use $k_1 = 2$ henceforth.

Impact Range

Another issue to consider is the validity of selecting $k_\omega = 8$ as the maximum impact value. Anh and Moffat [2005] conclude that this value is the best for their model, but that conclusion cannot be automatically transferred to the BM25 methods. For the validation, we used the data set wt10g with the method Unit and the values of $k_\omega = 2, 4, 8, 16, 32$ and 64. The results, not reported here, show that Unit even gives reasonable levels of effectiveness when $k_\omega = 4$, and that there is no noticeable further improvement when moving beyond $k_\omega = 8$. In the detailed experiments reported below, $k_\omega = 8$ was used.

Table 2. Effectiveness of impact-based BM25 models relative to Std-BM25 and Imp-VSM. In all cases the performance of VSM (not shown) is inferior to that of Imp-VSM. Figures in bold indicate results significantly greater than both of the baselines using a t-test at $p < 0.05$. For each effectiveness measure, the row *Mean* shows the per-query average over all collections, and the next row shows the same information, but without including TREC23.

Data set	Std-BM25	Impact-based methods (as % of Std-BM25)					
		Imp-VSM	Unit	Avg	UnitLog	AvgLog	QueryImp
Mean Average Precision							
TREC12	0.2025	103.70	105.09	105.09	102.47	**108.20**	104.54
TREC23	0.2320	92.76	79.87	96.34	70.78	83.58	84.40
TREC24	0.1435	99.16	**116.72**	100.63	112.33	**116.72**	107.18
TREC45	0.2049	95.36	104.39	94.68	105.12	105.27	104.00
TREC45-CR	0.2156	100.51	**114.61**	**104.04**	**114.05**	**115.45**	**115.12**
wt10g	0.1778	103.21	110.35	103.99	108.10	**113.39**	**111.30**
.GOV2	0.2520	101.75	**118.21**	**103.41**	**117.50**	**119.92**	**117.66**
Mean, All	0.2111	100.79	**110.28**	102.83	**110.88**	**111.01**	**110.16**
Mean, no TREC23	0.2097	101.32	**112.31**	103.26	**112.22**	**111.99**	**111.88**
Mean Precision at 10							
TREC12	0.4693	104.69	97.32	105.41	95.18	100.58	97.74
TREC23	0.4800	94.58	72.08	95.42	64.17	75.42	74.58
TREC24	0.2660	101.50	**117.29**	103.76	**115.04**	**117.29**	**110.53**
TREC45	0.3540	103.95	105.08	102.82	105.08	104.52	105.08
TREC45-CR	0.4028	101.19	**105.98**	103.20	105.09	**106.68**	**106.28**
wt10g	0.2850	102.81	103.16	105.26	102.46	104.91	102.46
.GOV2	0.5087	107.78	**115.96**	109.10	**116.10**	**115.71**	**117.55**
Mean, All	0.4136	103.07	104.40	104.50	105.19	106.69	104.85
Mean, no TREC23	0.4092	103.63	106.56	105.11	106.62	**107.61**	106.87

Evaluation

With the parameters established, Table 2 reports the effectiveness of the main Imp-BM25 variants described in Section 4. The method Naive was excluded because of its clearly inferior performance in the previous experiment.

Table 2 shows a number of interesting trends. First, the figures for TREC23 are dramatically different from all others. As TREC23 is the only case when long queries are employed, it appears that all of the impact-based models are inadequate for long queries (and it may also be that the same problem arises with the Imp-VSM approach).

Second, of the Imp-BM25 variants, AvgLog seems to be the best, especially for mean average precision – it improves MAP by around 12%, and mean P@10 and mean RR (not shown in the table) by around 8% in comparison with the original Std-BM25. This level of improvement arises in the majority of the test environments, and AvgLog can be taken to represent the Imp-BM25 model.

Third, three other Imp-BM25 variants, namely Unit, UnitLog, and QueryImp, also perform very well, with the effectiveness scores being very close to those of AvgLog. The only remaining variant – Avg – is also competitive. These results confirm our hypothesis that the inclusion of document impacts in the BM25 computation can lead to improvements in effectiveness. However, we reiterate that the conclusion is, at this stage, applicable only for short queries.

Fourth, comparison in terms of effectiveness between Unit and its query impact version QueryImp shows that the normalization of $\text{QTF}(q,t) \cdot \text{IDF}(t,D)$ to the integral term impact $\omega_{q,t}$ for Std-BM25 does not bring any benefit. In fact, it slightly reduces the effectiveness scores.

Having a similarity formula that gives good retrieval effectiveness is always encouraging. However a similarity formula that is simple is also nice. In this regard, the Unit method is notable. Despite the fact that it is very simple, it gives good retrieval effectiveness.

We conclude our experiments by returning to the histograms of Figure 2 and the collection WSJ, with AvgLog now taken to be the "Imp-BM25" method we were seeking. Figure 5 presents cumulative percentages for two of the distributions in Figure 2, with the corresponding curve for Imp-BM25 added. The Imp-BM25 document retrieval pattern is close to that of the relevant judgments, and also matches the Imp-VSM distribution.

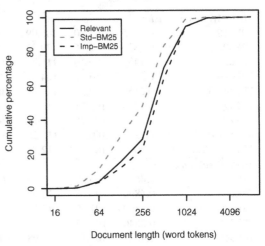

Fig. 5. Distribution of word lengths from the histograms of Figure 2(a) and Figure 2(c), represented as cumulative percentages, with Imp-BM25 added as a third line. The Imp-BM25 method has mean average precision of 0.2380.

Absolute Performance

Table 2 shows that the Imp-BM25 scheme (that is, the one listed in Table 1 as AvgLog) performs well relative to the baseline Std-BM25 scheme. It is interesting also to compare its performance with that of current language modeling approaches. To this end, we also compared Imp-BM25 against the Dirichlet-smoothed language modeling approach used by Metzler and Croft [2007], using the four document collections that were included both in their work and in our experiments. Note, however, that Metzler and Croft's need to train parameters against each collection has restricted the size of each of the test query sets. The data set description and comparison is provided in Table 3.

The results show that Imp-BM25 is competitive on the Web data (that is, the collections wt10g and .GOV2), which is pleasing given that there are no per-collection parameters required. On the other hand, the results for WSJ were disappointing, but it is not clear as how much the training factor contributes to the results of the LM in this case – where 100 queries were used for training, and only 50 for experiments.

Table 3. Effectiveness of Imp-BM25 relative to the language modeling results reported by Metzler and Croft [2007]. The choice of MAP is dictated by them. Significance testing was not possible.

Collection	Queries		MAP		
	Training (LM only)	Test	LM	Imp-BM25	Change
WSJ	051–150	151–200	0.3258	0.2976	–8.6%
TREC45-CR	301–450	601–700	0.2920	0.2814	–3.6%
wt10g	451–500	501–550	0.1861	0.1925	+3.4%
.GOV2	701–750	751–800	0.3234	0.3337	+3.2%

6 Conclusion and Discussion

This work has explored the use of document-centric impacts as defined by Anh and Moffat [2005] in combination with the BM25 similarity computation, and has shown that improved retrieval effectiveness for short queries can be attained. We based our comparison on both the original frequency-based BM25 and the vector space impact models. This conclusion is demonstrated using an extensive empirical study of most of the TREC ad-hoc data, and is cross-referenced against recent language model results.

A recent investigation by Metzler et al. [2008] showed that a probabilistic formulation of non-integral impacts using mean average precision as a guide for training achieved slightly (but significantly) better performance than the unsupervised integer-valued arrangement described by Anh and Moffat [2005], and used also in the experiments of this paper. Integer-valued impacts can thus be preferred if efficiency is a concern, or if insufficient data is available for training. (We also note that Metzler et al. give results that show that the alternative Language Modelling approach can lead to markedly improved retrieval effectiveness on all but large collections, albeit at the cost of further increased retrieval complexity.)

The impact-based BM25 models are based on simple formulations, and have a number of advantages in comparison to the original BM25. In particular, they do not require as many tuning parameters as the latter: the parameter b can be omitted entirely, and the parameter k_1 can be fixed more confidently than the original model. The new formulas also eliminate the use of individual document lengths, which speeds up the ranking process. Impacts can be stored in the indexes instead of term frequencies, and this may mean that the index can be organized in an impact-sorted manner that supports fast query processing.

The success of the Imp-BM25 methods show that it is possible and desirable to consider only frequency ranks inside each document, rather than their precise values. This is at least correct for the vector space model and the probabilistic BM25 model. It can now be suggested that, whenever a similarity measure contains term frequency and/or document length normalization, it is worth trying impacts instead.

Acknowledgments: The first and third authors were supported by the Australian Research Council, and the second by the Japan Society for the Promotion of Science. We thank the reviewers for their constructive comments.

References

Anh, V.N., Moffat, A.: Simplified similarity scoring using term ranks. In: Marchionini, G., Moffat, A., Tait, J., Baeza-Yates, R., Ziviani, N. (eds.) Proc. 28th Annual International ACM SIGIR Conference on Research and Development in Information Retrieval, August 2005, pp. 226–233. ACM Press, New York (2005)

Buckley, C., Salton, G., Allan, J.: Automatic retrieval with locality information using SMART. In: Harman, D.K. (ed.) Proceedings of the First Text REtrieval Conference (TREC-1), November 1993, pp. 59–72. National Institute of Standards and Technology (Special Publication 500-251), Gaithersburg (1993)

Clarke, C.L.A., Fuhr, N., Kando, N., Kraaij, W., de Vries, A.P. (eds.): Proc. 30th Annual International ACM SIGIR Conference on Research and Development in Information Retrieval, Amsterdam, The Netherlands, July 2007. ACM Press, New York (2007)

Cleveland, W.S., Grosse, E., Shyu, W.M.: Local regression models. In: Chambers, J.M., Hastie, T.J. (eds.) Statistical Models in S, ch. 8. Chapman & Hall/CRC, Boca Raton (1992)

Fuhr, N.: Models in information retrieval. Lectures on information retrieval, pp. 21–50. Springer, Heidelberg (2001)

Metzler, D., Croft, W.B.: Latent concept expansion using Markov random field. In: Clarke, et al. (eds.), pp. 311–318 (2007)

Metzler, D., Strohman, T., Croft, W.B.: A statistical view of binned retrieval models. In: Macdonald, C., Ounis, I., Plachouras, V., Ruthven, I., White, R.W. (eds.) ECIR 2008. LNCS, vol. 4956, pp. 175–186. Springer, Heidelberg (2008)

Robertson, S., Spärck Jones, K.: Relevance weighting of search terms. Journal of the American Society for Information Science 27(3), 129–146 (1976)

Robertson, S., Walker, S., Jones, S., Hancock-Beaulieu, M., Gatford, M.: Okapi at TREC–3. In: Harman, D. (ed.) Proc. Third Text REtrieval Conference (TREC–3), pp. 109–126. National Institute of Standards and Technology (Special Publication 500-225), Gaithersburg (1994)

Robertson, S., Walker, S., Beaulieu, M.: Okapi at TREC–7: automatic ad hoc, filtering, VLC and filtering tracks. In: Voorhees, E., Harman, D. (eds.) Proc. Seventh Text REtrieval Conference (TREC–7), November 1998, pp. 253–261. National Institute of Standards and Technology (Special Publication 500-242) (1998)

Salton, G., Buckley, C.: Term-weighting approaches in automatic text retrieval. Information Processing and Management 24(5), 513–523 (1988)

Singhal, A., Buckley, C., Mitra, M.: Pivoted document length normalization. In: Frei, H., Harman, D., Schäuble, P., Wilkinson, R. (eds.) Proc. 19th Annual International ACM SIGIR Conference on Research and Development in Information Retrieval, pp. 21–29. ACM Press, New York (1996)

Singhal, A., Salton, G., Mitra, M., Buckley, C.: Document length normalization. Information Processing and Management 32(5), 619–633 (1996)

Spärck Jones, K., Walker, S., Robertson, S.: A probabilistic model of information retrieval: Development and comparative experiments (Parts 1 and 2). Information Processing and Management 36, 779–808, 809–840 (2000)

Tao, T., Zhai, C.: An exploration of proximity measures in information retrieval. In: Clarke, et al. (eds.), pp. 295–302 (2007)

Zobel, J., Moffat, A.: Exploring the similarity space. ACM SIGIR Forum 32(1), 18–34 (1998)

The Effect of Weighted Term Frequencies on Probabilistic Latent Semantic Term Relationships

Laurence A.F. Park and Kotagiri Ramamohanarao

ARC Centre for Perceptive and Intelligent Manchines in Complex Environments,
Department of Computer Science and Software Engineering,
The University of Melbourne, Australia

Abstract. Probabilistic latent semantic analysis (PLSA) is a method of calculating term relationships within a document set using term frequencies. It is well known within the information retrieval community that raw term frequencies contain various biases that affect the precision of the retrieval system. Weighting schemes, such as BM25, have been developed in order to remove such biases and hence improve the overall quality of results from the retrieval system. We hypothesised that the biases found within raw term frequencies also affect the calculation of term relationships performed during PLSA. By using portions of the BM25 probabilistic weighting scheme, we have shown that applying weights to the raw term frequencies before performing PLSA leads to a significant increase in precision at 10 documents and average reciprocal rank. When using the BM25 weighted PLSA information in the form of a thesaurus, we achieved an average 8% increase in precision. Our thesaurus method was also compared to pseudo-relevance feedback and a co-occurrence thesaurus, both using BM25 weights. Precision results showed that the probabilistic latent semantic thesaurus using BM25 weights outperformed each method in terms of precision at 10 documents and average reciprocal rank.

Keywords: probabilistic latent semantic analysis, probabilistic model, information retrieval.

1   Introduction

For most information retrieval systems, a text document is a sequence of independent terms. Through further analysis of the document set, we are able to find clusters of terms that are related to each other; this process is considered to be the discovery of hidden topics. When given a collection of text documents, latent semantic analysis (LSA) [2] or probabilistic latent semantic analysis (PLSA) [3] are used to discover term relationships to hidden topics within the document set and hence relationships to other terms within the document set. The term relationships are calculated using the term frequencies found within a set of documents. Therefore, the term relationships are document set specific and are used to assist the increase of precision during the information retrieval process.

A. Amir, A. Turpin, and A. Moffat (Eds.): SPIRE 2008, LNCS 5280, pp. 63–74, 2008.

Latent semantic indexing, uses latent semantic analysis to construct an index based on relationships between the documents, terms and calculated topics. The process involves representing each document and term as a set of topics; when a query is provided, the documents with the most related topics to the query topics are considered the most relevant. It can be shown that this process is a mixture of term expansion using the latent semantic term relationships and document retrieval using a document-term frequency index [5, 7].

Recent experiments have shown that we are able to store probabilistic latent semantic information in a thesaurus and hence separate it from the document index [6, 8] This separation was shown to provide many benefits, including faster query times and using much less storage space when compared to a latent semantic index.

So far, probabilistic latent semantic term relationships have only been calculated using the raw frequency counts of each term in each document. It is well known that there are many biasing factors found with raw term frequency counts and there have been many research experiments performed by the information retrieval community in order to understand and remove these biases [1, 4]; the state of the art being BM25. This method is a term frequency weighting scheme that tries to remove any biases using probabilistic analysis of the document set.

We believe that the biasing factors found in raw term frequencies that disrupt the information retrieval process, also affect the term relationship calculations when using probabilistic latent semantic analysis. Therefore, we hypothesise that the term relationships obtained using PLSA will be more effective if calculated using weighted term frequencies rather than raw term frequencies. To examine this hypothesis, we will use the probabilistic latent semantic thesaurus, since it is able to isolate the term relationships and the effect the term weighting has on them.

This paper provides the following important contributions:

- An analysis of the effects of document and term weights on the PLSA term relationships through examination of retrieval results.
- A comparison of weighted PLSA to BM25 pseudo-relevence feedback and co-occurrence thesaurus term expansions methods.

In this document we will analyse the effectiveness of PLSA calculated term relationships when using the BM25 weighing scheme to weight our term frequencies. This will be compared to PLSA term relationships using raw term frequencies. The article will proceed as follows: section 2 will review the concept of PLSA and how it is used to discover hidden term relationships. Section 3 examines the bias found in term frequencies, how we can reduce their effect using BM25 and how to apply these effects to PLSA. Finally, we will examine the experiments performed and discuss the results in section 4.

2   Latent Semantic Analysis

Before we can begin our analysis of the effect of term weights on the PLSA term relationships, we must explain how the term relationships are calculated and how we can extract them from the latent semantic analysis process. In this section we will examine the latent semantic analysis concept.

2.1   Document Retrieval

The process in which the idea is transferred from the author's mind to the written article and then to the reader's mind, is a very lossy process. If we were able to model this process, then we would be able develop better methods of transferring our ideas to paper and also better methods of transferring ideas from paper to our own minds. Information retrieval systems try to model the former process in order to calculate which ideas are present in a document. Once the content of a document is known, the retrieval system can calculate better relevance judgements when given a query.

A basic document retrieval system comprises an inverted index containing the terms that are found in each document and an application to extract these values and compute document scores based on a provided query. When a query is given, the lists of documents associated to each query term are extracted from the index and combined using a document score function such as:

$$s(d, Q) = \sum_{t \in Q} w_{d,t} w_t w_{q,t} \tag{1}$$

where $s(d, Q)$ is the document score of document d given the set of query terms Q, $w_{d,t}$ is the document-term weight, w_t is the term weight, and $w_{q,t}$ is the query-term weight. Each of the weight values $w_{d,t}$, w_t and $w_{q,t}$ are based on $f_{d,t}$, f_t and $f_{q,t}$ respectively, where $f_{d,t}$ is the frequency of term t in document d, f_t is the number of documents term t appears in, and $f_{q,t}$ is the frequency of term t in query q.

Equation 1 shows us that document retrieval methods, which use a document-term index containing term frequencies, base their document score calculation on the occurrence of the user supplied query terms in each document. This allows the retrieval system to provide fast query times and use a conservative amount of storage, but the model suggests that all of the terms in the document set are independent of each other. For example, a search for "baby clothes" will return documents containing the terms "baby" or "clothes", but not provide documents containing related terms such as "infant", or "suits". This model assumes that authors write documents in the following manner:

1. the idea is constructed in the author's mind
2. specific terms are chosen from the term pool to express the idea on paper.

This model is shown in figure 1. Note that in this model, if other terms are chosen for the document, it would express a different idea because each of the terms are assumed independent of each other. We can see that this model does not reflect the actual process that an author does use to write a document.

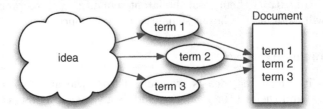

Fig. 1. A naïve document creation model. The author chooses specific terms for the document. If different terms are chosen the document will not convey the same message. This is the model used by retrieval systems that assume all terms are independent of each other (such as an inverted index of terms).

2.2   Latent Topics

We have seen in the previous section that the document retrieval model implies a poor document creation model. To make the model more realistic, we introduce an intermediate stage where the author chooses topics from a set of independent topics, to represent the document. Each of the topics contains a set of associated terms which are then chosen to include in the document. The process becomes:

1. the idea is constructed in the authors mind
2. specific topics are chosen from the topic pool to express the author's idea
3. for each topic, terms are chosen from the associated topic term pool to express the idea on paper

where the topic term pool is a set of terms that are related to the associated topic. Note that although the topics are independent, the associated terms may appear in many topics due to the synonymy found in many terms. This model is shown in figure 2. The final step allows the author to choose any of the terms associated to the selected topic to use within the document. This process suggests that as long as two documents contain the same topics, they can convey the same idea even though they contain different terms. The chosen topics must be the same in each document, but they are not written in the document; they are hidden from the reader and expressed in the terms that have been written.

Latent semantic analysis is the process of discovering these hidden topics and their relationship to the term and document set.

2.3   Probabilistic Latent Semantic Analysis

Probabilistic latent semantic analysis (PLSA) [3] is the process of calculating the term, topic and document relationships using probabilistic means. In this section, we will explain the basic concepts behind the method.

Consider the document set as being a bag filled with tokens; one token for every occurrence of a term in the document set. Each token has an associated term and document label attached. We can say that $P(d, t)$ is the probability that we put our hand in the bag and take out a token with the document label

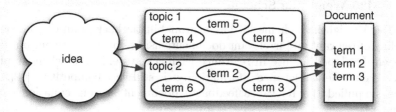

Fig. 2. The LSA document model. The author chooses specific topics for the document and then chooses a term from the topic to place in the document. This model implies that documents containing different terms can convey the same message, as long as the replacement terms are associated with the same topic.

d and term label t associated with it. Therefore if $f_{d,t}$ tokens are in the bag with labels d and t, implying that term t appeared $f_{d,t}$ times in document d, we obtain the sample probability:

$$\hat{P}(d,t) = \frac{f_{d,t}}{\sum_{\delta \in D} \sum_{\tau \in T} f_{\delta,\tau}} \qquad (2)$$

where D and T are the set of document and terms respectively and $\hat{P}(d,t)$ is the sample probability of document d and term t. PLSA attempts to model these sampled document-term probabilities as the sum of hidden topic distributions:

$$P(d,t) = \sum_{z \in Z} P(d|z)P(t|z)P(z) \qquad (3)$$

where Z is the set of hidden topics, $P(d,t)$ is the probability of term t being related to document d, $P(d|z)$ is the probability of document d given topic z, $P(t|z)$ is the probability of term t given topic z, and $P(z)$ is the probability of topic z. Using this model, we must fit our $|D| \times |T|$ samples using $|D| \times |Z| + |T| \times |Z| + |Z|$ parameters, where $|Z|$ is much smaller than $|D|$ and $|T|$.

3   Removing Bias in PLSA

Many weighting schemes have been developed for document retrieval systems to remove the bias found in non-homogeneous document collections [1, 4, 10]. Factors such as document length and term rarity can lead to the favour of certain irrelevant documents if not normalised.

   We would expect that these biases also exist when calculating term relationships. We have seen that our samples $\hat{P}(d,t)$ are crucial in the calculation of the unknown probabilities based on z. This leaves us with the question, what do we base our document-term sample probabilities on? We hypothesise that the biases found within raw term frequencies also affect the calculation of term relationships performed during PLSA.

   In this section, we will examine the popular BM25 weighting scheme and how we can apply it to PLSA.

3.1   BM25 Weighting Scheme

The BM25 weighting scheme [4] has a probabilistic background based on the modeling of relevant and irrelevant documents using Poisson distributions [9]. It has been developed for use in relevance feedback systems, but when simplified to use no document relevance information, it is still very competitive [12].

The simplified (no relevance feedback) document scoring equation can be shown as:

$$s(d, Q) = \sum_{t \in Q} w_{d,t} w_t \qquad (4)$$

where d is the document to be scored, Q is the set of query terms, $w_{d,t}$ and w_t are the document-term and term weights respectively.

The term weight is calculated as either the log odds of the term appearing in a document:

$$w_t = \log \left(\frac{N - f_t + 0.5}{f_t + 0.5} \right) \qquad (5)$$

or the negative log of the probability of the term appearing in a document:

$$w_{t+} = \log \left(\frac{N}{f_t} \right) \qquad (6)$$

where N is the number of documents and f_t is the number of documents containing term t. The term weight is used to reflect the importance of the term due to its rarity. For example a term that appears in all documents is not useful as a query term, since it will return all documents, therefore its weight is low. A term that appears in one document is very useful as a query term, therefore its weight should be high.

The document-term weight is the function:

$$w_{d,t} = \frac{(k_1 + 1) f_{d,t}}{K + f_{d,t}} \qquad (7)$$

where $f_{d,t}$ is the frequency of term t in document d, k_1 is a positive constant, and K is the pivoted document normalisation value. This function has two purposes. The first is to reduce the effect of large $f_{d,t}$ values. When searching for documents, one that contains twenty occurrences of a query term is not twice as relevant as one that contains ten occurrences of the same query term. In fact, they would both be considered just as relevant as each other. This function achieves this by reducing the increase in weight due to an increase in the term frequency. The second is to normalise the weight due to document length. A document that contains the query terms once in ten pages is not as relevant as one that contains the query terms in one page. The K value achieves this by normalising the documents based on their length.

3.2   Applying the Weights

Probabilistic latent semantic analysis calculates the maximum likelihood fit of the raw term frequencies (shown in equation 2). We want to perform a maximum likelihood fit of the term frequencies with biases removed, therefore we will perform PLSA on weighted term frequencies rather than raw term frequencies.

To use the weighted term frequencies, we simply substitute the weighted value where raw term frequencies are found. Therefore our new PLSA relationship becomes:

$$\hat{P}(d,t) = \frac{\omega_{d,t}}{\sum_{\delta \in D} \sum_{\tau \in T} \omega_{\delta,\tau}} \tag{8}$$

where $\omega_{d,t}$ is the weighted term frequency $f_{d,t}$.

PLSA uses the weighted term frequencies to construct a probabilistic model of the document set, therefore it is a requirement that the weight associated with each term in a document is positive. If we examine equation 5, we find that the log function returns negative values when applied to values less than one, which would occur when term t appears in over half of the documents. This property makes w_t unsuitable for use as an estimate of $P(d,t)$. The term weighting in equation 6 and the document-term weighting in equation 7 can never be negative, which make these weighting equations more suitable for our needs. Therefore we have the choice of using either of $\omega_{d,t} = w_{d,t}$, $\omega_{d,t} = w_t$ or $\omega_{d,t} = w_{d,t}w_t$. Once the weights are applied to every frequency value, we use the PLSA method to obtain the value of $P(d,t)$ and each of its components.

4   Experiments

We wish to analyse the effect of using weighted terms during the calculation of the PLSA term relationships. In this section, we describe the experiments performed and examine the data they produce.

We assumed that an increase in document retrieval precision implies that the term expansion is producing better terms for the query. Hence the probabilistic latent semantic analysis has established better relationships between the terms. Therefore, we will measure the effectiveness of the term relationships by examining the quality of the documents retrieved from a set of queries.

To store the weighted PLSA values, we will use a probabilistic latent semantic thesaurus (PLST), rather than a probabilistic latent semantic index (PLSI). The PLST has shown to provide greater precision, faster query times, and smaller storage space than the PLSI [8]. The PLST stores the probabilities $P(t_x|t_y)$ based on the computed $P(d|z)$, $P(z)$ and $P(t|z)$. The $P(t_x|t_y)$ values are used as a query expansion.

Our experimental environment was an information retrieval system consisting of a document-term frequency index using the BM25 weighting scheme and a probabilistic latent semantic thesaurus. Our experiments will examine the effect that weighting has on the the PLSA term relationships by performing one set of experiments with raw term frequencies ($\omega_{d,t} = f_{d,t}$) to calculate $P(t_x|t_y)$ and another set of experiments using BM25 weights ($\omega_{d,t} = w_{d,t}$, $\omega_{d,t} = w_{t+}$, and $\omega_{d,t} = w_{d,t}w_{t+}$) to calculate $P(t_x|t_y)$.

Previous work [6, 8] lead us to select the following constants for each latent semantic thesaurus: Only terms that appeared in more than 50 documents were

Table 1. Statistics of the four document sets used in the weighted latent semantic thesaurus experiments

document set	ZIFF1	ZIFF2	AP1	AP2
documents	75180	56920	84678	79919
median document length	181	167	353	346
avg. document length	412	394	375	370
unique terms	98206	82276	101708	95666
terms in 50 documents	7930	6781	10937	10498

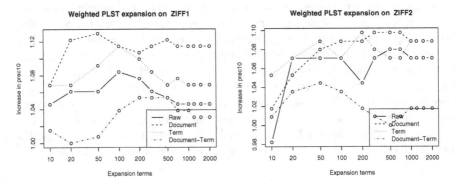

Fig. 3. A comparison of the increase in precision at 10 documents due to query expansion for PLST using $f_{d,t}$ (Raw), $w_{d,t}$ (Document), w_{t+} (Term), and $w_{d,t}w_{t+}$ (Document-Term) weights on the ZIFF1 and ZIFF2 document sets. The baseline BM25 precision at 10 documents with no expansion is 0.1985 and 0.1527 respectively.

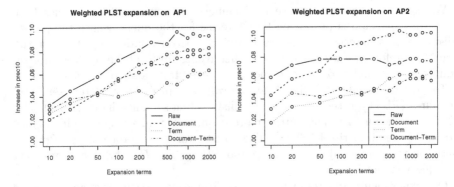

Fig. 4. A comparison of the increase in precision at 10 documents due to query expansion for PLST using $f_{d,t}$ (Raw), $w_{d,t}$ (Document), w_{t+} (Term), and $w_{d,t}w_{t+}$ (Document-Term) weights on the AP1 and AP2 document set. The baseline BM25 precision at 10 documents with no expansion is 0.3781 and 0.3554 respectively.

included in the thesaurus; the expansion terms were mixed with the query terms at a ratio of 0.6 to 0.4 respectively. Experiments were run on four separate document sets from TREC disks 1 and 2 named ZIFF1, ZIFF2, AP1 and AP2 (shown in table 1). On each document set, query expansions were performed using expansion sizes 10, 20 50, 100, 200, 300, 500, 700, 1000, 1200, 1500 and 2000. The expansion of size zero (implying no query expansion is performed) was used as a baseline to examine the precision of the retrieval system without using the probabilistic latent semantic term relationships. Therefore the results are presented in terms of increase in precision relative to this baseline. By setting the term expansion size to zero, we are switching off the PLST and thus our system becomes a BM25 document-term frequency index.

The experimental results showing precision at 10 documents are shown in figures 3 and 4. We can see in these plots that the $w_{d,t}$ weighted PLSA term relationships provide higher precision for most of our query expansion sizes for the ZIFF1 and ZIFF2 document sets. For the AP2 document set, PLSA using the raw term frequencies ($f_{d,t}$) provides higher precision than the $w_{d,t}$ weighted PLSA term expansion for 10, 20 and 50 terms. For all other expansion sizes the BM25 weighted PLSA expansion provides higher precision. It is interesting to note that the PLSA using the raw term frequencies ($f_{d,t}$) is generally flat for the three mentioned document sets. For the AP1 document set, we can see that PLSA using the raw term frequencies ($f_{d,t}$) provides higher precision for all levels of query expansion and is followed closely by the document weighted ($w_{d,t}$) expansion.

Significance testing using Wilcoxon's signed rank test was performed for three measures and is shown in table 2. The three measures used are mean average precision (MAP), precision after 10 documents (Prec10), and average reciprocal rank (ARR). MAP is used to judge the precision where many documents are required from the retrieval system, Prec10 is used to judge a system where a few documents are wanted, and ARR is used to judge a system where one document is wanted. The measures Prec10 and ARR are more useful for systems such as

Table 2. P-values from the Wilcoxon signed rank test. A P-value < 0.05 (marked with *) implies that using the associated weighting caused a significant increase in the associated measure. The measures shown are mean average precision (MAP), precision at 10 documents (Prec10) and average reciprocal rank (ARR).

Method	MAP	Prec10	ARR
$w_{d,t}$	0.618	0.011*	$2.97 \times 10^{-06*}$
w_{t+}	0.995	0.989	0.227
$w_{d,t}w_{t+}$	1	1	0.9999182

Table 3. Storage sizes in megabytes for each of the thesauruses using different weighting schemes. We can see that there is a clear drop in storage size for each of the four document sets, when weights are applied during the thesaurus construction.

Weight	AP1	AP2	ZIFF1	ZIFF2
$f_{d,t}$	99.25	92.56	55.93	43.93
$w_{d,t}$	86.75	82.68	41.68	31.87
w_t	78.50	73.50	40.56	32.93
$w_{d,t}w_t$	82.31	76.68	34.06	27.37

Web search engines, where the user does not require specific documents, but only a few documents to satisfy their information need.

We can see from the P-values that there is no significant increase in either of MAP, Prec10 and ARR when using w_{t+} and $w_{d,t}w_{t+}$ values for the PLST. There is, however, a very significant increase in Prec10 and ARR when using $w_{d,t}$ weights for the PLST.

We have also provided the storage required for each of the probabilistic latent semantic thesauruses in table 3. It is interesting to see that the storage required for each of the weighted thesauruses was much less than that needed by the thesaurus using raw term frequencies ($f_{d,t}$). This is probably due to the weighted values having a smaller range and thus requiring less bits for each level in each range.

4.1   Comparison to BM25 Pseudo-Relevance Feedback

To obtain an understanding of how well our weighted PLSA query expansion method performs, we have provided a comparison to the results obtained when using BM25 pseudo-relevance feedback [4] and a BM25 co-occurrence thesaurus.

Relevance feedback, unlike our static thesaurus method, is the dynamic process of supplying the retrieval system with a set of documents relevant to the query. The retrieval system then extracts a set of terms from the relevant documents to use as a query expansion. Pseudo-relevance feedback, unlike relevance feedback, does not obtain any relevance information from the user; it chooses the top ranking documents to the query as the set of pseudo-relevant documents. This set of documents is then used to obtain the term expansion. Pseudo-relevance feedback using BM25 weighting has been a very successful query expansion method at TREC, therefore it is a useful benchmark.

A co-occurrence thesaurus is a table of term to term relationships obtained by calculating:

$$P(t_x|t_y) = \frac{\sum_{d \in D} f_{d,t_x} f_{d,t_y}}{\sum_{t_z \in T} \sum_{d \in D} f_{d,t_z} f_{d,t_y}} \tag{9}$$

The co-occurrence thesaurus is used just as the PLST is used.

Previous experiments comparing PLSA using raw term frequencies to pseudo-relevance feedback and co-occurrence thesaurus using BM25 weights showed that PLSA provided significant increases in ARR and Prec10, but pseudo-relevance feedback provided greater MAP. We have shown that the BM25 weighted PLSA provides significant increases in ARR and Prec10 over PLSA, but there is no significant increase in MAP. Therefore we will observe the difference in ARR and Prec10 between BM25 weighted PLSA, pseudo-relevance feedback and the co-occurrence thesaurus. The prior results suggest the pseudo-relevance feedback will produce the greatest MAP.

We have produced plots in figure 5, comparing each of the mentioned method for various levels of query expansion.

We can see from these plots that the PLSA query expansion using BM25 document weights is far superior in terms of average reciprocal rank, achieving an average 8% increase. We can see that our PLST method using document

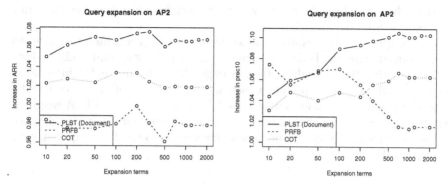

Fig. 5. Comparison of term expansion results on the AP2 document set using the average reciprocal rank (ARR) and precision at 10 documents (Prec10) measures. PLST (Document) is our probabilistic latent semantic thesaurus using document weights $(w_{d,t})$, PRFB is pseudo-relevance feedback, and COT is a co-occurrence thesaurus expansion method.

weights obtains a higher precision after 10 documents if 100 or more terms are chosen. Unfortunately, the relevance-feedback produces a greater mean average precision when using only a few terms. These results are similar to those of the PLST using unweighted term frequencies [8].

From these results we can see that our system would benefit a user who is searching for a few relevant documents, due to its high average reciprocal rank values. An example of this type of use would be found in typical Web searching. The pseudo-relevance feedback method would be more beneficial to a user who would want many or all relevant documents.

5   Conclusion

This article contains an analysis of the effect of using weighted terms during the probabilistic latent semantic analysis calculations and the impact it provides on probabilistic latent semantic term relationships.

We hypothesised that the term relationships obtained using PLSA will be more effective if calculated using weighted term frequencies rather than raw term frequencies. Raw term frequencies contain many forms of bias; weighted term frequencies are used to remove this bias during the query process, therefore weighted term frequencies should also be using when calculating probabilistic latent semantic term relationships.

Our hypothesis was tested by running precision experiments on a collection of document sets. We compared the precision from using a probabilistic latent semantic thesaurus built using raw term frequencies and a probabilistic latent semantic thesaurus built from weighted term frequencies. We found that using the thesaurus built from document weighted term frequencies provided a significant increase in precision at 10 document and average reciprocal rank. These results suggest that term relationships obtained using PLSA will be more effective when based on document weighted term frequencies rather than raw term frequencies.

We also compared the results obtained from the PLSA weighted thesaurus to those obtained using the BM25 pseudo-relevance feedback system. This analysis showed that the PLSA weighted thesaurus provided an average 8% increase in reciprocal rank and an increasing significance in precision after 10 documents, as the size of the term expansion increased. This implies that a PLSA weighted thesaurus retrieval system would be more useful than the BM25 pseudo-relevance feedback when found in an environment where a few document are required, such as a typical Web search.

References

1. Buckley, C., Walz, J.: SMART in TREC 8. In: Voorhees, Harman (eds.) [11], pp. 577–582
2. Dumais, S.T.: Improving the retrieval of information from external sources. Behaviour Research Methods, Instruments & Computers 23(2), 229–236 (1991)
3. Hofmann, T.: Probabilistic latent semantic indexing. In: Proceedings of the 22nd annual international ACM SIGIR conference on Research and development in information retrieval, pp. 50–57. ACM Press, New York (1999)
4. Sparck Jones, K., Walker, S., Robertson, S.E.: A probabilistic model of information retrieval: development and comparative experiments, part 2. Information Processing and Management 36(6), 809–840 (2000)
5. Park, L.A.F., Ramamohanarao, K.: Hybrid pre-query term expansion using latent semantic analysis. In: The Fourth IEEE International Conference on Data Mining, November 2004, pp. 178–185. IEEE Computer Society, Los Alamitos (2004)
6. Park, L.A.F., Ramamohanarao, K.: Query expansion using a collection dependent probabilistic latent semantic thesaurus. In: Zhou, Z.-H., Li, H., Yang, Q. (eds.) PAKDD 2007. LNCS (LNAI), vol. 4426, pp. 224–235. Springer, Heidelberg (2007)
7. Park, L.A.F., Ramamohanarao, K.: An analysis of latent semantic indexing term self preservation. ACM Transactions on Information Systems (to appear, 2008)
8. Park, L.A.F., Ramamohanarao, K.: Efficient storage and retrieval of probabilistic latent semantic information for information retrieval. The International Journal on Very Large Data Bases (to appear, 2008)
9. Robertson, S.E., Walker, S.: Some simple effective approximations to the 2-poisson model for probabilistic weighted retrieval. In: Proceedings of the 17th International Conference on Research and Development in Information Retrieval, London, pp. 232–241. Association of Computing Machinary, Inc., Springer, Heidelberg (1994)
10. Robertson, S.E., Walker, S.: Okapi/keenbow at TREC-8. In: Voorhees, Harman (eds.) [11], pp. 151–162
11. Voorhees, E.M., Harman, D.K. (eds.): The Eighth Text REtrieval Conference (TREC-8), Gaithersburg, Md. 20899, National Institute of Standards and Technology Special Publication 500-246, Department of Commerce, National Institute of Standards and Technology (November 1999)
12. Voorhees, E.M., Harman, D.K.: Overview of the eighth text retrieval conference (TREC-8). In: The Eighth Text REtrieval Conference (TREC-8) [11], pp. 1–23

Comparison of *s*-gram Proximity Measures in Out-of-Vocabulary Word Translation

Anni Järvelin[1] and Antti Järvelin[2]

[1] University of Tampere, Department of Information Studies,
FIN-33014 University of Tampere, Finland
anni.jarvelin@uta.fi
[2] University of Tampere, Department of Computer Sciences,
FIN-33014 University of Tampere, Finland
antti.jarvelin@cs.uta.fi

Abstract. Classified *s*-grams have been successfully used in cross-language information retrieval (CLIR) as an approximate string matching technique for translating out-of-vocabulary (OOV) words. For example, *s*-grams have consistently outperformed other approximate string matching techniques, like edit distance or *n*-grams. The Jaccard coefficient has traditionally been used as an *s*-gram based string proximity measure. However, other proximity measures for *s*-gram matching have not been tested. In the current study the performance of seven proximity measures for classified *s*-grams in CLIR context was evaluated using eleven language pairs. The binary proximity measures performed generally better than their non-binary counterparts, but the difference depended mainly on the padding used with *s*-grams. When no padding was used, the binary and non-binary proximity measures were nearly equal, though the performance at large deteriorated.

1 Introduction

Cross-Language Information Retrieval (CLIR) refers to retrieval of documents written in a language other than that of the user's request. The document collection's language is called the *target language* and the query language the *source language*. A typical approach to CLIR is automatically translating the query into the target language. For an overview of CLIR, see [1]. Out-of-vocabulary (OOV) words constitute a major problem in query translation in CLIR. Due to the terminology missing from dictionaries, untranslatable keys appear in queries. Many typical OOV words, like proper names and technical terms, are often important query keys [2]. Therefore their translation is essential for query performance. In European languages, technical terms often share a common Greek or Latin root but are rendered with different spelling. This provides a good basis for the use of approximate string matching in translating the OOV words, as the words similar to a query's OOV words can be found from the target document collection and recognized as the translations of the query words.

The classified *s*-gram matching technique is a generalization of the well-known *n*-gram matching technique developed as a solution to the OOV word problem

A. Amir, A. Turpin, and A. Moffat (Eds.): SPIRE 2008, LNCS 5280, pp. 75–86, 2008.
© Springer-Verlag Berlin Heidelberg 2008

in dictionary-based CLIR [3]. In s-gram matching the strings compared are decomposed into shorter substrings, called s-grams. Skipping characters is allowed when forming the s-grams and the degree of similarity between the strings is computed by comparing their s-gram sets. s-grams, or gapped q-grams, have also been described e.g. in [4] where they were applied for fast and efficient filtering for approximate string matching. The classified s-grams differ from the other gapped q-grams in that several different s-grams are grouped together into sets of s-grams prior to calculating the similarity. The classified s-grams have been developed with CLIR and natural language processing in mind, i.e., for relatively short strings including relatively little repetition of s-grams. In CLIR applications, the technique has outperformed several other established approximate string matching techniques, such as the edit distance, the longest common subsequence and n-grams [3,5].

There are several ways of calculating the s-gram proximity between two strings. In the context of n-gram matching the L_1 distance [6], its binary version Hamming distance [7], the Dice coefficient [8], and the Jaccard coefficient [9] among others have been used. Robertson and Willett [10] mention that any proximity measure could be used, while Zobel and Dart [7] propose that measures used in IR, such as the cosine measure [8], should not be appropriate for phonetic n-gram matching as they factor out the document length.

Only similarity measures based on the Jaccard coefficient have previously been tested with classified s-grams [3,5]. Clearly, other proximity measures could also be applied, but it is not obvious which might be the best suited ones. Järvelin et al. [11] formalized a few proximity measures for s-gram matching, e.g., the L_1 distance. They argued that, theoretically, the Jaccard coefficient may not be the choice proximity measure to be used in the s-gram matching, as it is binary and thus insensitive to the counts of each s-gram in the strings to be compared. The non-binary L_1 distance should be a more sensitive proximity measure, as it takes both the types of s-grams and their number in the strings compared into account. Järvelin et al. [11] did not test their claim empirically, but their definitions enable the comparison of different s-gram proximity measures.

As the choice of the proximity measure used with the s-grams may affect the performance of the technique, testing the different proximity measures is needed. This article contributes to the issue by reporting the results of an evaluation of several proximity measures for s-gram matching of cross-lingual spelling variants. Especially the differences between the binary and the non-binary proximity measures are considered. The binary proximity measures Jaccard coefficient, binary cosine similarity and Hamming distance were compared to their non-binary counterparts Tanimoto coefficient, cosine similarity and L_1 distance respectively. Also, the binary Dice coefficient was tested. Cross-lingual spelling variants in seven languages (English, Finnish, French, German, Italian, Spanish and Swedish) were used as source words that were translated into four target languages, English, German, Swedish and Finnish, using classified s-gram matching. In total eleven language pairs were used. The proximity measures' performance was evaluated as average translation precision.

Next, Section 2 provides an introduction to the *s*-grams and their proximity measures. Section 3 presents the materials and methods and Section 4 the results. Finally, section 5 contains a brief discussion and the conclusions.

2   *s*-gram Definitions

2.1   Introduction to *s*-grams

Word variation, where a language pair shares words written differently but having the same origin, is called cross-lingual spelling variation. Pirkola et al. [3] and Keskustalo et al. [5] showed that this kind of variation can advantageously be modeled with the *s*-grams. In *s*-gram matching the text strings to be compared are decomposed into substrings and the similarity between the strings is calculated as the overlap of their common substrings. Unlike in *n*-gram matching, skipping some characters is allowed when forming the *s*-grams. In CLIR applications substring length of two has been used. It has been found beneficial in IR applications to use padding spaces around the strings when forming *s*-grams [5,10]. This helps to get the characters at the beginning and at the end of a string properly presented in string comparison.

In *classified* *s*-gram matching technique [3] the *s*-grams originating from the same string are classified into sets based on the number of characters skipped prior to calculating the similarity. Only the *s*-grams belonging to the same set are compared to each other. *Gram class* indicates the skip length(s) used when generating a set of *s*-grams. The largest value in a gram class is called the *spanning length* of the gram class [5], e.g., for gram class $\{0, 1\}$, the spanning length is one. Two or more gram classes may also be combined into more general gram classes. The *character combination index (CCI)* then indicates the set of all the gram classes to be formed from a string, e.g. CCI $\{\{0\}, \{1, 2\}\}$ means that two gram classes are formed from a string: one with conventional *n*-grams formed of adjacent characters ($\{0\}$) and one with *s*-grams formed both by skipping one and two characters ($\{1, 2\}$). For the string "abracadabra", the *s*-gram set produced by the CCI $\{\{1, 2\}\}$ is $\{ar, ba, rc, aa, cd, db, bc, ra, ad, ca, ab, dr\}$, when duplicate *s*-grams are not listed.

2.2   *s*-gram Profiles and Their Proximities

s-gram-based string proximity measures are based on strings' *s-gram profiles*. The *s*-gram profile definitions given in this paper are extended from Ukkonen's [6] *n*-gram profile definition. Next strings' *s*-gram profiles are defined, which are then generalized for gram classes. Then various gram class based proximity measures are given, because the strings' CCI based proximity measures are calculated as the average gram class distance of the CCI's gram classes.

Definition 1. *Let $w = a_1 a_2 \ldots a_m$ be a string over a finite alphabet Σ, $n \in \mathbb{N}^+$ be a gram length, $k \in \mathbb{N}$ a skip length and let $x \in \Sigma^n$ be an s-gram. If $a_i a_{i+k+1} \ldots a_{i+(k+1)(n-1)} = x$ for some i, then w has a $s_{n,k}$-gram occurrence of*

x. Let $G_k(w)[x]$ denote the total number of $s_{n,k}$-gram occurrences of x in w. The $s_{n,k}$-gram profile of w is the vector $G_{n,k}(w) = (G_k(w)[x]), x \in \Sigma^n$.

s-gram profile can easily be generalized for gram classes. The gram class profiles are formed by summing up the s-gram profiles in a given gram class.

Definition 2. Let $w \in \Sigma^*$, $C \in \mathcal{P}(\mathbb{N})$ a gram class and $x \in \Sigma^n$. Let $G_C(w)[x] = \sum_{k \in C} G_k(w)[x]$. The gram class profile of w is the vector $G_{n,C}(w) = (G_C(w)[x])$, $x \in \Sigma^n$. In other words, $G_{n,C}(w) = \sum_{k \in C} G_{n,k}(w)$.

Sometimes the exact number of the occurrences of s-grams in the string is irrelevant, but merely the information if a specific s-gram occurs at all in the string is needed. This leads to the notion of binary gram class profile.

Definition 3. Let $w \in \Sigma^*$, and $C \in \mathcal{P}(\mathbb{N})$ a gram class and $x \in \Sigma^n$. Let

$$B_C(w)[x] = \begin{cases} 1 \text{ if } G_C(w)[x] > 0 \\ 0 \text{ otherwise} \end{cases}.$$

The binary gram class profile of w is the vector $B_{n,C}(w) = (B_C(w)[x]), x \in \Sigma^n$.

Various proximity measures can now be used to calculate string proximities based on the general and binary gram class profiles. Next, only the proximity measures using the general gram class profile of Definition 2 are given, because the corresponding proximity measures using binary profiles are defined by substituting the general s-gram profiles with binary ones in the following equations.

Let v and w be strings in Σ^*, $n \in \mathbb{N}^+$ be a gram length and $C \in \mathcal{P}(\mathbb{N})$ a gram class. L_1 distance for gram classes of strings v and w is

$$L1_{n,C}(v,w) = \sum_{x \in \Sigma^n} |G_C(v)[x] - G_C(w)[x]|. \tag{1}$$

The L_1 distance has been used with n-grams by Ukkonen [6] and its binary version, the Hamming distance, was proposed by Zobel and Dart [7]. Therefore its performance was investigated in s-gram based OOV word translation.

The cosine gram class similarity between v and w is defined as

$$Cos_{n,C}(v,w) = \frac{G_C(v)^T G_C(w)}{\|G_C(v)\|\|G_C(w)\|}, \tag{2}$$

where $\|\cdot\|$ denotes the Euclidean norm and T the transpose of a vector. Cosine similarity (or normalized dot product) is a widely utilized proximity measure in text retrieval applications [12] and therefore its performance in s-gram matching was also investigated along its binarized counterpart.

The Tanimoto coefficient [13] between gram classes of v and w is given by

$$T_{n,C}(v,w) = \frac{G_C(v)^T G_C(w)}{\|G_C(v)\|^2 - G_C(v)^T G_C(w) + \|G_C(w)\|^2}. \tag{3}$$

The Tanimoto coefficient was tested, because its binary counter part, the Jaccard coefficient, has traditionally been used in s-gram matching [3,5,11].

Turning to the binary profile based proximity measures, the Hamming distance $H_{n,C}(v,w)$ between v and w is derived by substituting the general gram class profile with binary profile in (1), binary cosine similarity $BinCos_{n,C}(v,w)$ by substituting with binary profiles in (2), and Jaccard coefficient $J_{n,C}(v,w)$ by doing the same substitution in (3).

Lastly, the Dice's coefficient was investigated, because it has been used in n-gram matching [10]. It is closely related to the Jaccard coefficient, but weights more the matching profile positions between the gram class profiles than the mismatching ones [12]. The Dice coefficient between v and w is given by

$$D_{n,C}(v,w) = \frac{2B_C(v)^T B_C(w)}{\|B_C(v)\|^2 + B_C(v)^T B_C(w) + \|B_C(w)\|^2}. \tag{4}$$

The character combination index based string proximity measures tested in this paper are defined as the average of strings' gram class proximities. For example, for a CCI $\mathscr{C} \in \mathcal{P}(\mathcal{P}(\mathbb{N}))$, and a gram length n, the CCI-distance corresponding to L_1 distance is

$$L1_{n,\mathscr{C}}(v,w) = \frac{1}{|\mathscr{C}|} \sum_{C \in \mathscr{C}} L1_{n,C}(v,w). \tag{5}$$

All CCI-based proximity measures tested below were defined analogously.

One problem that might arise when using the s-gram profiles in approximate string matching is the length of the profiles. With $s_{n,k}$-grams, the profile length is $|\Sigma|^n$ where Σ is the specified alphabet. For example, the standard English alphabet consists of 26 letters, and thus even the di-gram profiles are quite long. However, with natural languages, the s-gram and the gram class profiles are typically very sparse, and well suited for sparse vector implementations. Therefore, the proximities between the s-gram profiles can be evaluated efficiently.

3   Materials and Methods

3.1   Materials

The test data consisted of three parts: the search keys, the target words and the set of correct translations, i.e., the relevance judgments. 271 search keys were expressed in seven languages, which were English, Finnish, French, German, Italian, Spanish and Swedish. The search keys were mostly technical terms from the domains of biology, medicine, economics and technology, but also a list of geographical names obtained from [5] was included. These are typical cases of cross-lingual spelling variants that tend to be OOV words and thus problematic in CLIR. In total, 11 language pairs were used in the study, with four target languages: English, German, Finnish and Swedish. English was combined with all of the other languages as a target language and was also used as a source language with Finnish, German and Swedish. Translation was also done both ways between Swedish and German.

Target word lists (TWLs) consisted of CLEF 2003 [14] document collection's indices for the target languages. The collections are full-text newspaper document collections from 1994–1995. The size of the collections, and thus the TWLs, varies between languages. The English TWL consisted of ca 257,000, the Swedish TWL of ca 388,000 and the Finnish TWL of ca 535,000 unique word forms. The German CLEF03 collection was considerably bigger and thus only a part of it was used for creating a TWL including ca 391,000 unique word forms.

All the TWLs were lemmatized (i.e. the index words were returned into their basic forms) with the TWOL morphological analyzer by Lingsoft Ltd. The words not recognized by the morphological analyzer were indexed in the word forms they appeared in the text. Compounds were split and both the original compounds and their constituents were indexed. The missing translation equivalents of the search keys were added to the TWLs, and there was only one correct translation for each search key in the TWLs.

3.2 Methods

The performance of the proximity measures was tested as follows. The s-gram length was set to two, because earlier research [3,5] suggests it to be the most appropriate gram length for CLIR. Padding was used at both ends of the strings and the length of the padding string was $(n - 1)(k + 1)$, where n is the gram and k the skip length. Also, s-grams with no padding and padding only at the beginning of strings were tested for two language pairs (English-German and German-English) with CCI $\{\{0\}, \{1, 2\}\}$ to see how the padding affects the results. For each search key 100 best translations were produced, with exception of ties in the last place when all translations within the cohort of equal proximity values were included into the result set. Translations found later were not taken into consideration, as taking more than 2-4 s-gram translation candidates into a query deteriorates its performance [15].

This study concentrates on comparing the proximity measures. Exhaustive testing of all possible CCIs, proximity measures and language pairs was not sensible or even possible within this study. If skip lengths 0 – 4 were considered, there would be $2^5 - 1 = 31$ possible gram classes, and thus about $2^{31} - 1$, about two billion, combinations as possible CCIs. To be able to restrict the scope of the study to some evidently useful CCIs, statistics on typical cross-lingual spelling variation between French and English and German and English were used. Pirkola et al. [16] generated statistical transformation rules that model typical character changes and correspondences between several language pairs. The rules were generated from over 10,000 term pairs of medical words. They model the same cross-lingual spelling variation phenomenon as the s-grams, but are based on an independent method and character correspondence statistics from an independent large dataset. We mapped a subset of ca 200 most frequent transformation rules to the corresponding gram classes for both language pairs and calculated the frequency of each gram class in the data.

Table 1. The number of transformation rules corresponding to each gram class for French to English and German to English cross-lingual spelling variants

Gram class	{1}	{0,1}	{1,2}	{0,2}	{2}	{1,3}	{0,3}	{2,3}	{3}
Fr-En	56	65	44	11	12	7	3	5	3
Ge-En	117	37	36	20	7	3	4	1	0
Total	173	102	80	31	19	10	7	6	3

Table 2. The twelve CCIs used for the comparison of the proximity measures. Note that CCI_0 corresponds to the traditional n-grams. For CCI_0 gram length of two and three was experimented, for the remaining CCIs only gram length of two was used.

CCI_0	$\{\{0\}\}$	CCI_4	$\{\{0\},\{1\}\}$	CCI_8	$\{\{0,1\}\}$
CCI_1	$\{\{0\},\{0,1\}\}$	CCI_5	$\{\{0\},\{0,1\},\{1\}\}$	CCI_9	$\{\{0,1,2\}\}$
CCI_2	$\{\{0\},\{0,1\},\{1\},\{1,2\}\}$	CCI_6	$\{\{0\},\{1\},\{1,2\}\}$	CCI_{10}	$\{\{1\}\}$
CCI_3	$\{\{0\},\{0,1\},\{1,2\}\}$	CCI_7	$\{\{0\},\{1,2\}\}$	CCI_{11}	$\{\{1,2\}\}$

There were some differences between the language pairs, but the transformation rules that model character changes corresponding to the gram classes $\{1\}$, $\{0,1\}$ and $\{1,2\}$ were clearly the most common ones in the data. Table 1 summarizes the results for both languages. Based on this it seemed reasonable to use only gram classes with spanning length of two or less when matching cross-lingual spelling variants. Changes corresponding to the remaining clearly less frequent gram classes were thus discarded. Keskustalo et al. [5] reached an equal conclusion, when deciding which gram classes they should use.

Based on the results of Table 1, the gram classes $\{1\}$, $\{0,1\}$, and $\{1,2\}$ and gram class $\{0\}$ corresponding to the n-grams were selected as the base gram classes for the tested CCIs. In total, the twelve CCIs of Table 2 were used in the tests. For CCI_0, in addition to the gram length of two (di-grams), also gram length of three (tri-grams) was used. This set of CCI_0 - CCI_{11} is a representative set on effective s-grams, and by using this set a reliable picture of various s-gram proximity measures in s-gram matching can be obtained.

To compare the performances of the proximity measures, the average precision (AP, or reciprocal rank - as there is only one correct translation, these are the same) was calculated for each proximity measure at three different levels: among top 2, top 5 and top 100 highest ranked translation candidates. If the correct translation was in a cohort of words sharing the same proximity value with the target word, the average rank of the cohort was used. The top 2 and top 5 levels were the most interesting ones, as more translation candidates would deteriorate the query performance. The Friedman test [17] was used to test the statistical significance of the differences between the proximity measures. Below, statistically significant difference corresponds to α-level $\alpha = 0.01$, statistically highly significant difference to α-level $\alpha = 0.001$, and statistically almost significant α-level $\alpha = 0.05$.

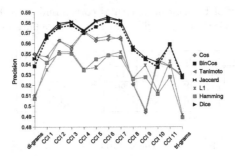

Fig. 1. The medians of APs of the proximity measures at top 5 translation candidates for all CCIs over all language pairs, zoomed in for clarity

4   Results

4.1   CCIs and Proximity Measures over All Languages

The results for all proximity measures and CCIs over all language pairs are presented in Fig. 1 and in Table 3 as the medians of AP when the top 5 translation candidates are considered. The results in top 2 and top 100 followed the same trends and are not presented due to the lack of space. The results divide the s-grams into two groups: the s-grams with CCIs that combine several s-gram types into a gram class and the s-grams where only one s-gram type is present in each gram class. In the former group (CCIs 1, 2, 3, 5, 6, 7, 8, 9, 11), the binary proximity measures performed clearly better than their non-binary counterparts, i.e., Jaccard performed better than Tanimoto, binary cosine better than cosine and Hamming better than L_1. The differences between Jaccard and Tanimoto and binary cosine and cosine were statistically significant for 7 of these 9 CCIs for 8 language pairs out of 11. For CCI_5 the differences were statistically significant only for five language pairs (EN-FI, EN-GE, FI-EN, FR-EN, IT-EN) of which two (EN-GE, FR-EN) were only almost significant. For CCI_6 the differences were statistically significant for seven language pairs (EN-FI, EN-GE, FR-EN, GE-EN, IT-EN, SP-EN, SW-EN), two of these (IT-EN, SW-EN) being almost significant. For the two closest related language pairs (GE-SW and SW-GE) the differences were not statistically significant. Also, for EN-SW the differences were statistically significant only for CCIs 1, 2, 8, and 9. The differences between Hamming and L_1 were typically not statistically significant. The three best measures Dice, Jaccard and binary cosine performed similarly and clearly better than the rest of the proximity measures. L_1 and Hamming were the worst proximity measures. The performance difference between them and the other proximity measures was statistically significant for all language pairs and CCIs.

In the latter group, including the adjacent di-grams and tri-grams (CCI_0) and the s-grams with CCI_4 and CCI_{10}, the difference between the binary and non-binary proximity measures was smaller and always to the advantage of the non-binary measures. These differences were nevertheless never statistically significant. Tanimoto was the best proximity measure, while L_1 and Hamming were

Table 3. The medians of the APs of the proximity measures among top 5 translation candidates for all CCIs over all language pairs. The best proximity measures for each CCI are in bold. Tanimoto coefficient performs best for n-grams and s-grams with CCI_{10} and Dice coefficient performs best for the s-grams with CCI's that combine several s-gram types into more general gram classes.

CCI	Cos	BinCos	Tanimoto	Jaccard	L_1	Hamming	Dice
			Proximity measure				
di-grams	0.5490	0.5382	**0.5493**	0.5454	0.5100	0.5070	0.5454
CCI_1	0.5417	0.5655	0.5475	0.5678	0.5349	0.5418	**0.5683**
CCI_2	0.5627	0.5755	0.5629	0.5774	0.5522	0.5506	**0.5795**
CCI_3	0.5549	0.5774	0.5573	0.5807	0.5522	0.5504	**0.5810**
CCI_4	0.5708	0.5699	0.5715	0.5715	0.5355	0.5343	**0.5726**
CCI_5	0.5624	0.5760	0.5647	0.5811	0.5371	0.5449	**0.5819**
CCI_6	0.5638	0.5816	0.5670	0.5839	0.5490	0.5486	**0.5855**
CCI_7	0.5656	0.5781	0.5637	0.5818	0.5518	0.5469	**0.5821**
CCI_8	0.5208	0.5539	0.5214	**0.5567**	0.5265	0.5258	**0.5567**
CCI_9	0.4939	0.5448	0.4958	**0.5465**	0.5392	0.5276	**0.5465**
CCI_{10}	0.5417	0.5373	**0.5446**	0.5418	0.5142	0.5114	0.5418
CCI_{11}	0.5380	**0.5592**	0.5382	0.5585	0.5429	0.5279	0.5585
tri-grams	0.5280	0.5272	**0.5296**	**0.5296**	0.4913	0.4891	**0.5296**
MEDIAN	0.5490	0.5655	0.5493	0.5678	0.5371	0.5343	**0.5683**

the worst ones the difference being always statistically significant. n-grams performed clearly worse than the s-grams with CCIs combining s-gram types into more general gram classes. The s-grams with CCI_4, combining two gram classes of a single s-gram type, performed better. This suggests that the s-gram technique benefits from combining gram classes into one CCI. It also seems that the more s-gram types were combined into a gram class, the more the performance of Tanimoto and cosine suffered. The CCI_9 is an example of this, showing a notable fall in the performance of Tanimoto and cosine in Fig. 1.

4.2 Results for Each Language Pair

The results in Fig. 1 and in Table 3 are medians over all the language pairs tested. To give a better picture of the results for the different language pairs, a typical CCI was selected to represent each group. CCI_6 represents the s-grams that combine several s-gram types in the gram classes. The results are presented for all language pairs in Fig. 2 (a) as AP among the top 5 translation candidates. The binary proximity measures performed better than their non-binary counterparts. Differences between Jaccard and Tanimoto and binary cosine and cosine were statistically significant for 7 language pairs, as mentioned above (not for FI-EN, SW-GE, GE-SW, EN-SW). Dice, Jaccard and binary cosine were the best proximity measures, while L_1 and Hamming were the worst ones. The differences between these were consistently statistically significant. Fig. 2 (b) presents the results for CCI_0 di-grams representing the other class of s-grams as AP among

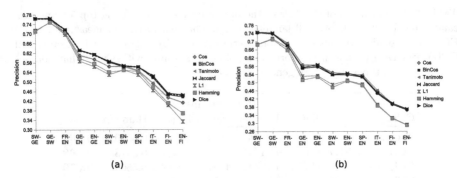

Fig. 2. (a) The AP of the proximity measures at top 5 for all language pairs for the s-grams with CCI_6. (b) The AP of the proximity measures at top 5 for all language pairs for traditional di-grams (CCI_0). Both figures are zoomed in for clarity.

the top 5 translation candidates. The results were scattered depending on the language pair, though still in line with the median results presented in Table 3 and Fig. 1. The non-binary proximity measures (Tanimoto and cosine) performed on average better than their binary counterparts, but the differences were not statistically significant. Hamming distance and L_1 were the worst measures, with statistically significant difference to the other proximity measures. Tri-grams performed generally worse than di-grams.

4.3   Padding

The differences between the binary and non-binary proximity measures were clearly reduced when no padding or padding only at the beginning of the strings were used. When no padding at all was used, the results deteriorated for all proximity measures and more for the binary than the non-binary proximity measures. For cosine and Tanimoto, the results even improved slightly for one of the two language pairs (GE-EN). Thus the differences between corresponding binary and non-binary proximity measures were reduced and were not statistically significant. When only the left-side padding was used, the overall effect on results was a little unclear: for English to German matching the best results deteriorated slightly, but for German to English the top results improved slightly. The non-binary proximity measures improved in comparison to their binary counterparts and the differences between them were not statistically significant. L_1 and Hamming suffered both from not using padding and also from using padding only at the beginning of the strings. They were always clearly the worst proximity measures with a statistically highly significant difference between them and the other proximity measures.

5   Discussion

To sum up the results, the binary proximity measures performed better than their non-binary counterparts in s-gram based matching of OOV words. Dice,

Jaccard and binary cosine performed best and any of these measures could be beneficially used. The difference between the binary and non-binary proximity measures seems to depend on the CCI used: when a number of different s-gram types were combined into a more general gram class (such as $\{\{1, 2\}\}$), the binary proximity measures clearly outperformed their non-binary counterparts. For the CCIs where only one s-gram type was present in each gram class (the traditional n-grams, CCI_4, and CCI_{10}), the differences between the binary and non-binary proximity measures vanished. Also, the more s-gram types were combined into a gram class, the more the performance of Tanimoto and cosine suffered.

This seems to be linked to the padding used with s-grams: When several s-gram types are combined into one gram class and padding was used, identical s-grams from both ends of strings are formed repeatedly and become overweighted when using non-binary proximity measures. As character changes are especially common at the ends of cross-lingual spelling variants (e.g. *antiseptic - antiseptique*), this damages the performance of the non-binary proximity measures. Removing the padding is nevertheless not a guarantee of success as it may affect the overall performance of the s-gram matching negatively. Keskustalo et al. [5] have found earlier that whether the padding on both sides of strings or only at the beginning performs best depends on the language pair at hand. For s-gram matching implementations using non-binary s-gram profiles, the repetitive occurrences of s-grams including padding characters should be ignored.

L_1 and its binary counterpart Hamming distance did not perform well and they do not seem suitable proximity measures for this application area. With these proximity measures the distance between two strings is calculated as the mean value of the different s-grams in the gram classes. This causes the measures to favor short words as no s-grams can be formed of one letter words (without padding) and none or very few of two or three letter words. Therefore, L_1 and Hamming give more non-relevant short words at the top ranks in the result lists than the other proximity measures. This is also reflected in the fact that the results for L_1 and Hamming deteriorated when the padding was removed.

Non-binary proximity measures are suitable for applications where a lot of repetition of s-grams occur (e.g. gene matching). In cross-lingual OOV word matching the alphabet used is rather large and the strings processed quite short. Consequently the repetition of s-grams is not extensive and therefore the binary and non-binary s-gram profiles approach each other. Therefore, no advantage is achieved with the use of the non-binary proximity measures.

Acknowledgments

The authors wish to thank Academy Professor Kalervo Järvelin, Ph.D., Docent Ari Pirkola, Ph.D., and Mr. Heikki Keskustalo, M.Sc. from University of Tampere for their support and comments on the paper. The first author was funded by Tampere Graduate School in Information Science and Engineering (TISE) and Academy of Finland under grant # 1209960.

References

1. Kishida, K.: Technical issues of cross-language infromation retrieval: a review. Inf. Process. Manage 41(3), 433–455 (2005)
2. Pirkola, A., Järvelin, K.: Employing the resolution power of search keys. JA-SIST 52(7), 575–583 (2001)
3. Pirkola, A., Keskustalo, H., Leppänen, E., Känsälä, A.P., Järvelin, K.: Targeted s-gram matching: a novel n-gram maching technique for cross- and monolingual word form variants. Information Research 7(2) (2002), http://InformationR.net/ir/7-2/paper126.html
4. Burkhardt, S., Kärkkäinen, J.: Better filtering with gapped q-grams. Fundamenta Informaticae 56(1–2), 51–70 (2003)
5. Keskustalo, H., Pirkola, A., Visala, K., Leppänen, E., Järvelin, K.: Non-adjacent digrams improve matching of cross-lingual spelling variants. In: Nascimento, M.A., de Moura, E.S., Oliveira, A.L. (eds.) SPIRE 2003. LNCS, vol. 2857, pp. 252–265. Springer, Heidelberg (2003)
6. Ukkonen, E.: Approximate string-matching with q-grams and maximal matches. Theor. Comput. Sci. 92(1), 191–211 (1992)
7. Zobel, J., Dart, P.: Phonetic string matching: lessons from information retrieval. In: SIGIR 1996: Proceedings of the 19th ACM SIGIR Conference, pp. 166–172. ACM Press, New York (1996)
8. Salton, G., McGill, M.J.: Introduction to Modern Information Retrieval, 1st edn. McGraw-Hill, New York (1983)
9. Pfeiffer, U., Poersch, T., Fuhr, N.: Retrieval effectiveness of proper name search methods. Inf. Process. Manage 32(6), 667–679 (1996)
10. Robertson, A.M., Willett, P.: Applications of n-grams in textual information systems. J. Doc. 54(1), 48–69 (1998)
11. Järvelin, A., Järvelin, A., Järvelin, K.: s-grams: defining generalized n-grams for information retrieval. Inf. Process. Manage. 43(4), 1005–1019 (2007)
12. Hand, D.J., Mannila, H., Smyth, P.: Principles of Data Mining, 1st edn. MIT Press, Cambridge (2001)
13. Theodoridis, S., Koutroumbas, K.: Pattern Recognition, 2nd edn. Academic Press, London (2003)
14. Gonzalo, J., Peters, C.: Introduction. In: Peters, C., Gonzalo, J., Braschler, M., Kluck, M. (eds.) CLEF 2003. LNCS, vol. 3237, pp. 1–6. Springer, Heidelberg (2004), http://clef.iei.pi.cnr.it/
15. Hedlund, T., Airio, E., Keskustalo, H., Lehtokangas, R., Pirkola, A., Järvelin, K.: Dictionary-based cross-language information retrieval: Learning experiences from CLEF 2000 – 2002. Information Retrieval - Special Issue on CLEF Cross-Language IR 7(1–2), 99–119 (2004)
16. Pirkola, A., Toivonen, J., Keskustalo, H., Visala, K., Järvelin, K.: Fuzzy translation of cross-lingual spelling variants. In: SIGIR 2003: Proceedings of the 26th ACM SIGIR Conference, pp. 345–352. ACM Press, New York (2003)
17. Conover, W.J.: Practical Nonparametric Statistics, 3rd edn. Wiley, New York (1999)

Speeding Up Pattern Matching by Text Sampling

Francisco Claude[1,*], Gonzalo Navarro[1,*], Hannu Peltola[2],
Leena Salmela[2], and Jorma Tarhio[2]

[1] Department of Computer Science, University of Chile
{fclaude,gnavarro}@dcc.uchile.cl
[2] Department of Computer Science and Engineering
Helsinki University of Technology
{hpeltola,lsalmela,tarhio}@cs.hut.fi

Abstract. We introduce a novel alphabet sampling technique for speeding up both online and indexed string matching. We choose a subset of the alphabet and select the corresponding subsequence of the text. Online or indexed searching is then carried out on that subsequence, and candidate matches are verified in the full text. We show that this speeds up online searching, especially for moderate to long patterns, by a factor of up to 5. For indexed searching we achieve indexes that are as fast as the classical suffix array, yet occupy space less than 0.5 times the text size (instead of 4) plus text. Our experiments show no competitive alternatives in a wide space/time range.

1 Introduction

The string matching problem is to find all the occurrences of a given pattern $P = p_0 p_1 \ldots p_{m-1}$ in a large text $T = t_0 t_1 \ldots t_{n-1}$, both being sequences of characters drawn from an alphabet Σ of size σ.

One approach to string matching is *online* searching, which means the text is not preprocessed. Thus these algorithms need to scan the text when searching and their time cost is of the form $\mathcal{O}(n \cdot f(m))$. The worst-case complexity of the problem is $\Theta(n)$, first achieved by the Knuth-Morris-Pratt algorithm [9]. The average complexity of the problem is $\Theta(n \log_\sigma m/m)$, achieved for example by the BDM algorithm [3]. Other non-optimal algorithms such as the Boyer-Moore-Horspool (BMH) algorithm [7] are very competitive in practice.

The second approach, *indexed searching*, tries to speed up searching by preprocessing the text and building a data structure that allows searching in $\mathcal{O}(m \cdot g(n) + occ \cdot h(n))$ time, where occ is the number of occurrences of the pattern in the text. Popular solutions to this approach are suffix trees and suffix arrays [10]. The first gives an $\mathcal{O}(m + occ)$ time solution, while the suffix array gives an $\mathcal{O}(m \log n + occ)$ time complexity which can be improved to $\mathcal{O}(m + occ)$ using extra space [1]. The problem of these approaches is that the space needed is too large for many practical situations (4–20 times the text size). Recently, a lot

---

* Partially funded by Millennium Nucleus Center for Web Research, Grant P04-067-F, Mideplan, Chile.

A. Amir, A. Turpin, and A. Moffat (Eds.): SPIRE 2008, LNCS 5280, pp. 87–98, 2008.

of effort has been spent to compress these indexes [13] obtaining a significant reduction in space, but requiring considerable implementation effort [5].

In this work we explore sampling the text by removing a set of characters from the alphabet. We first apply an online algorithm to this sampled text, obtaining an approach in between online searching and indexed searching. We call this kind of structure a *semi-index*. This is a data structure built on top of a text, which permits searching faster than any online algorithm, yet its search complexity is still of the form $\mathcal{O}(n \cdot f(m))$. To be interesting, a semi-index should be easy to implement and require little extra space. Several other semi-indexes exist in the literature, even without using that name. For example, q-gram indexes [12], directly searchable compression formats [11], and other sampling approaches.

We also consider indexing the sampled text. We build a suffix array indexing the sampled positions of the text, and get a sampled suffix array. This approach is similar to the sparse suffix array [8] as both index a subset of the suffixes, but the different sampling properties induce rather different search algorithms.

A challenge in our method is how to choose the best alphabet subset to sample. We present analytical results, supported by experiments, that simplify this process by drastically reducing the number of combinations to try. We show that it is sufficient in practice to sample the least frequent characters up to some limit. In both cases, online and indexed, our sampling technique significantly improves upon the state of the art, especially for relatively long search patterns. For example, online searching is speeded up by a factor of up to 5 on English text. For indexed searching we achieve indexes that are as fast as the classical suffix array, yet occupy less than 0.5 times the text size (instead of 4) plus text.

2   Text Sampling

The main idea of our online approach is to choose a subset of the alphabet to be the sampled alphabet and then to build a subsequence of the text by omitting all characters not in the sampled alphabet. At regular intervals we map the positions of the sampled text to their corresponding positions in the original text. When searching, we build the sampled pattern from the pattern by omitting all characters not in the sampled alphabet and then search for this sampled pattern in the sampled text. For each candidate returned by this search we verify a short range of the original text with the help of the position mapping.

Let $T = t_0 t_1 \ldots t_{n-1}$ be the text over the alphabet Σ and $\tilde{\Sigma} \subset \Sigma$ the sampled alphabet. The proposed semi-index is composed of the following items:

- Sampled text $\tilde{T}$: Let $\tilde{T} = t_{i_0} t_{i_1} \ldots t_{i_{\tilde{n}-1}}$ be the sequence of the t_i's that belong to the sampled alphabet $\tilde{\Sigma}$. The length of the sampled text is $\tilde{n}$.
- The position mapping M: A table of size $\lfloor \tilde{n}/q \rfloor$ where $M[i]$ maps the $q \cdot i$'th character of $\tilde{T}$ to its corresponding position in T so $\tilde{T}[q \cdot i] = T[M[i]]$.

Given a pattern $P = p_0 p_1 \ldots p_{m-1}$, search on this semi-index is carried out as follows. Let $\tilde{P} = p_{j_0} p_{j_1} \ldots p_{j_{\tilde{m}-1}}$ be the subsequence of p_i's that belong to the sampled alphabet $\tilde{\Sigma}$. The length of the sampled pattern is thus $\tilde{m}$. The sampled

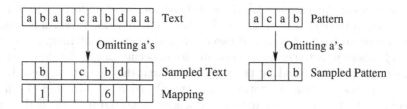

Fig. 1. Example of preprocessing

---

search $(\tilde{T} = \tilde{t}_0\tilde{t}_1 \ldots \tilde{t}_{\tilde{n}-1}, \tilde{P} = \tilde{p}_0\tilde{p}_1 \ldots \tilde{p}_{\tilde{m}-1}, T = t_0t_1 \ldots t_{n-1},$
        $P = p_0p_1 \ldots p_{m-1}, j_0, q, M[0 \ldots \tilde{n}/q])$

1.   for $(i \leftarrow 0$ to $\sigma - 1)$ $d[i] \leftarrow \tilde{m}$
2.   for $(i \leftarrow 0$ to $\tilde{m} - 2)$ $d[\tilde{p}_i] \leftarrow \tilde{m} - 1 - i$
3. $pos \leftarrow 0$
4.   while $(pos < \tilde{n} - \tilde{m})$
5. $j \leftarrow \tilde{m} - 1$
6.   while $(j \geq 0$ and $\tilde{t}_{pos+j} = \tilde{p}_j)$ $j \leftarrow j - 1$
7.   if $(j = -1)$
8.   Check for occurrence from $M[pos/q] + (pos \bmod q) - j_0$
9.   to $M[pos/q + 1] - (q - pos \bmod q) - j_0$
10. $pos \leftarrow pos + d[\tilde{t}_{pos+\tilde{m}-1}]$

Fig. 2. Searching the sampled text for a sampled pattern with the BMH algorithm

---

text $\tilde{T}$ is then searched for $\tilde{P}$, and for every occurrence, the positions to check in the original text are delimited by the position mapping M. If the sampled pattern is found in position i_r in $\tilde{T}$, the area $T[M[i_r/q] + (i_r \bmod q) - j_0 \ldots M[i_r/q + 1] - (q - i_r \bmod q) - j_0]$ is checked for possible startings of real occurrences.

For example, if the text is $T = abaacabdaa$, the sampled text built omitting the a's ($\tilde{\Sigma} = \{b, c, d\}$) is $\tilde{T} = t_1t_4t_6t_7 = bcbd$. If we map every other position in the sampled text, the position mapping M is $\{1, 6\}$. For searching the pattern $acab$ we omit the a's and get $\tilde{P} = p_1p_3 = cb$. We search for $\tilde{P} = cb$ in $\tilde{T} = bcbd$, finding an occurrence at position 1. The previous mapped position is $M[0] = 1$, so $\tilde{t}_0$ corresponds to t_1, and the next mapped position is $M[1] = 6$, so $\tilde{t}_2$ corresponds to t_6. Because the first sampled character in P is in position 1, we verify the area $1 \ldots 4$ in the original text finding the match at position 3. Preprocessing for the text and pattern of the previous example is shown in Fig. 1.

Because the sampled patterns tend to be quite short, we implemented the search phase with the BMH algorithm [7], which has been found to be fast in such settings [14]. Figure 2 shows the algorithm for this basic method.

Although the above scheme works well for most of the patterns, it is obvious that there are some bad patterns which would be searched faster in the original text. The average complexity of the BMH algorithm is $\mathcal{O}(n(1/m + 1/\sigma)) = \mathcal{O}(n/\min(m, \sigma))$ assuming a uniform and independent distribution of the characters of the alphabet [2]. If the distribution is not uniform, a better approximation is to replace σ by the the effective alphabet size $\bar{\sigma}$, which is defined as the inverse of the probability of two random characters matching, i.e. $1/\bar{\sigma} = \sum_{c \in \Sigma} p_c^2$, where p_c is the empirical probability of occurrence of the character c.

To determine if it would be faster to just search the pattern in the original text we tried calculating the ratios $n/\min(m, \bar{\sigma})$ and $n \cdot (1/m + 1/\bar{\sigma})$ both for the sampled text and pattern and for the original text and pattern. If the ratio is lower for the original text and pattern, we search only in the original text. The results were better using the ratio $n/\min(m, \bar{\sigma})$.

3   Optimal Sampling for Online Search

A question arises from the previous description of our sampling method: How to form the sampled alphabet $\tilde{\Sigma}$? We will first analyze how the average running time of the BMH algorithm changes when we sample the text and then, based on this, we will develop a method to find the optimal sampled alphabet. Throughout this section we assume that the characters are independent and we analyze the approach for a general pattern not known when preprocessing the text.

Let us define $b_A = \sum_{c \in A} p_c$ and $a_A = \sum_{c \in A} p_c^2$ where $A \subset \Sigma$. Now the length of the sampled text will be $b_{\tilde{\Sigma}} n$, the average length of the sampled pattern $b_{\tilde{\Sigma}} m$ (assuming it distributes similarly to the text) and the probability of two random characters matching in the sampled text $a_{\tilde{\Sigma}}/b_{\tilde{\Sigma}}^2$. Given the average complexity of the BMH algorithm, $\mathcal{O}(n(1/m + 1/\bar{\sigma}))$, the average search cost in the sampled text is

$$\mathcal{O}\left(b_{\tilde{\Sigma}} n \left(\frac{1}{b_{\tilde{\Sigma}} m} + \frac{a_{\tilde{\Sigma}}}{b_{\tilde{\Sigma}}^2}\right)\right) = \mathcal{O}\left(n \left(\frac{1}{m} + \frac{a_{\tilde{\Sigma}}}{b_{\tilde{\Sigma}}}\right)\right) .$$

When considering the verification cost we assume for simplicity that the mapping M contains the position of each sampled character in the original text, i.e. $q = 1$. The probability that a position has to be verified is then

$$p_{\text{ver}} = \sum_{i=0}^{m} \binom{m}{i} b_{\tilde{\Sigma}}^i (1 - b_{\tilde{\Sigma}})^{m-i} \left(\frac{a_{\tilde{\Sigma}}}{b_{\tilde{\Sigma}}^2}\right)^i = \left(\frac{a_{\tilde{\Sigma}}}{b_{\tilde{\Sigma}}} + 1 - b_{\tilde{\Sigma}}\right)^m .$$

If we assume that each verification costs $\mathcal{O}(m)$ then the cost of verification is

$$n \cdot p_{\text{ver}} \cdot \mathcal{O}(m) = n \cdot \left(\frac{a_{\tilde{\Sigma}}}{b_{\tilde{\Sigma}}} + 1 - b_{\tilde{\Sigma}}\right)^m \cdot \mathcal{O}(m) .$$

The total cost of searching in our scheme is thus

$$\mathcal{O}\left(n \cdot \left(\frac{1}{m} + \frac{a_{\tilde{\Sigma}}}{b_{\tilde{\Sigma}}} + \left(\frac{a_{\tilde{\Sigma}}}{b_{\tilde{\Sigma}}} + 1 - b_{\tilde{\Sigma}}\right)^m \cdot m\right)\right)$$

and hence the optimal sampled alphabet $\tilde{\Sigma}$ minimizes the cost per text character

$$E(\tilde{\Sigma}) = \frac{1}{m} + \frac{a_{\tilde{\Sigma}}}{b_{\tilde{\Sigma}}} + \left(\frac{a_{\tilde{\Sigma}}}{b_{\tilde{\Sigma}}} + 1 - b_{\tilde{\Sigma}}\right)^m \cdot m$$

which can be divided into the search cost in the sampled text $1/m + a_{\tilde{\Sigma}}/b_{\tilde{\Sigma}}$ and the verification cost $(a_{\tilde{\Sigma}}/b_{\tilde{\Sigma}} + 1 - b_{\tilde{\Sigma}})^m \cdot m$.

The verification cost always increases when a character is removed from the alphabet so the search cost in the sampled text must decrease for the combined cost to decrease. If $R = \Sigma \backslash \tilde{\Sigma}$ is the set of removed characters, the function

$$h_R(p) = \frac{1}{m} + \frac{a_\Sigma - a_R - p^2}{1 - b_R - p}$$

gives the search cost in the sampled text, per text character, if an additional character with probability p is removed. The derivative of $h_R(p)$ is

$$h'_R(p) = 1 - \frac{(1 - b_R)^2 - (a_\Sigma - a_R)}{(1 - b_R - p)^2}$$

which has exactly one zero $p_z = (1 - b_R) - \sqrt{(1 - b_R)^2 - (a_\Sigma - a_R)}$ in the interval $[0, 1 - b_R]$. We can see that the function $h_R(p)$ is increasing until p_z and decreasing after that. Solving the equation $h_R(p_R) = h_R(0)$ we get $p_R = (a_\Sigma - a_R)/(1 - b_R)$. So removing a single additional character decreases the search cost in the sampled text only if the probability of occurrence for that character is larger than p_R. Otherwise both the search cost in the sampled text and the verification cost will increase and thus removing the character is not beneficial.

Suppose now that we have already fixed whether we are going to keep or remove each character with probability of occurrence higher than p_c and now we need to decide if we should remove the character c. If $p_c > p_R$, we will need to explore both options as removing the character will decrease search cost in the sampled text and increase verification cost. However, if $p_c < p_R$ we know that if we added only c to R the searching time in the sampled text would also increase and therefore we should not remove c. But could it be beneficial to remove c together with a set of other characters with probabilities of occurrence less than p_R? In fact it cannot be. Suppose that we remove a character c with probability $p_c < p_R$. Now the new removed set will be $R' = R \cup \{c\}$ so we get $a_{R'} = a_R + p_c^2$ and $b_{R'} = b_R + p_c$. Now the new critical probability will be

$$p_{R'} = \frac{a_\Sigma - a_{R'}}{1 - b_{R'}} = \frac{a_\Sigma - a_R - p_c^2}{1 - b_R - p_c} \ .$$

We know that $h_R(p_c) > h_R(p_R) = h_R(0)$ because $p_c < p_R$. Therefore

$$\frac{1}{m} + \frac{a_\Sigma - a_R - p_c^2}{1 - b_R - p_c} > \frac{1}{m} + \frac{a_\Sigma - a_R}{1 - b_R}$$

and so

$$p_{R'} = \frac{a_\Sigma - a_R - p_c^2}{1 - b_R - p_c} > \frac{a_\Sigma - a_R}{1 - b_R} = p_R \ .$$

Thus even now it is not good to remove a character with probability less than the critical value p_R for the previous set and this will again hold if another character with a small probability is removed. Therefore we do not need to consider

$R_{opt} = \{\}$
sort characters of Σ in descending order
find_opt$(0, \{\})$
return R_{opt}

find_opt(i, R)
1. if $(i = \sigma)$
2.   if $(E(\Sigma \backslash R) < E(\Sigma \backslash R_{opt}))$
3. $R_{opt} = R$
4. else
5. $p_R = \frac{a_\Sigma - a_R}{1 - b_R}$
6.   if $(p_i > p_R)$
7.   find_opt$(i + 1, R \cup \{i\})$
8.   find_opt$(i + 1, R)$
9.   else
10.   find_opt(σ, R)

Fig. 3. Pseudo code for searching for the optimal set of removed characters

removing characters with probabilities less than p_R. Note however that removing a character with a higher probability will decrease the critical probability p_R and after this it can be beneficial to remove a previously unbeneficial character. In fact, if the sampled alphabet contains two characters with different probabilities of occurrence, the probability of occurrence for the most frequent character in the sampled alphabet is always larger than p_R. Thus it is always beneficial for searching in the sampled text to remove the most frequent character.

The above can be applied to prune the exhaustive search for the optimal set of removed characters. First we sort the characters of the alphabet in the decreasing order of frequency. We then figure out if it is beneficial for searching in the sampled text to remove the most frequent character not considered yet. If it is, we try both removing and not removing that character and proceed recursively for both cases. If it is not, we prune the search here because none of the remaining characters should be removed. Figure 3 gives the pseudo code.

In practice when using this pruning technique the number of examined sets drops drastically as compared to the exhaustive search, although the worst case is still exponential. For example, the number of examined sets drops from 2^{61} to 2,810 when considering the King James Bible as the text.

Table 1. Predicted and observed optimal number of removed characters for the King James Bible. The predicted optima are computed with the algorithm suggested by the analysis, which in our experiments always returned a set of most frequent characters.

m	10	20	30	40	50	60	70	80	90	100
Predicted optimal number of removed characters	3	7	9	11	12	13	14	15	16	16
Observed optimal number of removed characters	3	7	11	13	14	15	17	17	16	18

In our experiments, the optimal set of removed characters always contained the most frequent characters up to some limit depending on the length of the pattern, as shown in Table 1. Therefore a simpler heuristic is to remove the k most frequent characters for varying k and choose the set that predicts the best overall time. However, if the verification cost is very high for some reason (e.g. going to disk to retrieve the text, or uncompressing part of it) it is possible that the optimal set of removed characters is not a set of most frequent characters.

4 Sampled Suffix Array

To turn the sampling approach into an index, we use a suffix array to index the sampled positions of the text. When constructing the suffix array, only suffixes starting with a sampled character will be considered, but the sorting will still be done considering the full suffixes. The resulting sampled suffix array is like the suffix array of the original text where suffixes starting with unsampled characters have been omitted. The construction of the sampled suffix array can be done in $\mathcal{O}(n)$ time using $\mathcal{O}(\tilde{n})$ words of space if we apply the construction technique of the word suffix array [4]. The sampled suffix array for the text $T = abaacabdaa$ is shown in Fig. 4, where the sampled alphabet is $\tilde{\Sigma} = \{b, c, d\}$.

Search on the sampled suffix array is carried out as follows. Given a pattern $P = p_0 p_1 \ldots p_{m-1}$ we first find the first sampled character of the pattern. Let this be at index j. The pattern is now divided into the unsampled prefix $p_0 \ldots p_{j-1}$ and the suffix starting with the first sampled character $p_j \ldots p_{m-1}$. We search the sampled suffix array for this suffix of the pattern like in an ordinary suffix array. Each candidate match returned by this search will then be verified by comparing the unsampled prefix against the text.

We could also construct the suffix array directly for the sampled text but this would entail more verifications as the unsampled characters of the pattern suffix would not be required to match. We would also need to store the sampled text, or to skip the unsampled characters in the original text each time we read a suffix.

The sampled suffix array resembles a sparse suffix array [8], which indexes regularly sampled text positions. However, we only need to make one search on

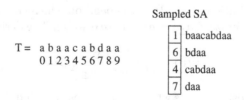

Fig. 4. The sampled suffix array for the text $T = abaacabdaa$ with the sampled alphabet $\tilde{\Sigma} = \{b, c, d\}$. The sorted suffixes are only shown for convenience. They are not part of the structure.

the sampled suffix array, while using a sparse suffix array one would need to make q searches if the sparse suffix array indexes every q'th position. On the other hand, the sampled suffix array can only be used for patterns that contain at least one sampled character whereas the sparse suffix array can be used if the pattern length is at least q. The variance of the search time when using the sampled suffix array is also larger than when using a sparse suffix array because in the sampled suffix array we have much less control over the length of the string that is used in the suffix array search.

5　Optimal Sampling for Suffix Array

Suppose that we have enough space to create the sampled suffix array for $b \cdot n$ suffixes where $0 < b < 1$. How should we now choose the sampled alphabet $\tilde{\Sigma}$ so that the search time would be optimal? Obviously $b_{\tilde{\Sigma}} = b$ but we still have a number of possible sampled alphabets to choose from. The search on the suffix array will compare the suffix of the pattern starting with the first sampled character against a text string $\mathcal{O}(\log n)$ times. The comparison time is minimized when the probability of matching for the first sampled character is minimized. Thus the sampled alphabet $\tilde{\Sigma}$ should be a set of least frequent characters.

Let us then consider the verification. The probability that two random characters are unsampled and match is $a_R = a_\Sigma - a_{\tilde{\Sigma}}$ where R is the set of removed characters. Thus the average cost of a single verification is $1/(1 - a_\Sigma + a_{\tilde{\Sigma}})$.

The probability that the suffix of the pattern starting with the first sampled character matches a random string of equal length is

$$b_{\tilde{\Sigma}} \frac{a_{\tilde{\Sigma}}}{b_{\tilde{\Sigma}}^2} (a_\Sigma)^{m_s - 1} = \frac{a_{\tilde{\Sigma}}}{b_{\tilde{\Sigma}}} (a_\Sigma)^{m_s - 1}$$

where m_s is the length of the suffix starting with the first sampled character. This is also the probability of verification per character in the original text. The average cost of verification per text character is then

$$\frac{a_{\tilde{\Sigma}}}{b_{\tilde{\Sigma}}} (a_\Sigma)^{m_s - 1} \cdot \frac{1}{1 - a_\Sigma + a_{\tilde{\Sigma}}} = \frac{a_{\tilde{\Sigma}}}{1 - a_\Sigma + a_{\tilde{\Sigma}}} \cdot \frac{(a_\Sigma)^{m_s - 1}}{b_{\tilde{\Sigma}}} .$$

Because we attempt to determine the optimal sampled alphabet such that $b_{\tilde{\Sigma}} = b$, $b_{\tilde{\Sigma}}$ and the distribution of m_s do not depend on which characters we remove. Thus we should minimize $f(a_{\tilde{\Sigma}}) = a_{\tilde{\Sigma}}/(1 - a_\Sigma + a_{\tilde{\Sigma}})$. The derivative of $f(a_{\tilde{\Sigma}})$ is

$$f'(a_{\tilde{\Sigma}}) = \frac{1 - a_\Sigma}{(1 - a_\Sigma + a_{\tilde{\Sigma}})^2} > 0$$

so the verification cost increases when $a_{\tilde{\Sigma}}$ increases. To minimize $a_{\tilde{\Sigma}}$ the sampled alphabet $\tilde{\Sigma}$ should be a set of least frequent characters. This also minimizes the total cost because also the suffix array search cost is minimized by this choice. Interestingly, this corresponds to the simplified heuristic we proposed in Sect. 3.

6   Experiments

6.1   Semi-index

To determine the sampled alphabet, we ran the exact algorithm of Sect. 3 for different pattern lengths to choose the sampled alphabet that produces the smallest estimated cost $E(\tilde{\Sigma})$. For all pattern lengths the algorithm recommended removing a set of most frequent characters. To see how well these results correspond to practice, we tested the semi-index approach by removing the k most frequent characters from the text for varying k. We used a 2 MB prefix of the King James Bible as the text, and the patterns are random substrings of the text. For each pattern length 500 patterns were generated, and the reported running times are averages over 200 runs with each of the patterns. The most frequent characters in the decreasing order of frequency were "␣ethaonsirdlfum,wycgbp" where ␣ is

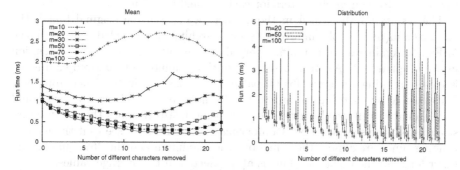

Fig. 5. The running time for various pattern lengths for the basic method. The left figure shows the mean running time; the right shows the median, minimum, maximum, and 25% and 75% quartiles.

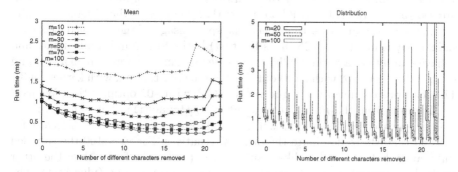

Fig. 6. The running time for various pattern lengths for the tuned version where searching in the sampled text is skipped if it looks like searching in the original text is faster. The left figure shows the mean running time; the right figure shows the median, minimum, maximum, and 25% and 75% quartiles.

the space character. The tests were run on a 1.0 GHz AMD Athlon dual core processor with 2 GB of memory, 64 kB L1 cache and 512 kB L2 cache, running Linux 2.6.23. The code is in C and compiled with gcc using -O3 optimization.

Figure 5 shows the results of these experiments with the basic method mapping every 64'th sampled character to its position in the original text. If we make the mapping sparser the running time will start to increase a little earlier, but the effect is quite mild. The results for zero removed characters correspond to the original BMH algorithm. As we can see, the semi-index is up to 5 times faster, especially when the patterns are long. Figure 5 also shows that, for each pattern length, there is an optimal number of characters to remove. A comparison of these optima and those given by the analysis is shown in Table 1. As we can see, the analysis gives reasonably good results although it recommends removing too few characters with long patterns, because we estimated the verification time quite pessimistically. When more characters are removed it is unlikely that we would need to scan m characters for each verified position.

The results for the tuned method, where we search the original text if the ratio $n/\min(m, \bar{\sigma})$ looks unfavorable for searching the sampled text, is shown in Fig. 6. Again we are mapping every 64'th sampled character to its position in the original text. As we can see, the optimal number of removed characters is closer to being the same for all pattern lengths than in the basic approach. For example by choosing to remove the 13 most frequent characters, we would do reasonably well for all pattern lengths using just 0.18 times the original text size to store the sampled text. Comparing Figs. 5 and 6 we see that the median running times are almost the same, but the maximum and the 75% quartile are lower for the tuned method. This is also reflected in the average values.

6.2 Sampled Suffix Array

Figure 7 shows the results obtained by comparing our sampled suffix array against our implementation of the sparse suffix array [8] and the locally compressed suffix array (LCSA) [6], an index that compresses the differential suffix array using Re-Pair. Note that when the space usage of the sampled or sparse suffix array is maximal (3.25 times the text) both of them index all suffixes and behave exactly like a normal suffix array. The experiments were run on a Pentium IV 2.0 GHz processor with 2 GB of RAM running SuSE Linux with kernel 2.4.31. The code was compiled using gcc version 3.3.6 with -O9 optimization. We used 50 MB texts from the *PizzaChili* site, http://pizzachili.dcc.uchile.cl.

Our approach performs very well for moderate to long patterns. Already for $m = 50$ it starts to dominate the other alternatives. For $m = 100$ the sampled suffix array behaves almost like a suffix array (and much faster than the other methods), even when using less than 0.5 times the text size (plus text). The novel compressed self-indexes [5,13] are designed to use much less space (e.g. 0.8 times the text size including the text) but take much more time, and thus are inappropriate for this comparison. We chose the LCSA as an alternative that compresses less but is much faster than the other self-indexes [6]. Its compression performance varies widely with the text type, and is not particularly good on

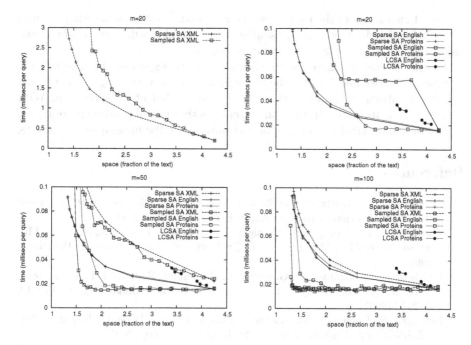

Fig. 7. Search times for the sampled and sparse suffix arrays and LCSA for XML, English and protein data. LCSA uses little space for XML data but it is much slower than the other approaches, so these results are not shown. The top figures show results for pattern length 20 and the bottom figures show the results for pattern lengths 50 and 100. The space fraction includes that of the text, so it is of the form $1 + \frac{\text{index size}}{\text{text size}}$.

English and Proteins. On XML it requires extra space equal to the size of the text, yet its times are much higher and fall well outside the plot (and this is still much faster than the other self-indexes!). The LCSA, on the other hand, would perform better on shorter patterns, where our index is not competitive.

7   Conclusions and Further Work

We have presented two sampling approaches to speed up string matching with long patterns. The sampled semi-index profits from nonuniform character distribution to gain a speedup over online searching, while the sampled suffix array works also with a uniform distribution. It is also worth noting that in the semi-index approach the sampled text is an internal structure of the semi-index so any transform, like compression or code splitting [15], could be applied to it.

The current approach is not applicable to small alphabets. To extend the approach to smaller alphabets we could use q-grams. In the semi-index approach we would then define a sampled alphabet for each $(q - 1)$-long context and the sampled text would contain those characters that are sampled in the context where they occur. When searching for a pattern, we must always discard the first $q - 1$ characters of the pattern as their context is not known. Using q-grams with the sampled suffix array is simpler. The sampled suffix array would just index all suffixes starting with a sampled q-gram.

Another interesting direction to minimize the extra space of the semi-index is to replace the original text by the subsequence of the non-sampled characters, and use a bitmap to indicate the subset each symbol of T belongs to. With *rank/select* capabilities [13] this bitmap replaces the current position mapping for verification and permits searching on the sampled or the unsampled characters.

References

1. Abouelhoda, M., Kurtz, S., Ohlebusch, E.: Replacing suffix trees with enchanced suffix arrays. Journal of Discrete Algorithms 2(1), 53–86 (2004)
2. Baeza-Yates, R.: String searching algorithms revisited. In: Dehne, F., Sack, J.R., Santoro, N. (eds.) WADS 1989. LNCS, vol. 382, pp. 75–96. Springer, Heidelberg (1989)
3. Crochemore, M., Czumaj, A., Gąsieniec, L., Jarominek, S., Lecroq, T., Plandowski, W., Rytter, W.: Speeding up two string-matching algorithms. Algorithmica 12, 247–267 (1994)
4. Ferragina, P., Fischer, J.: Suffix arrays on words. In: Ma, B., Zhang, K. (eds.) CPM 2007. LNCS, vol. 4580, pp. 328–339. Springer, Heidelberg (2007)
5. Ferragina, P., González, R., Navarro, G., Venturini, R.: Compressed text indexes: From theory to practice (manuscript 2007), http://pizzachili.dcc.uchile.cl
6. González, R., Navarro, G.: Compressed text indexes with fast locate. In: Ma, B., Zhang, K. (eds.) CPM 2007. LNCS, vol. 4580, pp. 216–227. Springer, Heidelberg (2007)
7. Horspool, R.N.: Practical fast searching in strings. Software – Practise & Experience 10, 501–506 (1980)
8. Kärkkäinen, J., Ukkonen, E.: Sparse suffix trees. In: Cai, J., Wong, C.K. (eds.) COCOON 1996. LNCS, vol. 1090, pp. 219–230. Springer, Heidelberg (1996)
9. Knuth, D.E., Morris, J.H., Pratt, V.R.: Fast pattern matching in strings. SIAM Journal on Computing 6, 323–350 (1977)
10. Manber, U., Myers, G.: Suffix arrays: A new method for online string searches. SIAM Journal on Computing 22(5), 935–948 (1993)
11. Moura, E., Navarro, G., Ziviani, N., Baeza-Yates, R.: Fast and flexible word searching on compressed text. ACM Trans. on Information Systems 18(2), 113–139 (2000)
12. Navarro, G., Baeza-Yates, R., Sutinen, E., Tarhio, J.: Indexing methods for approximate string matching. IEEE Data Engineering Bulletin 24(4), 19–27 (2001)
13. Navarro, G., Mäkinen, V.: Compressed full-text indexes. ACM Computing Surveys 39(1), 1–61 (2007)
14. Navarro, G., Raffinot, M.: Flexible Pattern Matching in Strings – Practical on-line search algorithms for texts and biological sequences. Cambridge University Press, Cambridge (2002)
15. Rautio, J., Tanninen, J., Tarhio, J.: String matching with stopper encoding and code splitting. In: Apostolico, A., Takeda, M. (eds.) CPM 2002. LNCS, vol. 2373, pp. 45–52. Springer, Heidelberg (2002)

Mismatch Sampling

Raphaël Clifford[1], Klim Efremenko[2], Benny Porat[3], Ely Porat[3,*],
and Amir Rothschild[4]

[1] University of Bristol, Dept. of Computer Science, Bristol, BS8 1UB, UK
clifford@cs.bris.ac.uk
[2] Bar-Ilan University, Dept. of Computer Science, 52900 Ramat-Gan,
Israel and Weizman institute, Dept. of Computer Science and Applied Mathematics,
Rehovot, Israel
klimefrem@gmail.com
[3] Bar-Ilan University, Dept. of Computer Science, 52900 Ramat-Gan, Israel
bennyporat@gmail.com, porately@cs.biu.ac.il
[4] Tel-Aviv University, Dept. of computer science, Tel-Aviv, Israel and Bar-Ilan
University, Dept. of Computer Science, 52900 Ramat-Gan, Israel
rotshch@post.tau.ac.il

Abstract. We consider the well known problem of pattern matching
under the Hamming distance. Previous approaches have shown how to
count the number of mismatches efficiently, especially when a bound is
known for the maximum Hamming distance. Our interest is different in
that we wish collect a random sample of mismatches of fixed size at
each position in the text. Given a pattern p of length m and a text t
of length n, we show how to sample with high probability c mismatches
where possible from every alignment of p and t in $O((c + \log n)(n +
m \log m) \log m)$ time. Further, we guarantee that the mismatches are
sampled uniformly and can therefore be seen as representative of the
types of mismatches that occur.

1 Introduction

Approximate pattern matching is one of the most studied problems in computer
science. Numerous measures of approximation have been developed over the years
with wide ranging applications from computer vision to bioinformatics. The challenge of approximate matching is that with every different way of measure the distance between two strings comes the need to develop often entirely novel techniques
to cope with the need for ever larger amounts of data to be processed efficiently.

One of the most common and simplest measures of approximation is the
Hamming distance. Given a pattern p of length m and a text t of length n, the
task is to return the number of mismatches between p and every substring of t
of length m. Much work has gone into fast solutions to this general problem as
well as a restricted version called k-mismatch where only distances up to k are
to be reported. However, in many situations it is desirable to know not only how

---

* Research supported in part by the Binational Science Foundation (BSF) and the
Israel Science Foundation (ISF).

A. Amir, A. Turpin, and A. Moffat (Eds.): SPIRE 2008, LNCS 5280, pp. 99–108, 2008.
© Springer-Verlag Berlin Heidelberg 2008

many mismatches occur but also to have some idea what the mismatches are. Unfortunately, in the worst case no algorithm that returns all mismatches can run in less than $\Theta(nm)$ time.

In order to be able to understand which mismatches occur it will therefore be necessary to return a fixed sample of the mismatches at each alignment. We call the problem we consider Mismatch Sampling and define it as follows. Given an integer c, sample uniformly at random c distinct mismatches that occur between the pattern and text at each possible alignment. Where the Hamming distance is less than c, all mismatches are to be reported. Such samples will have a variety of interpretations depending on the context but can be seen as representing typical spelling errors when searching text or for example, common DNA mutations in the context of bioinformatics. In the general case where the maximum Hamming may be as large as m, we are not aware of any existing techniques which improve on the naive $\Theta(nm)$ time algorithm in the worst case.

In the process of tackling the Mismatch Sampling problem, we also give a faster randomised algorithm for the 1-mismatch problem which improves the time complexity from $O(n \log m)$ to $O(n + m \log m)$. The 1-mismatch algorithm is a powerful tool in its own right which has been used to develop related algorithms for the k-mismatch with don't cares problem [3] and generalised pattern matching [9], for example. In the latter case, the new 1-mismatch algorithm can be used as a direct replacement and in the former the same speedup applies if the don't cares occur only in the pattern.

2   Preliminaries

Let Σ be a set of characters which we term the *alphabet* and let $t = t_1 t_2 \dots t_n \in \Sigma^n$ be the text and $p = p_1 p_2 \dots p_m \in \Sigma^m$ the pattern. The terms *symbol* and *character* are used interchangeably throughout. Similarly, we will sometimes refer to a *location* in a string and synonymously at other times a *position*. We will also refer to an *alignment* of the pattern and text which is to be understood as the location in the text where the pattern starts to be compared.

Definition 1. *Define $HD(i)$ to be the Hamming distance between p and $t[i, \dots, i + m - 1]$.*

Our algorithms make extensive use of the fast Fourier transform (FFT). An important property of the FFT is that in the RAM model, the cross-correlation,

$$(t \otimes p)[i] \stackrel{\text{def}}{=} \sum_{j=1}^{m} p_j t_{i+j-1}, \ \ 0 \leq i \leq n - m + 1,$$

can be calculated accurately and efficiently in $\mathcal{O}(n \log n)$ time both over the integers $\mathbb{Z}$, and a finite field F_q (see e.g. [4], Chapter 32). By a standard trick of splitting the text into overlapping substrings of length $2m$, the running time can be further reduced to $\mathcal{O}(n \log m)$.

We will often assume that the text is of length $2m$ in the presentation of an algorithm or analysis in this paper and that the reader is familiar with this splitting technique.

In order to fix terminology we give a definition of the term *with high probability* which is also abbreviated to w.h.p. Our definition is at the stricter end of those found in the literature when analysing randomised algorithms and has as one consequence that the bounds still hold even if the algorithm is repeated a polynomial number of times.

Definition 2. *We say that an algorithm outputs the correct answer with high probability or* w.h.p. *in time* $\Theta(f(n))$ *if for every* $\alpha \geq 1$, *there exist a value* $c_\alpha > 0$ *depending on* α, *such that after* $\Theta(f(n))$ *time, the algorithm outputs the correct answer with probability at least* $1 - \frac{c_\alpha}{n^\alpha}$. *Note that the constant in the* Θ *notation may also depend on* α.

3   Related Work and Previous Results

Much progress has been made in finding fast algorithms for the Hamming distance problem over the last 20 years. $O(n\sqrt{m \log m})$ time solutions based on repeated applications of the FFT were given independently by both Abrahamson and Kosaraju in 1987 [1, 7]. The major improvements have concentrated on a bounded version of the problem called k-mismatch. In this problem an integer bound k is given in advance and only Hamming distances less than or equal to k need be reported. In 1985 Landau and Vishkin gave a beautiful $O(nk)$ algorithm that is not FFT based which uses constant time lowest common ancestor (LCA) operations on the suffix tree of p and t [8]. This was subsequently improved in [2] to $O(n\sqrt{k \log k})$ time by a method based on filtering and FFTs again. Approximations within a multiplicative factor of $(1 + \epsilon)$ to the Hamming distance can also be found in $O(n/\epsilon^2 \log m)$ time   [6]. More recently, a randomised algorithm for the k-mismatch problem with *don't cares* was given which runs in $O(n(k + \log m \log k) \log n)$ time and returns all the mismatches found [3].

The existing fast k-mismatch algorithms do also return the mismatches that have been found and so might appear to be helpful for our problem of mismatch sampling. However, in our case k can be as large as m which would give an algorithm whose time complexity is no better than a naive $O(nm)$ solution. Despite this limitation, some the techniques we will employ are related to those given in the previous work of [3]. The most important similarity is the idea of sampling single mismatches using masked versions of the patterns. However the solution we present is different and more efficient than those given before and as we will show, a number of further technical obstacles need to be overcome before we are able to provide a full solution for the Mismatch Sampling problem.

4   Results and New Techniques

We summarise the main results and discuss the techniques that were developed.

- The first result in Section 5 is a randomised algorithm that solves a problem called 1-mismatch with constant probability in $O(n + m \log m)$ time. The

1-mismatch problem is to find a single mismatch, where at least one occurs, at every alignment. The algorithm gives the correct answer with constant probability as long as the true Hamming distance $HD(i)$, can be estimated to within a constant factor.

The main new technique here is a sampling trick which masks out enough of the pattern so that only one mismatch is likely to remain. This enables us to sample single mismatches even when the true Hamming distance may be considerably larger. In order to perform this sampling efficiently, we show that the expensive cross-correlation calculations using FFTs need only be performed over a constant number of arrays of length $2m$. The only computations that have to be performed on an array of the length of the text run in linear time. This saves a log factor in the overall time complexity as well as being practically more efficient due to the overheads inherent in the FFT calculations.

We also show that by performing all calculations modulo a large prime q, we can ensure that the single mismatches found are chosen uniformly at random from the set of mismatches at each alignment. The overall time complexity of the algorithms is maintained by performing the FFT calculations and hence the cross-correlations, over the field F_q.

– In Section 6 we present the first solution for the Mismatch Sampling problem which samples $\min(c, HD(i))$ mismatches w.h.p. at every alignment. The 1-mismatch algorithm is repeatedly run over $O(\log m)$ stages with each stage providing a different estimate of the Hamming distances between pattern and text. This gives an $O(c \log n(n + m \log m) \log m)$ time algorithm. It is important to note that the probabilistic bounds we give are particularly strong and hold for every position in the text simultaneously.

– Finally we show how by using a k-mismatch algorithm as a preprocessing step, we are able to speed up the Mismatch Sampling algorithm and still find the correct answer w.h.p. The main idea is quickly to eliminate all positions where the Hamming distance is less than $2c$ and then concentrate only on those remaining positions. The overall time complexity is therefore reduced to $O((c + \log n)(n + m \log m) \log m)$ time. We also show that the algorithm can easily be made Las Vegas while maintaining the same running time w.h.p.

5   Randomised 1-Mismatch

We first present the main algorithmic tool that will be used to sample distinct mismatches. The 1-mismatch problem is to find a single mismatch, where at least one occurs, between the pattern and every alignment of the text. The overall strategy for Mismatch Sampling will be to repeatedly sample single mismatches from each alignment of the pattern and text using an algorithm for the 1-mismatch problem.

Our solution to the 1-mismatch problem is randomised and requires $O(n + m \log m)$ time per iteration, returning a single sampled mismatch for each alignment where $HD(i) \geq 1$, with constant probability. However, in order to find the mismatches with constant probability we will require an estimate within a constant factor of the Hamming distance at each alignment of the pattern and text. For the time being we assume that such an estimate is available and in Section 6 we show that only $O(\log m)$ distinct estimates will be required overall.

In order to be able to sample individual mismatches we must find a way to eliminate all other mismatches that could have occurred. We first create masked versions of the pattern, so that an alignment of the masked pattern and the text is likely to only contain one mismatch. To perform this efficiently we create a random array r of length $2m$. A *sampling rate* s is then defined which determines the probability that a given r_j will be set to zero. For a given sampling rate s, the aim is for 1-mismatch to find single mismatches for every alignment i for which $s \leq HD(i) \leq 2s$. The aim will be to mask out all but one mismatch at each alignment by multiplying the difference $(p_j - t_{i+j-1})$ by the random element r_{i+j-1}.

We define the random array r such that each $r_j = 0$ with probability $((s-1)/s)$ and 1 with probability $1/s$, r_j is chosen independently and uniformly at random from $[1, \ldots, q-1]$. We set q to be a prime which is larger than $\max((\max_{i,j} |(p_j - t_i)|), n))$. We then compute the cross-correlation between the pattern and r and an array r' of the same length as r such that $r'_i = ir_i$. This gives two arrays $A = p \otimes r$ and $C = p \otimes r'$. In this way, any values set to zero in r will effectively eliminate the contribution from corresponding values in p. The cross-correlation calculations over arrays of length $2m$, need only to be performed once per iteration of the 1-mismatch algorithm and do not require the input text. In this way it can be seen as a preprocessing step.

For each position i in the text we then calculate $\sum_{j=1}^{m} r_{i+j-1}(i+j-1)(p_j - t_{i+j-1})/\sum_{j=1}^{m} r_{i+j-1}(p_j - t_{i+j-1})$ with all calculations performed over the finite field F_q. This calculation is the main body of the 1-mismatch algorithm and Algorithm 1 describes the main steps assuming the text of length $2m$. Using the standard method of splitting the text into segments of length $2m$ with overlap m described in Section 2 the algorithm can then be applied to the whole text.

For any i where there is exactly one mismatch between p and $t[i, \ldots, i+m-1]$, $E[i]$ is the location in t of the mismatch and $E[i] - i + 1$ is the location in p. We can check for all the positions where there are no mismatches in linear time using any of the well known exact matching algorithms. As we have the location of the proposed mismatch in both the pattern and text a simple constant time check per alignment will tell us if we have indeed found a mismatch.

Input: Pattern p, text t, random array r and prime q
Output: $E[i]$ contains a single mismatch location with probability at least $1/2e$
Compute A s.t. $A[i] = \sum_{j=1}^{m} r_{i+j-1}p_j$ for each $1 \leq i \leq m$;
Compute B s.t. $B[i] = \sum_{j=1}^{m} r_{i+j-1}t_{i+j-1}$ for each $1 \leq i \leq m$;
Compute C s.t. $C[i] = \sum_{j=1}^{m} (i+j-1)r_{i+j-1}p_j$ for each $1 \leq i \leq m$;
Compute D s.t. $D[i] = \sum_{j=1}^{m} (i+j-1)r_{i+j-1}t_{i+j-1}$ for each $1 \leq i \leq m$;
Compute $E = (C - D)/(A - B)$;

Algorithm 1. Randomised 1-MISMATCH for text of length $2m$

Lemma 1. *Algorithm 1 run over a text of length n takes $O(n + m \log m)$ time.*

Proof. Calculating B, D and E for every partition of the text into sections of length $2m$ takes $O(n)$ in total. The time required to compute A and C is dominated by the running time of the FFT on an array of length $O(m)$ and is therefore $O(m \log m)$. A and C do not need to be recalculated for each segment of the text of length $2m$. The total running time of Algorithm 1 is therefore $\Theta(n + m \log m)$

Lemma 2. *For a given alignment i and sampling rate $s \le HD(i) \le 2s$, Algorithm 1 samples a single mismatch uniformly from the set of mismatches at alignment i with probability at least $1/2e$.*

Proof. Suppose $k = HD(i) \ge 1$, and let $i_1, ..., i_k$ be the locations of the mismatches. Algorithm 1 will find a single mismatch i_1 if $r_{i+i_j-1} = 0$ for all $1 < j \le k$ and $r_{i+i_1-1} > 0$. Therefore the probability of a single mismatch being found is $\frac{k}{s}(1 - \frac{1}{s})^k$ which is bounded below by $1/2e$. It is also possible that the algorithm will accidentally find a mismatch even when there are two or more mismatch positions available. However this can only increase the probability of a mismatch being found. Finally, it is also possible that the denominator $(A - B) = \sum_{j=1}^{m} r_{i+j-1}(p_j - t_{i+j-1}) = 0$ in which case we know that there can't be a single mismatch at the relevant alignment and so we can simply discard the result.

We now show that any mismatch found is selected uniformly at random. If there is only one mismatch then the result follows immediately from the observation that the non-zero elements of r are chosen with equal probability. We also know that each non-zero r_{i+j-1} is chosen uniformly from $[1, \ldots, q]$ and that all calculations are being performed in F_q. As the sum of two uniform random numbers in F_q is also a uniform random number, then it follows that if there is in fact more than one mismatch from which $(C - D)/(A - B)$ returns a valid mismatch position, the mismatch will also be chosen uniformly at random.

Non-uniform Sampling

In our application we require that the 1-mismatch algorithm returns mismatches chosen uniformly at random in order that the full Mismatch Sampling algorithm will return uniformly random subsets of the possible mismatches. However, 1-mismatch is also useful as a tool in its own right and there can be situations where it is not required that the mismatches are chosen at random. In this case a simplification to Algorithm 1 is possible which may provide a practical speedup. Instead of choosing r_j from a large range, we can simply make r a binary array with r_j set to 0 with probability $(s - 1)/s$ and 1 with probability $1/s$. The calculations, including the FFTs can then be performed over the integers instead of the finite field F_q.

In this case, the mismatches found will no longer be chosen uniformly at random from the possible mismatches at each alignment. Why this is true is

---

Input: Pattern p, text t and an integer c
Output: $O[i]$ =sample of up to $\min(HD(i), c)$ distinct mismatches
/* Iterate over $O(\log m)$ `sample rates`   */
for $\ell = 1$ *to* $\log_2 m$ **do**
    for $O(c \log n)$ *times* **do**
        Create random array r with sampling rate $s = 2^{\ell-1}$;
        Perform 1-MISMATCH(p, t, r);
        Add new mismatches to output O;
    end
end

Algorithm 2. Simple Mismatch Sampling

best shown with an example. Consider $p = 4, 4, 4$ and $t = 3, 2, 3$ so that $E = (r_1 + 4r_2 + 3r_3)/(r_1 + 2r_2 + r_3)$. In this case, if r has two or more non-zero positions then the only mismatch that will be found is at position 2. However, single mismatches will still be found with probability at least $1/2e$ although no longer uniformly at random.

6 The Mismatch Sampling Algorithm

In this Section we will present the main Mismatch Sampling algorithm. This will be done in two phases. In the first we will give a simple algorithm based on repeated applications of 1-mismatch which runs in $O((c \log n)(n + m \log m) \log m)$ time and samples c mismatches w.h.p. wherever possible w.h.p. We will then show how to speed up the approach using k-mismatch as a preprocessing stage resulting in the final $O((c + \log n)(n + m \log m) \log m)$ Mismatch Sampling algorithm. We also discuss how this algorithm can be made Las Vegas without increasing the time complexity w.h.p.

To start, recall from Section 5 that the 1-mismatch algorithm requires an estimate of the Hamming distance. In order to apply it to the full problem where the Hamming distance at each alignment is not known, we will require $O(\log m)$ stages overall. At each stage ℓ we set the sampling rate s set to $2^{\ell-1}$. The algorithm which is set out in Algorithm 2 will repeat 1-mismatch a sufficient number of times at each sampling rate so that when the correct sampling rate is found, we will find c mismatches w.h.p., assuming the true Hamming distance is at least c.

The following Lemma shows that when the correct sampling rate is found for a particular alignment i, all $\min(c, HD(i))$ mismatches will be found w.h.p. The proof is an application of the *coupon collector's problem* (see e.g. [5]). Although the usual analysis of the coupon collector's problem would require only $O(c \log c)$ iterations we set the number of iterations to $O(c \log n)$ in order to ensure that the probabilistic bound holds at all alignments in the text simultaneously.

Lemma 3. *For all alignments i such that $s \leq HD(i) \leq 2s$, at least $\min(c, HD(i))$ distinct mismatches will be found after $O(c \log n)$ iterations of 1-mis-match w.h.p. and they will be chosen uniformly at random from the set of mismatches alignment i.*

We can now give the running time of the first Mismatch Sampling algorithm.

Theorem 3. *For each $1 \leq i \leq n$, Algorithm 2 samples $\min(c, HD(i))$ mismatches w.h.p. uniformly at random in $O((c \log n)(n + m \log m) \log m)$ time.*

Proof. From Lemma 3 we know that after $O(c \log n)$ iterations we will find $\min(c, HD(i))$ mismatches w.h.p. for $s \leq HD(i) \leq 2s$. Therefore, by repeating this process for each of $O(\log m)$ sampling rates, s, we will find $\min(c, HD(i))$ mismatches w.h.p. at every alignment i. Our algorithm performs $O(c \log n \log m)$ 1-mismatch procedures. Therefore the overall running time is $O(c \log n(n + m \log m) \log m)$.

Mismatch Sampling in $O((c + \log n)(n + m \log m) \log m)$ time

We now show the final speedup and give the full Mismatch Sampling algorithm. First, we observe that some positions are easier to sample c mismatches from than others. In particular, the following Lemma shows that if there are more than $2c$ mismatches, we can sample c mismatches more quickly than before.

Lemma 4. *For all alignments i such that $s \leq HD(i) \leq 2s$, and $HD(i) \geq 2c$, c distinct mismatches will be found after $O(c + \log n)$ iterations of 1-mismatch w.h.p. and will be chosen uniformly at random from the set of mismatches at alignment i.*

Proof. By Lemma 2 we will find one mismatch at every iteration of 1-mismatch algorithm with constant probability. Because $HD(i) \geq 2c$, if we have found fewer than c mismatches then probability that a discovered mismatch will be new is at least $1/2$. So at every iteration of the 1-mismatch algorithm we will found a *new* mismatch with constant probability. So after $O(c + \log n)$ iterations we will have found at least c mismatches w.h.p. As before, this bound holds at every alignment in the text simultaneously.

The second part of the improvement is to eliminate all the alignments where fewer than $2c$ mismatches occur. This can be done by using the k-mismatch algorithm of Landau and Vishkin [8] and setting $k = 2c$. After this preprocessing step we will have found $\min(c, HD(i))$ mismatches for all alignments where $HD(i) \leq 2c$ in $O(nc)$ time. We can now concentrate only on those alignments where $HD(i) > 2c$.

Algorithm 3 sets out the main steps and Theorem 4 sets out the final running time.

```
Input: Pattern p, text t and an integer c
Output: O[i] =sample of up to min(HD(i), c) distinct mismatches
/* Eliminate alignments with few mismatches              */
Run 2c-mismatch(t,p) ;
/* Many mismatches stage                                 */
for ℓ = ⌊log 2c⌋ to log m do
    for O(c + log n) times do
        Create random array r with sampling rate s = 2^{ℓ-1};
        Perform 1-MISMATCH(p, t, r);
        Add new mismatches to output O;
    end
end
```

Algorithm 3. Mismatch sampling

Theorem 4. *For each* $1 \leq i \leq n$, *Algorithm 3 samples* $\min(c, HD(i))$ *mismatches w.h.p. uniformly at randomly in* $O((c + \log n)(n + m \log m) \log m)$ *time.*

Proof. The 2c-mismatch algorithm handles the cases with at most $2c$ mismatches and runs in $O(nc)$ time. Then we choose random array r and perform the 1-mismatch algorithm $O((\log \frac{m}{c})(c + \log n)$ times taking $O((n + m \log m)(\log \frac{m}{c})(c + \log m)$ time overall. After performing these two stages, at each alignment i we have found $\min(c, HD(i))$ mismatches w.h.p. Therefore we will also find $\min(c, HD(i))$ mismatches at all alignments w.h.p.

By repeating the 1-mismatch step in Algorithm 3 until $\min(c, HD(i))$ mismatches are found at each alignment, the algorithm can straightforwardly be made Las Vegas and it follows from Theorem 4 that the running time will be $O((c + \log n)(n + m \log m) \log m)$ w.h.p.

7   Discussion

The motivation for Mismatch Sampling can be applied to any number of approximate pattern matching problems and it is of interest to know which will allow a fixed size sample to be given efficiently. The most obvious direct extension is to Mismatch Sampling for the Hamming distance with don't cares problem considered in [3], where a deterministic 1-mismatch algorithm allowing don't cares is developed. The sampling rate s from Section 6 can now be used to choose positions from the patterns with the remaining positions replaced with don't care characters. Following the same overall strategy of sampling single mismatches over $O(\log m)$ stages will now give a Mismatch Sampling algorithm allowing don't cares that runs in $O((c + \log n)(n \log^2 m))$ time. The problem of sampling errors where pattern matching is to be performed over more sophisticated approximation measures, such as the edit distance for example, appears to be considerably more challenging.

References

[1] Abrahamson, K.: Generalized string matching. SIAM journal on Computing 16(6), 1039–1051 (1987)

[2] Amir, A., Lewenstein, M., Porat, E.: Faster algorithms for string matching with k mismatches. J. Algorithms 50(2), 257–275 (2004)

[3] Clifford, R., Efremenko, K., Porat, E., Rothschild, A.: k-mismatch with don't cares. In: Arge, L., Hoffmann, M., Welzl, E. (eds.) ESA 2007. LNCS, vol. 4698, pp. 151–162. Springer, Heidelberg (2007)

[4] Cormen, T.H., Leiserson, C.E., Rivest, R.L.: Introduction to Algorithms. MIT Press, Cambridge (1990)

[5] Feller, W.: An introduction to probability theory and its applications, vol. 1. Wiley, Chichester (1968)

[6] Indyk, P.: Faster algorithms for string matching problems: Matching the convolution bound. In: Proceedings of the 38th Annual Symposium on Foundations of Computer Science, FOCS 1998, pp. 166–173 (1998)

[7] Kosaraju, S.R.: Efficient string matching (manuscript, 1987)

[8] Landau, G.M., Vishkin, U.: Efficient string matching with k mismatches. Theoretical Computer Science 43, 239–249 (1986)

[9] Porat, E., Efremenko, K.: Approximating general metric distances between a pattern and a text. In: Huang, S.-T. (ed.) SODA, pp. 419–427. SIAM, Philadelphia (2008)

Sliding CDAWG Perfection*

Martin Senft and Tomáš Dvořák

Faculty of Mathematics and Physics
Charles University, Prague, Czech Republic
{Martin.Senft,Tomas.Dvorak}@mff.cuni.cz

Abstract. The Compact Directed Acyclic Word Graph (CDAWG) is a well-known suffix data structure designed for an efficient solution to problems on strings. Some applications, especially those from the data compression field, require maintaining a CDAWG over a sliding window. The fastest known solution to this problem is an approximation algorithm that slides a CDAWG in an amortized constant time. However, the existence of an exact algorithm performing within the same complexity bounds has been an open question so far. We show that the answer to this question is negative and present an on-line algorithm with the best asymptotic complexity possible.

1 Introduction

The Compact Directed Acyclic Word Graph (CDAWG) [1,2,3] is a well known suffix data structure which allows an efficient solution to a number of problems on strings. Some applications, especially those from the data compression field [4,5], require maintaining a CDAWG over a sliding window. The sliding could be implemented in a naive way by building a new CDAWG from scratch for each subsequent window, which requires time and space proportional to the window size. However, it would be desirable to design a solution that does the change incrementally to preserve the additional information (e.g. branching statistics [4,5]) added to the CDAWG by the algorithm that uses it.

The process of incremental sliding of the CDAWG over some string can be split into two operations: deletion of the leftmost character of the string and addition of a new character to the right side. The addition is solved by an on-line CDAWG construction algorithm [3] that works in an amortized constant time. However, the deletion appears to be a bigger challenge. The best existing solution is an approximation algorithm that works in an amortized constant time [2]. But this algorithm may sometimes delete up to the half of the characters from the underlying string instead of one.

CDAWG can be obtained by a node minimization of a suffix tree as well as by a path compression of a Directed Acyclic Word Graph (DAWG). While suffix tree allows perfect deletion in a constant time [6,4], DAWG requires a time proportional to the size of the window [7]. Does a CDAWG inherit this property from a suffix tree, or from a DAWG? We close this issue as follows.

---

* Supported in part by the GAUK Grant 69408.

A. Amir, A. Turpin, and A. Moffat (Eds.): SPIRE 2008, LNCS 5280, pp. 109–120, 2008.

Theorem 1 (Main Result). *The problem to maintain a CDAWG for a perfect sliding window of size k over a string on an alphabet Σ, has the following amortized time complexity per one sliding window move*

$$\Theta(k) \quad \textit{if } |\Sigma| \geq 2,$$
$$\Theta(1) \quad \textit{if } |\Sigma| = 1.$$

2   Notation and Definitions

The concepts used in this paper but not defined below can be found e.g. in [8]. Let Σ be a finite alphabet. A sequence σ of n elements of Σ is called a *string* of length $|\sigma| = n$. For every $1 \leq i \leq j \leq |\sigma|$, $\sigma[i]$ denotes the i-th character of σ while $\sigma[i..j] = \sigma[i]\sigma[i+1]\ldots\sigma[j]$. The set of all strings over Σ is denoted by Σ^*, while λ denotes the empty string. A *concatenation* of strings α and β is denoted by $\alpha\beta$. Strings α, β and γ (each possibly empty) are called a *prefix*, *factor*, and *suffix* of a string $\sigma = \alpha\beta\gamma$. Moreover, $\beta = \sigma[i..j]$ is called an *occurrence* of β in σ at position i. If $\sigma[i-1]$ ($\sigma[j+1]$) exists, it is called the *left* (*right*) *context* of this occurrence. The sets of all prefixes, factors, and suffixes of σ are denoted by $\mathrm{Pref}(\sigma)$, $\mathrm{Fact}(\sigma)$, and $\mathrm{Suf}(\sigma)$, respectively.

A factor β of σ is called *right* (*left*) *branching* if it occurs in σ in at least two distinct right (left) contexts, and *unique* if it occurs in σ exactly once. Let $\mathrm{Bran}^{\mathrm{R}}(\sigma)$ and $\mathrm{USuf}(\sigma)$ denote the sets of right branching factors and unique suffixes of σ, respectively, while $\mathrm{EBUS}(\sigma) = \{\lambda\} \cup \mathrm{Bran}^{\mathrm{R}}(\sigma) \cup \mathrm{USuf}(\sigma)$. The notation $A = B \dot\cup C$ means that $A = B \cup C$ and $B \cap C = \emptyset$. Finally, the set of all vertices of a graph G is denoted by $\mathrm{V}(G)$.

2.1   CDAWG Definition and Properties

We begin by introducing two concepts that simplify the CDAWG definition.

Definition 2 (Right End Equivalence). *For every $\alpha, \beta \in \Sigma^*$,*

$$\alpha \equiv_\sigma^R \beta \quad \textit{if } \{\gamma\alpha \mid \gamma\alpha \in \mathrm{Pref}(\sigma)\} = \{\delta\beta \mid \delta\beta \in \mathrm{Pref}(\sigma)\}.$$

It is straightforward to verify that $\equiv_\sigma^R$ is an equivalence relation and hence all strings can be partitioned into classes with respect to $\equiv_\sigma^R$. The equivalence class where string α belongs and the longest member of the same class are denoted by $[\alpha]_\sigma^R$ and $(\alpha)_\sigma^R$, respectively. Less formally, $[\alpha]_\sigma^R$ contains all strings whose occurrences in σ end on the same positions as α. Note that there always is a *degenerated class* DC_σ^R consisting of all strings of $\Sigma^* \setminus \mathrm{Fact}(\sigma)$.

Definition 3 (Right Extension). *The* right extension *of $\alpha \in \mathrm{Fact}(\sigma)$, denoted by $\langle\alpha\rangle_\sigma^R$, is the shortest string $\beta \in \mathrm{EBUS}(\sigma)$ such that $\alpha \in \mathrm{Pref}(\beta)$.*

Informally, $\langle\alpha\rangle_\sigma^R$ is created by adding characters to the right of α until a member of $\mathrm{EBUS}(\sigma)$ is reached.

Definition 4 (Compact Directed Acyclic Word Graph). *The* Compact Directed Acyclic Word Graph *for a string* $\sigma \in \Sigma^*$, *denoted by* $\mathrm{CDAWG}(\sigma)$, *is a directed acyclic graph whose vertices are equivalence classes* $[\alpha]_\sigma^R$ *for every* $\alpha \in \mathrm{EBUS}(\sigma)$. *For every pair of distinct vertices* $[\alpha]_\sigma^R, [\gamma]_\sigma^R$, *there is a*

- directed edge *from* $[\alpha]_\sigma^R$ *to* $[\gamma]_\sigma^R$ *labelled with string* $a\beta$ *for some* $a \in \Sigma$ *and* $\beta \in \Sigma^*$, *if* $\langle \alpha a \rangle_\sigma^R = \alpha a \beta = \gamma$,
- suffix link *from* $[\alpha]_\sigma^R$ *to* $[\gamma]_\sigma^R$, *if* $(\gamma)_\sigma^R$ *equals the longest suffix of* α *which is not a member of* $[\alpha]_\sigma^R$.

Note that CDAWG was originally introduced by Blumer et al. [1] as a graph with vertex set based on $\mathrm{EBUS}(\sigma) \cup \mathrm{Suf}(\sigma)$. The definition provided above, which follows the approach of Inenaga et al. [3,2], is more suitable for our purposes, as it allows an efficient realization of operation Update described below.

Before proceeding further, we inspect some basic properties of a CDAWG. Note that some simpler proofs must have been omitted due to the space constraints.

Lemma 5 (Classification of Vertices). *For every non-empty* $\sigma \in \Sigma^*$,

$$V(\mathrm{CDAWG}(\sigma)) = \left\{[\lambda]_\sigma^R\right\} \dot{\cup} \left\{[\sigma]_\sigma^R\right\} \dot{\cup} \left\{[\alpha]_\sigma^R \mid \alpha \neq \lambda, \alpha \in \mathrm{Bran}^R(\sigma)\right\}.$$

Consequently, $\mathrm{CDAWG}(\sigma)$ has a unique *source* $[\lambda]_\sigma^R$ with no incoming edges and a unique *sink* $[\sigma]_\sigma^R$ with no outgoing edges. The path from the source to the sink of $\mathrm{CDAWG}(\sigma)$ that spells out σ is called *backbone*.

It is sometimes convenient to treat a string $\alpha \in \mathrm{Fact}(\sigma) \setminus \mathrm{EBUS}(\sigma)$ as an *implicit node* of $\mathrm{CDAWG}(\sigma)$, located on the edge leading from $[\beta]_\sigma^R$ to $[\gamma]_\sigma^R$, where β is the longest member of $\mathrm{Pref}(\alpha) \cap \mathrm{EBUS}(\sigma)$, and $\gamma = \langle \alpha \rangle_\sigma^R$. In this context, strings $\alpha \in \mathrm{EBUS}(\sigma)$, which are represented by corresponding vertices $[\alpha]_\sigma^R$ of $\mathrm{CDAWG}(\sigma)$, are called *explicit nodes*. We define the *active point* of a $\mathrm{CDAWG}(\sigma)$ as the node α such that α is the longest non-unique suffix of σ.

2.2   CDAWG Operations

In this paper, we study CDAWG with respect to the following two operations.

Definition 6 (CDAWG Operations).
Operation Delete *receives* $\mathrm{CDAWG}(a\beta)$ *for some* $a \in \Sigma, \beta \in \Sigma^*$ *and returns* $\mathrm{CDAWG}(\beta)$.
Operation Update *receives* $\mathrm{CDAWG}(\alpha)$ *for some* $\alpha \in \Sigma^*$ *and a character* $b \in \Sigma$ *and returns* $\mathrm{CDAWG}(\alpha b)$.

Now we are ready to formulate the main problem studied in this paper.

Definition 7 (CDAWG for a Perfect Sliding Window).
INPUT: *A string* $\sigma \in \Sigma^*$ *and an integer* k *such that* $0 < k < |\sigma|$.
OUTPUT: *A sequence* $C_1, C_2, \ldots, C_{|\sigma|-k+1}$, *such that* $C_1 = \mathrm{CDAWG}(\sigma[1..k])$ *and* $C_{i+1} = \mathrm{Update}(\mathrm{Delete}(C_i), \sigma[i+k])$.

It is known that the worst-case time complexity of the same problem is $\Theta(n)$ for a suffix tree [6,4] and $\Theta((n-k)k)$ for a DAWG [7]. The purpose of this paper is to determine the complexity for a CDAWG.

3   Perfect Sliding in Linear Time Is Impossible

In this section we show that the lower bound on the time required to maintain CDAWG for a perfect sliding window is the same as that for a DAWG [7].

Lemma 8 (Adversary String). *Let $l \geq 2$, $m \geq 1$, $\omega = (ab)^m a$ for distinct $a, b \in \Sigma$ and $\sigma \in \text{Fact}(\omega^l)$ such that $\omega a \in \text{Fact}(\sigma)$. Let $v = |\,V(\text{CDAWG}(\sigma))|$.*

$$v \begin{cases} = m + 2 & \text{if } \sigma[1] = a, \\ \geq m + p + 2, \text{ where } 0 \leq p < m \text{ and } b(ab)^p a\omega a \in \text{Pref}(\sigma) & \text{if } \sigma[1] = b. \end{cases}$$

Proof. In both cases $\text{Bran}^R(\sigma) \setminus \{\lambda\} = \{(ab)^i a \mid 0 \leq i \leq m - 1\} \cup \{b(ab)^i a \mid 0 \leq i \leq m - 2\}$. Also, $(ab)^i a \equiv_\sigma^R (ab)^j a$ iff $i = j$ and $b(ab)^i a \equiv_\sigma^R b(ab)^j a$ iff $i = j$.

If $\sigma[1] = a$, then $b(ab)^i a \equiv_\sigma^R (ab)^{i+1} a$ for every $0 \leq i \leq m - 2$ and therefore $\{[\alpha]_\sigma^R \mid \alpha \neq \lambda, \alpha \in \text{Bran}^R(\sigma)\} = \{[(ab)^i a]_\sigma^R \mid 0 \leq i \leq m-1\}$. Hence, by Lemma 5, $\text{CDAWG}(\sigma)$ contains $m + 2$ vertices. This settles part $\sigma[1] = a$.

However, if $\sigma[1] = b$, then $b(ab)^p a\omega a \in \text{Pref}(\sigma)$ for some $0 \leq p < m$. Hence $b(ab)^i a$ is a prefix of σ for each $0 \leq i \leq p$ and therefore it cannot be $\equiv_\sigma^R$ equivalent to $(ab)^j a$ for any $j > i$. Moreover, as $a(ab)^j a \in \text{Fact}(\sigma)$ for every $0 \leq j \leq m$ in this case, we also have $b(ab)^i a \not\equiv_\sigma^R (ab)^j a$ for any $j \leq i \leq p$. On the other hand, $b(ab)^i a \equiv_\sigma^R (ab)^{i+1} a$ still holds for every $p < i \leq m - 2$ as in the previous case.

It follows that $\{[\alpha]_\sigma^R \mid \alpha \neq \lambda, \alpha \in \text{Bran}^R(\sigma)\} = \{[(ab)^i a]_\sigma^R \mid 0 \leq i \leq m - 1\} \cup \{[b(ab)^i a]_\sigma^R \mid 0 \leq i \leq \min(p, m - 2)\}$ is a set of size $m + \min(p, m - 2) + 1 \geq m + p$. The conclusion of case $\sigma[1] = b$ now follows from Lemma 5.   □

Theorem 9 (Lower Bound). *Let $|\Sigma| \geq 2$. Then an arbitrary algorithm that maintains CDAWG for a perfect sliding window of size k over a string $\sigma \in \Sigma^*$, $|\sigma| = n$, requires $\Omega((n - k)k)$ time in the worst case.*

Proof. First note that at least $n - k$ steps of the algorithm are necessary to move the sliding window from the initial to the final position. In case $k \leq 5$ this gives the desired lower bound.

For $k \geq 6$, put $m = \lfloor \frac{k-2}{4} \rfloor$ and $\omega = (ab)^m a$ for distinct $a, b \in \Sigma$. Consider the input string $\sigma = \omega^n[1..n]$ where $n > k$. Then $k \geq 4m + 2$ and hence each string $\sigma[i..i + k - 1]$, $1 \leq i \leq n - k + 1$, contains ωa as a factor. Since σ is a power of a string of length $2m + 1$, it suffices to consider only $i \leq 2m + 1$. Let C_i denote $\text{CDAWG}(\sigma[i..i + k - 1])$. From Lemma 8 we obtain that for $i = 1, 2, \ldots, 2m + 1$

$$|V(C_i)| \begin{cases} = m + 2 & \text{if } i \text{ is odd}, \\ \geq 2m - \frac{i}{2} + 2 & \text{if } i \text{ is even}. \end{cases}$$

Hence for every even i, at least $m - \frac{i}{2}$ vertices must be added to construct C_i from C_{i-1}. Assuming that a creation of a vertex takes at least one step of the algorithm, the total number of steps required to construct C_i from C_{i-1} for $i = 2, 3, \ldots, 2m + 1$ is at least $\sum_{i=2, i \text{ even}}^{2m} (m - \frac{i}{2}) = \sum_{i=1}^m (m - i) = \frac{(m^2 - m)}{2}$. Due to the periodic nature of σ, this is also a lower bound on the number of steps to construct C_i for $i = q(2m + 1) + 2, q(2m + 1) + 3, \ldots, (q + 1)(2m + 1)$, where

$0 \leq q < \lfloor \frac{n-k+1}{2m+1} \rfloor$. Consequently, the total number of steps can be bounded from below by

$$\left\lfloor \frac{n-k+1}{2m+1} \right\rfloor \frac{m^2 - m}{2} = \Omega((n-k)\,m) = \Omega\left((n-k)\left\lfloor \frac{k-2}{4} \right\rfloor\right) = \Omega((n-k)\,k).\ \square$$

Recall that the naive algorithm can build CDAWG for each subsequent window from scratch in time $O(k)$ [3] and therefore its total running time is bounded by $O((n-k)k)$. This together with the above theorem verifies Theorem 1.

4 Change Analysis

Now that we have derived the complexity of sliding the CDAWG, we describe an incremental algorithm achieving that bound. Unlike the naive solution, it builds a CDAWG for each subsequent window by local modifications of the existing data structure, preserving the additional information [4,5]. Recall that sliding can be divided into two operations, described by Definition 6. As operation Update is done easily using the on-line CDAWG construction algorithm [3], we concentrate on operation Delete. First we need to describe changes made in a CDAWG by one application of this operation. To discover what happens when transforming $CDAWG(a\beta)$ into $CDAWG(\beta)$, we analyse the individual components of CDAWG definition. Beside the equivalence relation, it employs the set EBUS and the right extension. However, the definitions of both rely solely on λ, Bran^R and USuf. Thus we can concentrate only on changes to the equivalence classes and the sets of right branching factors and unique suffixes.

Lemma 10 (Observations).

1. Let $\alpha \in \text{Fact}(\sigma)$.
    (a) $[\alpha]_\sigma^R = \{\beta \in \text{Suf}((\alpha)_\sigma^R) \mid |\beta| \geq p\}$ for some $0 \leq p \leq |(\alpha)_\sigma^R|$.
    (b) $\alpha = (\alpha)_\sigma^R$ iff α is a prefix or a left branching factor of σ.
2. If $\alpha\beta \in \text{Fact}(\sigma)$, then
    (a) α has at least as many occurrences in σ as $\alpha\beta$,
    (b) α has at least as many left contexts in σ as $\alpha\beta$,
    (c) $|[\alpha]_\sigma^R| \leq |[\alpha\beta]_\sigma^R|$
3. $a\alpha \in \text{Pref}(a\beta) \cap \text{Pref}(\beta)$ iff $a\alpha = a^m \in \text{Pref}(\beta)$.
4. $\text{Bran}^R(a\beta) = \text{Bran}^R(\beta) \cup \{\textit{the longest right branching prefix of } a\beta\}$.
5. $\text{USuf}(a\beta) \setminus \{a\beta\} = \text{USuf}(\beta) \setminus \{\textit{the longest non-unique prefix of } a\beta\}$.
6. $|\{c\gamma \in \text{Fact}(\sigma) \mid c \in \Sigma, \gamma \in \text{Pref}(\sigma)\}| \leq |\{[\alpha]_\sigma^R \mid \alpha \in \text{Fact}(\sigma)\}|$.

Following the approach of Blumer [7], the next lemma describes what happens to (non-degenerated) equivalence classes when the first character of $a\beta$ is removed.

Lemma 11 (Modifications). Let $\gamma = (\gamma)_{a\beta}^R$.

1. If γ is not a prefix of $a\beta$, then either $[\gamma]_\beta^R = [\gamma]_{a\beta}^R$ or $[\gamma]_\beta^R = [\gamma]_{a\beta}^R \cup \{\delta\}$, where δ is the longest suffix of γ not in $[\gamma]_{a\beta}^R$ and $\delta \in \text{Pref}(a\beta)$.

2. If $\gamma = a\alpha$ is a prefix of $a\beta$, then $a\alpha \in [a\alpha]^R_\beta$, $[a\alpha]^R_{a\beta} \setminus \{a\alpha\} \subseteq [\alpha]^R_\beta$ and $\alpha = (\alpha)^R_\beta$. In particular,

   (a) if $[a\alpha]^R_{a\beta} = \{a\alpha\}$ and $a\alpha$ occurs exactly once in $a\beta$, then $a\alpha \in \mathrm{DC}^R_\beta$.

   (b) if $[a\alpha]^R_{a\beta} = \{a\alpha\}$ and $a\alpha$ occurs more than once in $a\beta$ and all occurrences, except the first, have the same left context b and $a\alpha$ is not a prefix of β, then $a\alpha \in [ba\alpha]^R_\beta$.

   (c) if $[a\alpha]^R_{a\beta} = \{a\alpha\}$, $a\alpha$ occurs more than once in $a\beta$ and all occurrences, except the first, have the same left context b and $a\alpha$ is a prefix of β, then $b = a$ and $[a\alpha]^R_{a\beta} = [a\alpha]^R_\beta = \{a\alpha\}$.

   (d) if $[a\alpha]^R_{a\beta} = \{a\alpha\}$ and $a\alpha$ occurs in $a\beta$ in at least two different left contexts, then $[a\alpha]^R_{a\beta} = [a\alpha]^R_\beta = \{a\alpha\}$.

   (e) if $[a\alpha]^R_{a\beta} \neq \{a\alpha\}$ and $a\alpha$ occurs exactly once in $a\beta$, then $a\alpha \in \mathrm{DC}^R_\beta$.

   (f) if $[a\alpha]^R_{a\beta} \neq \{a\alpha\}$, $a\alpha$ occurs more than once in $a\beta$ and all occurrences, except the first, have the same left context b, then $a\alpha \in [ba\alpha]^R_\beta$.

   (g) if $[a\alpha]^R_{a\beta} \neq \{a\alpha\}$ and $a\alpha$ occurs in $a\beta$ in at least two different left contexts, then $[a\alpha]^R_\beta = \{a\alpha\}$.

3. Let $a\alpha, a\alpha b \in \mathrm{Pref}(a\beta)$, then only the following combinations of cases from part 2 are possible for $[\alpha b]^R_{a\beta}$ and $[a\alpha b]^R_{a\beta}$.

		$[a\alpha b]^R_{a\beta}$						
		2a	2b	2c	2d	2e	2f	2g
	2a	+	−	−	−	+	−	−
	2b	+	+	−	−	+	+	−
	2c	+	+	+	−	+	+	−
$[a\alpha]^R_{a\beta}$	2d	+	+	+	+	+	+	+
	2e	−	−	−	−	+	−	−
	2f	−	−	−	−	+	+	−
	2g	−	−	−	−	+	+	+

Proof. 1. When γ is not a prefix of $a\beta$, then the same holds for the rest of $[\gamma]^R_{a\beta}$ by Lemma 10. Hence all occurrences of these strings in $a\beta$ lie in β and they are equivalent under $\equiv^R_\beta$ as they were under $\equiv^R_{a\beta}$. Also occurrence(s) separating members of $[\gamma]^R_{a\beta}$ from any longer strings still exists in β. However, if δ were a prefix of $a\beta$, it may have lost the only occurrence that separated it from members of $[\gamma]^R_{a\beta}$ and now possibly belongs to $[\gamma]^R_\beta$. But then the longest suffix of δ, different from δ, is still a prefix of β and thus cannot belong to $[\gamma]^R_\beta$. By Lemma 10, there are no more members in this class.

2. When $a\alpha$ is a prefix of $a\beta$, then strings from $[a\alpha]^R_{a\beta} \setminus \{a\alpha\}$ are not prefixes of $a\beta$ and so all their occurrences are in β. Hence $a\alpha \in [a\alpha]^R_\beta \neq [\alpha]^R_\beta$, while members of $[a\alpha]^R_{a\beta} \setminus \{a\alpha\}$, if any, stay together in $[\alpha]^R_\beta$. Moreover, by Lemma 10, $\alpha = (\alpha)^R_\beta$.

   (a) As $a\alpha$ has no occurrences in β, it falls into the degenerated class DC^R_β.

   (b) As all occurrences of $a\alpha$ in β are preceded by b and $a\alpha$ is not a prefix of β, $a\alpha$ is equivalent with $ba\alpha$.

(c) Because $a\alpha$ is a prefix of β, it is the longest string in $[a\alpha]_\beta^R$ by Lemma 10. Thus $[a\alpha]_\beta^R = \{a\alpha\}$.

(d) There exist two distinct characters b and c, such that $ba\alpha$ and $ca\alpha$ occur in $a\beta$. Either they both occur in β, or one of them only appears as a prefix of $a\beta$, which means that $a\alpha$ is a prefix of β. In both cases $a\alpha$ is not equivalent with any longer string in β. Hence $[a\alpha]_\beta^R = \{a\alpha\}$.

(e) As in 2a, $a\alpha \in [a\beta]_\beta^R$.

(f) There is no need to worry about whether or not is $a\alpha$ a prefix of β. If it were a prefix of β, then $a\alpha = a^m$ by Lemma 10. But a^m can not be equivalent with any of its suffixes as they occur both as a prefix and a suffix of a^m. Hence $a\alpha \in [ba\alpha]_\beta^R$ as in 2b.

(g) $[a\alpha]_\beta^R = \{a\alpha\}$ for the same reasons as in 2d.

3. Apply part 2 of Lemma 10. $\qquad\square$

We need to be able to detect which case of Lemma 11 applies to any given vertex $[a\alpha]_{a\beta}^R$ of CDAWG$(a\beta)$. The following lemma lists the properties needed and a description how the detection is done.

Lemma 12 (Detection). *Let $[a\alpha]_{a\beta}^R \in V(\text{CDAWG}(a\beta))$ and $a\alpha = (a\alpha)_{a\beta}^R$.*

1. *$a\alpha \in \text{Pref}(a\beta)$ iff $[a\alpha]_{a\beta}^R$ lies on the backbone.*
2. *If $a\alpha$ is a prefix of $a\beta$ then $[a\alpha]_{a\beta}^R = \{a\alpha\}$ iff none of the $[a\alpha]_{a\beta}^R$ ancestors, including $[a\alpha]_{a\beta}^R$, on the backbone is of in-degree at least two.*
3. *If $a\alpha$ is a prefix of $a\beta$, then $a\alpha$ occurs in $a\beta$ more than once iff there exists a node $ba\alpha$ for some $b \in \Sigma$.*
4. *If $a\alpha$ is a prefix of $a\beta$, then $a\alpha$ occurs in $a\beta$ in at least two distinct left contexts iff there exist nodes $ba\alpha$ and $ca\alpha$ for distinct $b, c \in \Sigma$.*
5. *If $a\alpha$ is a prefix of $a\beta$, then $a\alpha$ is also a prefix of β iff $\alpha = a^m$ and a is the next symbol on the backbone.*
6. *$a\alpha$ occurs more than once in $a\beta$ and all occurrences, except the first, are in the same left context b iff $ba\alpha$ is a node while $ca\alpha$ for any $c \neq b$ is not.*

Note that the suffix links that are useful for sliding DAWG [7] cannot be used for the detection in case 4, as most of them are only implicit due to the path compression. For example, string aa is both left and right branching in string $\sigma = aaabaacaa$, but vertex $[aa]_\sigma^R$ has no incoming suffix link in CDAWG(σ).

The analysis is completed by description of the actions needed to make the changes necessary to transform CDAWG$(a\beta)$ into CDAWG(β).

Lemma 13 (Sink Type). *Let $[a\alpha]_{a\beta}^R$ be a vertex of CDAWG$(a\beta)$. Then $[a\alpha]_{a\beta}^R$ is the sink of CDAWG$(a\beta)$ iff $[a\alpha]_{a\beta}^R$ is of type 2a or 2e of Lemma 11.*

Lemma 14 (Actions). *Let $a\alpha$ be a prefix of $a\beta$, $v = [a\alpha]_{a\beta}^R \in V(\text{CDAWG}(a\beta))$ a vertex on the backbone, e the backbone edge leading to v, and u the active point. Then the following actions need to be done for each v to change CDAWG$(a\beta)$ to CDAWG(β). The actions depend on the Lemma 11-type of the class $[a\alpha]_{a\beta}^R$ and are done for the shortest $a\alpha$ first. No other actions are necessary.*

2a or **2e.** *If u is an implicit node, then the label of edge e must be shortened at u. Otherwise, node u is explicit and edge e must be deleted. If u has only one out-edge f now, then its label must be appended to every incoming edge of u, the incoming edges redirected to point to the target of f and both u and f deleted. No more changes to the graph are necessary.*

2b or **2f.** *Redirect edge e to the vertex $[\langle ba\alpha \rangle_{a\beta}^R]_{a\beta}^R$, extending its label by γ, where $ba\alpha\gamma = \langle ba\alpha \rangle_{a\beta}^R$, if necessary. If the type is 2b, then remove vertex v and all edges and vertices below that are no longer on any path from the source. No more changes to the graph are necessary.*

2c or **2d.** *No changes to v or its edges are necessary. Move on to the next vertex lying on the backbone.*

2g. *Create a new vertex w and add out-edges to w that are duplicates of out-edges leading from v. Redirect e to w. Move on to the next vertex lying on the backbone.*

Proof. We start by examining implicit backbone node $a\alpha$ using Lemma 11 and Lemma 13. First, implicit node of type 2a or 2e may appear only on an edge leading to the sink. Any such node should be deleted. This happens either when cleaning up after edge redirection in case 2b or when removing or shortening the edge to the sink. Second, implicit node of type 2b or 2f should be joined with $ba\alpha$. This happens when redirecting edge in case 2b or 2f or when shortening the edge to the sink as only vertices of types 2a, 2b, 2f, or 2e are below. Third, implicit node of type 2c or 2d should be preserved as is. It could be influenced by edge redirection or shortening, but this does not result in a join to another class as in 2b or 2f due to the differing number of occurrences. Last, implicit node of type 2g can only appear on an edge leading to a branching vertex of type 2f or 2g and is split from its original class when this edge is redirected, but does not join another class due to differing occurrences.

2a or **2e:** By Lemma 13, vertex v is the sink. Lemma 11 tells us to remove the longest member from the class represented by vertex v. This is done by removing edge e. However, strings with more than one occurrence could be represented on this edge and would be lost after deletion. To preserve these strings, edge e is not removed, but shortened to save even the longest of these strings, which is represented by node u. Note that shortening effectively deletes the longest member from the class represented by v and adds u to the class of unique suffixes. After deleting edge e, there could be two issues with the resulting graph. First, the sink could have no more incoming edges and must be deleted. Second, explicit node u ceased to be branching and now has only one out-edge f. Thus node u must be made implicit by redirecting any incoming edges to the target of f while appending the label of f to their labels and deleting u and f afterwards. As we have reached the sink, all vertices were already handled.

2b or **2f:** As Lemma 11 tells us, we have to split $a\alpha$ from its class and add it to the class $[ba\alpha]_{a\beta}^R$. The split is done by redirecting edge e to another target. However, the target class might be represented only by implicit nodes, where $ba\alpha$ is represented by w lying on edge f leading to vertex x. To solve this issue, we have to redirect edge e to x, but also to append the part of the label of f below w to the label of e.

By redirecting the edge, we have split all strings of the $a\alpha\gamma$ type, with more than one occurrence, from their original classes and added them to classes containing $ba\alpha\gamma$, if different. This solves any nodes of type 2a, 2b, 2f, or 2e below. However, this also does not change the nodes not representing prefixes of $a\beta$, as $a\alpha\gamma$ and $ba\alpha\gamma$ were already equivalent. Hence the last thing needed is to remove all nodes below v that are now unreachable from the source.

2c or **2d:** This case calls for no changes to the class represented by v, but there may be other vertices below that need attention. So we move on to the next vertex lying on the backbone.

2g: Lemma 11 requires $a\alpha$ to be split from the rest of $[a\alpha]^R_{a\beta}$, which is exactly what happens when we create new vertex w, duplicate out-edges and redirect edge e. Note that this does no changes to the vertices below and classes they represent as all source-sink paths passing through v and w have exactly the same labels as before. However, some of these vertices may need our attention. So we move on to the next vertex lying on the backbone.   □

5 Algorithm

The preceding change analysis is materialized in Algorithm 1, described on pages 118–119. Note that for the sake of space and clarity we omit commands needed to keep suffix links, active point and edge labels (cf. [9]) valid.

To verify the upper bound stated in Theorem 1, we need to analyse the running time of $\text{Delete}(\text{CDAWG}(a\beta))$. Let e denote the number of edges of $\text{CDAWG}(a\beta)$ and $k = |a\beta|$. The initialization part (lines 1.1–1.11) is the only part outside of loops and takes only a constant amount of time.

We now turn our attention to loops. Note that in our analysis we ignore the time needed for branching in vertices, i.e. maximum time B needed to select the outgoing edge whose label starts with the desired character. When the implementation allows a binary search at each explicit node, $B = O(\log |\Sigma|)$. If $|\Sigma|$ is not considered to be a constant, the total running time of our algorithm derived below should be multiplied by B.

First, there are the two loops that initialize and update left contexts and begin at lines 1.15 and 1.21, respectively. These loops trace out $ca\beta$ for all $c \in \Sigma$ in parallel. The number of steps is proportional to the number of factors $c\gamma$ of $a\beta$ for which $\gamma \in \text{Pref}(a\beta)$. By part 6 of Lemma 10, this number can be bounded by the number of equivalence classes with respect to $\equiv^R_{a\beta}$, which is known to be $O(k)$, see e.g. [8, Theorem 5.3.5].

Second, there is the main loop starting at line 1.20. We have already covered the inner loop starting at line 1.21. The remaining code on lines 1.30 to 1.38 with the exception of the call to the procedure $\text{AdjustExplicitVertex}$ handling the vertices at line 1.37 takes a constant time in each iteration. In the procedure $\text{AdjustExplicitVertex}$, there are three lines that take other than a constant time per iteration. Lines 2.7 and 2.21 are calls to the Cleanup procedure, that

---

Algorithm 1. Delete(CDAWG($a\beta$))

1.1 $location := \uparrow [\lambda]_{a\beta}^R;$ `// Last node on the backbone`

1.2 $last\text{-}nonunique := location;$ `// Last node with two occurrences`

1.3 $active\text{-}edge :=$ NIL;   `// Last edge used to enter` $location$

1.4 $new\text{-}vertex :=$ NIL;   `// New explicit node`

1.5 $trivial\text{-}class :=$ true;   `// Does` $location$ `represent class of size 1?`

1.6 $simple\text{-}string :=$ true;   `// Does` $location$ `represent string` a^m`?`

1.7 $stop :=$ false;   `// Is the main loop done?`

1.8 $contexts[\,];$ `// Dictionary of (character` $c, \uparrow$`node` $a\beta$`)`

1.9 $last\text{-}context := (a\beta)[1];$ `// Last seen left context of` $location$

1.10 $count := 0;$ `// Number of left contexts for` $location$

1.11 $i := 1;$ `// Index to characters of string` $a\beta$

1.12   **if** *there is only one out-edge* e *leading from* $location$ **then**

1.13   shorten the label of edge e by one character or delete both edge e and its target if the label is only one character long;

1.14   **stop**

1.15   **forall** *out-edges* e *leading from* $location$ **do**

1.16   **if** c *is the first character on the edge label of* e **then**

1.17 $last\text{-}context :=$ c;

1.18 $contexts[c] := \uparrow$node representing c;

1.19 $count{+}{+}$;

1.20   **while not** $stop$ **and** $i \leq |a\beta|$ **do**

1.21   **forall** *entries* $contexts[c]$ *in the dictionary* **do**

1.22   **if** $(ca\beta)[1..i]$ *is a node of* CDAWG($a\beta$) **then**

1.23 $contexts[c] := \uparrow$node $(ca\beta)[1..i]$;

1.24   **else**

1.25   **if** $count = 1$ **then**

1.26 $last\text{-}nonunique := location$;

1.27 $last\text{-}context := c$;

1.28   delete $contexts[c]$;

1.29 $count{-}{-}$;

1.30 $location := \uparrow$node $(a\beta)[1..i]$;

1.31 $active\text{-}edge := \uparrow$last edge used to reach $location$;

1.32   **if** *simple-string* **and** $(a\beta)[i] \neq (a\beta)[1]$ **then**

1.33 $simple\text{-}string :=$ false;

1.34   **if** *trivial-class* **and** $location$ *has* ≥ 2 *incoming edges* **then**

1.35 $trivial\text{-}class :=$ false;

1.36   **if** $location$ *is an explicit node* **then**

1.37   AdjustExplicitVertex;

1.38 $i{+}{+}$;

---

are followed by instructions that stop the outer loop and prevent a second call to Cleanup. Procedure Cleanup deletes the subgraph that is no longer accessible from the source, using a depth first search, and so its complexity is bounded

---

Procedure AdjustExplicitVertex

2.1 **if** *trivial-class* **then**

2.2   **switch** *count* **do**

2.3   **case** 0 // Can not be reached - handled at line 1.12

2.4   **case** 1

2.5   **if** *simple-string* **and** $(\alpha\beta)[i+1] = (\alpha\beta)[1]$ **then**

2.6   redirect *active-edge* to point to the vertex representing $\langle contexts[last\text{-}context]\rangle^{R}_{\alpha\beta}$ and adjust its label accordingly;

2.7   Cleanup(*location*);

2.8   *stop* := true;

2.9   **else**

            // Nothing to do

2.10   **case** ≥ 2

         // Nothing to do

2.11 **else**

2.12   **switch** *count* **do**

2.13   **case** 0

2.14   **if** *last-nonunique is an implicit node on active-edge* **then**

2.15   shorten *active-edge* at implicit node *last-nonunique*;

2.16   redirect *active-edge* to *location*;

2.17   **else**

2.18   delete *active-edge*;

2.19   **case** 1

2.20   redirect *active-edge* to point to the vertex representing $\langle contexts[last\text{-}context]\rangle^{R}_{\alpha\beta}$ and adjust its label accordingly;

2.21   Cleanup(*location*);

2.22   *stop* := true;

2.23   **case** ≥ 2

2.24   *new-vertex* := ↑new vertex;

2.25   redirect *active-edge* to *new-vertex*;

2.26   duplicate all out-edges of *location* as out-edges of *new-vertex*;

2.27   *location* := *new-vertex*;

Procedure Cleanup(*location*)

3.1 **if** *location has no incoming edges* **then**

3.2   **foreach** *out-edge e from location to target* **do**

3.3   delete edge *e*;

3.4   Cleanup(*target*);

3.5   delete vertex *location*;

---

by $O(e)$. Line 2.26 duplicates all outgoing edges of the current vertex to a new vertex. This may be done many times. However, every edge of CDAWG($a\beta$) is duplicated at most once and thus the total work done by this line is bounded by $O(e)$. So we have shown that the total work of the main loop without the inner loop at line 1.21 depends on the number of iterations, which is limited by k, and the number of edges e, which can be also bounded by $O(k)$, see e.g. [1].

Consequently, the total running time of Algorithm 1 is bounded by $O(k)$.

6   Concluding Remarks

This paper closes the issue on the exact complexity of the perfectly sliding CDAWG. It also describes an on-line algorithm for perfect sliding with asymptotically optimal running time. This offers a freedom to choose the approximation [2] for speed or our solution for perfect sliding. In particular, perfect sliding may be useful for a CDAWG variant of the suffix tree based compression [5].

One may ask what is the expected time complexity of our algorithm. Note that the naive solution takes $\Theta(k)$ expected time per window move. On the other hand, Blumer [7] shows that her moving window algorithm for a DAWG requires $O(n \log k)$ expected time. Is there a similar bound for a CDAWG?

References

1. Blumer, A., Blumer, J., Haussler, D., McConnell, R.M., Ehrenfeucht, A.: Complete inverted files for efficient text retrieval and analysis. Journal of the ACM 34(3), 578–595 (1987)
2. Inenaga, S., Shinohara, A., Takeda, M., Arikawa, S.: Compact directed acyclic word graphs for a sliding window. Journal of Discrete Algorithms 2(1), 33–51 (2004)
3. Inenaga, S., Hoshino, H., Shinohara, A., Takeda, M., Arikawa, S., Mauri, G., Pavesi, G.: On-line construction of compact directed acyclic word graphs. Discrete Applied Mathematics 146(2), 156–179 (2005)
4. Larsson, N.J.: Structures of String Matching and Data Compression. Ph.D thesis, Department of Computer Science, Lund University, Sweden (1999)
5. Senft, M.: Compressed by the suffix tree. In: Storer, J.A., Cohn, M. (eds.) DCC, pp. 183–192. IEEE Computer Society, Los Alamitos (2006)
6. Fiala, E.R., Greene, D.H.: Data compression with finite windows. Communications of the ACM 32(4), 490–505 (1989)
7. Blumer, J.: How much is that DAWG in the window? Journal of Algorithms 8, 451–469 (1987)
8. Smyth, B.: Computing Patterns in Strings. Addison-Wesley, Reading (2003)
9. Senft, M.: Suffix tree for a sliding window: An overview. In: Šafránková, J. (ed.) WDS 2005, pp. 41–46. Matfyzpress, Praha (2005)

Self-indexing Natural Language[*]

Nieves R. Brisaboa[1], Antonio Fariña[1], Gonzalo Navarro[2], Angeles S. Places[1],
and Eduardo Rodríguez[1]

[1] Database Lab. Univ. da Coruña, Spain
{brisaboa,fari,asplaces,erodriguezl}@udc.es
[2] Dept. of Computer Science, Univ. of Chile
gnavarro@dcc.uchile.cl

Abstract. Self-indexing is a concept developed for indexing arbitrary
strings. It has been enormously successful to reduce the size of the large
indexes typically used on strings, namely suffix trees and arrays. Self-
indexes represent a string in a space close to its compressed size and
provide indexed searching on it. On natural language, a compressed inverted index over the compressed text already provides a reasonable alternative, in space and time, for indexed searching of words and phrases.
In this paper we explore the possibility of regarding natural language
text as a string of words and applying a self-index to it. There are several challenges involved, such as dealing with a very large alphabet and
detaching searchable content from non-searchable presentation aspects
in the text. As a result, we show that the self-index requires space very
close to that of the best word-based compressors, and that it obtains
better search time than inverted indexes (using the same overall space)
when searching for phrases.

1   Introduction and Related Work

Text indexing has become the only alternative to provide searching capabilities
on the extremely large collections of strings that arise from different fields, such
as bioinformatics (DNA and protein sequences), the Web and other natural
language collections, software development (source code), multimedia databases
and signal processing (music, audio, video and numeric streams), and so on.

For many years, the *inverted index* and its variants [4] have been a simple and
effective solution to index natural language text, and the base of the success of
Web search engines. We note that "natural language" is used to denote text that
is composed of an alternating sequence of "words" and "separators", which can
be easily distinguished syntactically; that the set of different words follows some
statistical laws such as growing sublinearly with the text size (Heaps' law [14]);
and especially that only whole words and sequences thereof (called "phrases")
can be searched for. These limitations have been widely accepted despite they
exclude many human languages (such as Chinese and Korean).

[*] Funded in part (for the Spanish group) by MEC grant (TIN2006-15071-C03-03), and
(for the third author) by Yahoo! Research grant "Compact Data Structures". We
also thank AECI grant A/8065/07.

A. Amir, A. Turpin, and A. Moffat (Eds.): SPIRE 2008, LNCS 5280, pp. 121–132, 2008.

On so-called natural language text, the basic inverted index consists of a *vocabulary*, that is, the set of different words in the text, and a *posting list* recording the text positions of each vocabulary word in increasing order. This simple data structure immediately answers single-word searches, and can handle phrase searches by essentially intersecting the corresponding posting lists. The way to carry out the intersections is still an active area of research [5, 6, 10, 28].

To save space, compression techniques have been applied to inverted indexes. In general the idea is to differentially encode each posting list (as its numbers are increasing) and encode those *gaps* with an encoding that favors small numbers. Some absolute samples are also inserted to allow fast intersections. The famous book *Managing Gigabytes* [32] describes this technology in detail. The text can be compressed as well, the preferred choice being Huffman coding [15] where source symbols are words and target symbols are bits (hence called "word-oriented bit-wise Huffman"). To further save space, the text can be divided into blocks, so that the postings point to the blocks where the word appears. This is called a *block-addressing* inverted index [3, 25]. At search time, the resulting blocks must be sequentially scanned to find the exact occurrences. The block size provides an obvious space/time tradeoff. In this tradeoff, it is advantageous to opt for a text compression method that permits much faster searches than bitwise Huffman [9, 25]. Nowadays, very efficient indexed searching can be obtained by occupying, with the compressed text plus the compressed index, 30% to 40% of the original text size (and removing the original text of course).

Other variants of inverted indexes, out of the scope of this paper, are oriented to document (rather than exact position) retrieval, or to relevance ranking [4].

The situation, up to the last decade, was far less satisfactory with other types of sequences. Without a concept of word, it is necessary to provide searching for *any* text substring. This was accomplished with powerful data structures called suffix trees [1, 31] and suffix arrays [17]. Those were able to locate the *occ* occurrences of a pattern of length m in $O(m + occ)$ time, regardless of the text size. However, they require 10–20 (suffix trees) or at best 4 (suffix arrays) times the text size, plus the text, and this rendered them unsuitable in many cases.

This changed drastically with the rise of *compressed self-indexes*, which were able to represent the text in space proportional to its empirical entropy [18], and within that space, offer indexed searching for any text substring [24]. For example, the smallest compressed self-index [12] offers searching in $O(m\lceil \frac{\log \sigma}{\log \log n}\rceil +$ $occ \cdot \log^{1+\epsilon} n)$ time, where n is the collection size, σ is the alphabet size, and ϵ is any positive constant. Another index, Sadakane's Compressed Suffix Array (CSA) [26], performs equally well in practice (despite not in theory) and has the interest for this paper of smoothly handling very large alphabets.

For example, on natural language texts, these indexes take around 60% to 70% of the original text size (and replace it). This is remarkable compared with the 400% plus text needed by suffix arrays, yet not competitive with the 30% to 40% achieved by compressed inverted indexes over compressed text. However, the comparison is not fair because self-indexes can search for any text substring whereas inverted indexes search only for whole words and phrases.

In this paper we explore the idea of applying a compressed self-index (as developed for general strings) over the sequence of *words* of a natural language text, that is, regarding the words as the basic symbols. This is promising because a self-index achieving high-order entropy should capture the dependence between consecutive words, which is significant in natural language [7, Chapter 4]. Moreover, even the slower CSA is able to locate the occurrences of a phrase of m *words* in $O(m \log n + occ \cdot \log^{1+\epsilon} n)$ time (and know the *number* of occurrences in just $O(m \log n)$ time). This compares favorably with inverted indexes, which need to carry out intersections. For example, for a phrase of 2 words appearing occ_1 and occ_2 times, an inverted index can take time $O(occ_1 + occ_2)$ or $O(occ_1 \log occ_2)$, where both occ_1 and occ_2 are (possibly much) larger than occ.

Applying a self-index to natural language words poses some challenges. A first one is that the alphabet is very large, and this rules out the theoretically best schemes [12, 13], which achieve k-th order entropy at the price of $\Omega(\sigma^k)$ extra space, where σ is the vocabulary size in our case. A text of n words is known to have a vocabulary of size $\sigma = O(n^\beta)$ [14], where $\beta \approx 0.5$ [4]. Thus σ^k may become $\Omega(n)$ already for $k = 2$! However, other self-indexes such as Sadakane's CSA [26] approach high-order entropy space without such a dependence on σ. Our first structure, the *Word CSA (WCSA)*, results from regarding the text as a sequence of word and separator identifiers and representing it with a CSA.

A second challenge is that, in many applications, we wish to have more flexible searching. For example, inverted indexes often permit to find phrases regardless of whether the words are separated by a space, two spaces, a tab, a newline, etc. This complicates the simple WCSA model where the self-index can reproduce the original text and thus the latter can be discarded. We must store some information on the separators in order to be able of exactly recreating the original text. Moreover, it is customary to apply some *filtering* on the text words to be searched [4], that is, users normally want to regard "preprocess", "pre-process", and "PRE-PROCESS" as occurrences of "preprocess", and even also "preprocessing" and "preprocessed" (the latter is achieved by *stemming*, that is, indexing/searching the roots of the words). It is also usual to disregard *stopwords* (articles, prepositions, etc.) in the searches. This shows that there should be a *presentation layer*, where the text is filtered into the *searchable sequence* of (possibly stemmed, lowercase, stopwords removed) bare words, and the *presentation sequence* containing the separators and all extra information on the bare words that permits recreating the original sequence. The searchable sequence is self-indexed, while the presentation sequence is just compressed with a technique that permits fast direct access for displaying purposes. Both sequences are compressed by different means, thus the choice of what is searchable is not a space/time tradeoff but depends on user's needs. We call *Flexible WCSA (FWCSA)* this second data structure.

Our resulting data structures achieve excellent compression results, close to many natural language text compressors (that do not provide any indexing). Texts are usually compressed to around 35-40% of their original size with the FWCSA (values up to 30% can be obtained depending on the parameters used, but the resulting index becomes slow). We compare FWCSA with a block addressing inverted index (II) over compressed text using the same amount of space and offering the same functionality (a full word-addressing inverted index requires much more space, around 60-70%). The results show that, with the same available space requirements, FWCSA overcomes II when we are interested in compression ratios below 40%. When more space is available, the FWCSA is still faster for locating occurrences on either single words or phrases, except on words with many occurrences, where the II becomes superior. Also FWCSA obtains better results in the extraction of snippets for phrases in most cases.

The WCSA requires even less space, around 1-2 percentage points less than FWCSA in compression ratio. We compare the WCSA with recent related works that offer similar functionality: (1) Compressing the text with a word-oriented bytewise Huffman-like compressor prior to applying a basic (character-oriented) self-index to the result [11]; (2) reordering the bytes of the output of a word-oriented dense-code compressor in a wavelet-tree-like [13] shape, to give search capabilities to the compressed text [8]; and finally (3) a block addressing inverted index with the same functionality. Again WCSA is the best choice when little memory is available. By increasing the size of the indexes until around 45% in compression ratio, the WCSA is still the best choice for dealing with searches on phrases composed of several words. However, the wavelet-tree-like index performs better when single-word patterns are searched for.

We note that the (F)WCSA operates in main memory, and therefore requires that the compressed text does not exceed the available RAM. Because of its access pattern, the (F)WCSA is not promising on secondary memory, whereas inverted indexes perform well. Recently, however, there has been much interest in inverted indexes that operate in RAM [28, 30], motivated by the large main memories available at reasonable prices (up to 4GB is standard) and the common distributed architectures where the text collection resides in the RAM of several computers (then the problem is how to integrate the results of several indexes across the slow network). Therefore, main memory data structures are of interest nowadays, unlike what was assumed 10 years ago.

2   Sadakane's Compressed Suffix Array (CSA)

Let $T[1, n]$ be a sequence over an alphabet Σ of size σ. The *suffix array* [17] $A[1, n]$ of T is a permutation of $[1, n]$ of all the suffixes $T[i, n]$ so that $T[A[i], n] \prec T[A[i+1], n]$ for all $1 \le i < n$, being $\prec$ the lexicographic ordering. Since every substring of T is the prefix of a suffix, and all suffixes prefixed by a search pattern $P[1, m]$ are contiguous in A, we can binary search A for the interval $A[sp, ep]$ of the pointers to all the occurrences of (i.e., suffixes starting with) P in T, in time $O(m \log n)$. Each step of the binary search needs to access $T[A[i], A[i] + m - 1]$ for some i, in order to compare that string with $P[1, m]$.

Let us now define another permutation $\Psi[1, n]$ such that $\Psi(i) = A^{-1}[A[i] + 1]$ (or $A^{-1}[1]$ if $A[i] = n$). Hence $\Psi(i)$ tells where in A is there the pointer following $T[A[i]]$. Assume one has computed $C[1, \sigma]$, so that $C[c]$ is the number of occurrences of symbols $\prec c$ in T. We show how can one obtain the successive letters of $T[A[i]...]$ (so as to carry out the binary search) with Ψ and C and without A and T. To extract the first letter, note that all the suffixes starting with c are in the area $A[C[c]+1, C[c+1]]$, and therefore a binary search on C for the c such that $C[c] < i \leq C[c+1]$ gives the desired first letter, $T[A[i]] = c$. To extract the next letter, we use the identity $T[A[i]+1] = T[A[\Psi(i)]]$, thus we simply have to move to $i' \leftarrow \Psi(i)$ and carry out the same process again to obtain $T[A[i']]$, and so on. This is sufficent to replace A and T.

The binary search on C can be implemented in constant time as follows. Set up a string $S[1, \sigma']$, $\sigma' \leq \sigma$, containing all the different symbols that actually occur in T, in increasing lexicographical ordering. Also, set up a bitmap $D[1, n]$ with all zeros except $D[C[c]+1] = 1$ for all $c \in \Sigma$. Now, the c corresponding to an i value is $c = S[rank(D, i)]$, where $rank(D, i)$ is the number of 1s in $D[1, i]$. This is (easily) computed in constant time using $o(n)$ bits on top of D [16, 22].

The description above is the essential idea of Sadakane's CSA [26], where we have removed several possible optimizations that are not promising for our particular application (backward searching, compressed bitmaps, etc.). One important remaining point is how to compress Ψ, as in principle it is as large as the suffix array A it replaces. Sadakane shows that Ψ is formed by σ increasing subsequences, and thus it can be compressed to around the zero-order entropy of T, more precisely $nH_0(T) + O(n \log \log \sigma)$, by gap encoding its differential values. Furthermore, as shown later [24], Ψ contains at most $nH_k + \sigma^k$ (for any k) runs of values, so that consecutive differences equal 1 within each run. Thus, by enriching the gap encoding with run-length compression of those runs one achieves higher-order compression. Absolute Ψ values at regular intervals d are retained to permit fast random access to Ψ (yielding constant time in theory).

Note that, since we do not have A anymore, determining the interval $A[sp, ep]$ is not sufficient to locate the occurrences, that is, to output the values $A[i]$ in the interval. For this sake, the text is sampled at regular intervals l, and the suffix array positions pointing to sampled text positions are recorded, in suffix array order, into an array $A_S[1, n/l]$. Those sampled positions in A are marked in a bitmap $B_A[1, n]$, thus if $B_A[i] = 1$ we know that $A[i] = A_S[rank(B_A, i)]$. Otherwise, we try $i \leftarrow \Psi(i)$ successively, as we are virtually moving forward in T by one position at each iteration. Hence, if we determine $A[i] = j$ after k applications of Ψ, then our original value was $j - k$. Due to the regular sampling in T we carry out at most l iterations until finding a sampled position in A.

Finally, in order to discard T, we need to be able to extract any substring $T[a, b]$. For the same sampled text positions $j \cdot l$ sampled above, we store $A^{-1}[j \cdot l]$ in text position order into an array $A_S^{-1}[1, n/l]$. Thus, we find the latest sampled position $j \cdot l$ preceding a, $j = \lfloor a/l \rfloor$, and know that $j \cdot l$ is pointed from $i = A_S^{-1}[j]$. From that i we use the mechanism we have described to extract a string using C and Ψ, to find out the substring $T[j \cdot l, b]$ which covers the one of interest to us. (This is not the way Sadakane's theoretical description handles this [26], but the way he implemented it in practice.)

3   A Word-Based CSA

In this section we present the simple word-based self-index (WCSA). It can be seen as the adaptation of Sadakane's CSA [26] to a large word-based alphabet.

To create the WCSA we first map each different word or separator[1] (let us call both "words") from the source text to an integer id. Then, an integer sequence Sid is formed with the identifiers of the consecutive text words and a vocabulary array V is created to store the word corresponding to each id. Finally, Sid is self-indexed by building an integer-based CSA (iCSA) on it. The algorithm to create iCSA first builds the suffix array A of Sid, as well as D, and can discard Sid. Then, arrays A^{-1} and ψ are created, as well as B_A, A_S and A_S^{-1}. Then A and A^{-1} can be discarded. Assuming that there are σ different words, the vocabulary used by the iCSA is $\{1, 2, \ldots, \sigma\}$, so it remains implicit and there is no need to store it (nor $S[1, \sigma']$). Finally, ψ is compressed by storing some absolute samples and Huffman-encoding the consecutive gaps, including a special encoding for the runs.[2] To sum up, WCSA consists of the vector of words V (sorted alphabetically) and a bottom layer composed of an iCSA built on Sid.

As any typical self-index, iCSA provides the following basic functions using the CSA algorithms described: $countiCSA(P')$ counts the number of occurrences of pattern P' in Sid; $locateiCSA(P')$ locates P''s positions in Sid; and $extractiCSA(l,r)$ retrieves the integers $Sid[l] \ldots Sid[r]$.

Searches for a pattern $P = \{w_1, w_2, \ldots w_m\}$ on the WCSA start by binary searching V for each word w_i of P to obtain its corresponding id_i (its position in V), hence obtaining a new pattern $P' = \{id_1, id_2, \ldots id_m\}$ to be searched in the iCSA. Operation $countWordsWCSA(P)$ is directly translated into $countiCSA(P')$, and $locateWordsWCSA(P)$ to $locateiCSA(P')$ (note this gives word offsets, not byte offsets, of the occurrences). Finally, $extractWordsWCSA(s,e)$ recovers the original text from the sth to the eth word: We obtain the word ids with $extractiCSA(s,e)$ and then retrieve the original words stored at those positions (ids) in array V. Notice that snippets composed of k words around the occurrences of P' can be obtained by applying $occs = locateWordsWCSA(P')$ followed by $extractWordsWCSA(occs[i - k], occs[i + k])$ for each $i \in [1..|occs|]$.

4   Flexible Word-Based CSA

We show how a more flexible index can be obtained based on WCSA. Our Flexible WCSA (FWCSA) can deal with many typical requirements of natural language searching, such as case-insensitive search, stemming, disregarding stopwords and/or separators, etc. The FWCSA does not index the original text as

---

[1] We parse the text using the spaceless model: If a word is followed by a single blank, that separator is not encoded but implicitly regenerated at snippet extraction time. This saves 70% of the separators [21].

[2] Further details were omitted for lack of space.

such, but rather a *normalized* version of it. Normalization is a user-defined function from (original) words and separators to (normalized) words or a null word. It can be used to express the requirements above[3]. We map the set of different normalized words to integer ids, then replace each word from the original text by the *id* of its normalized version (or ignore it if the normalization gives the null word), and finally build an iCSA on the resulting sequence of ids.

As we want FWCSA to be able to recover any part of the original text, some additional information has to be stored in what we call the *presentation layer*.

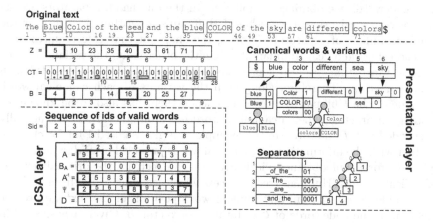

Fig. 1. General Structure of FWCSA

Fig. 1 shows the general structure of WCSA. A first pass over the original text is needed to gather some statistics from the source text. We split the original text into "valid words" and "separators". A "valid word" is a text word or separator[4] that normalization does not map to the null word. A "separator" in this context is all the text between valid words, that is, a maximal sequence of text words and separators mapped to the null word by normalization. Hence valid words and separators strictly alternate in the text[5]. A vocabulary of *canonical* (i.e., normalized) words is built, and kept sorted alphabetically. For each canonical word, a list with all the variants that the normalization process maps to it is stored (sorted by frequency). Similarly, a vocabulary containing all the "separators" in the source text is created and sorted by frequency.

A second pass over the original text permits to fill the structures from the presentation layer shown in Fig. 1, as well as array *Sid*. Notice that $Sid[1] = 2$

---

[3] For example, if one wishes a case-insensitive search ignoring stopwords and separators, a proper normalization could map all words to their lowercase version, and stopwords and separators to the null word.

[4] What is a text word can also be user-defined, being the typical definition a maximal sequence of letters and digits.

[5] If normalization wishes to keep separators as valid words, we insert dummy "separators" between valid words.

because the first valid word from the text, "Blue" is mapped via normalization to the second canonical word, "blue". Once the presentation layer is built, the iCSA structure is constructed over the sequence Sid.

In the presentation layer, bitmap CT keeps a compressed representation of the presentation aspect of the text. Based on the alternation between words and separators, CT will have a codeword belonging to a word, followed by the codeword of a separator, and so on. As an example, in Fig. 1, we can observe that $CT[1..3]$ ='001' is the codeword associated to the separator "The " and $CT[4]$ ='1' is the codeword of the variant "Blue" of the canonical word "blue". Those codewords are obtained as follows. On the one hand, separators are assigned a codeword using a word-based Huffman's algorithm [15, 19] over the whole vocabulary of separators (storing the shape of that tree requires little overhead using canonical Huffman [20]). On the other hand, the variants of each canonical word (that are also kept sorted by frequency) are also encoded with the same method. Therefore, along with the variants of each canonical word, the shape of the Huffman tree used to encode them has also to be known for decoding. In practice, when a canonical word has a unique variant it is actually not encoded in CT (however, in the example in Fig. 1 we used 1 bit for clarity). Together with the information on canonical words provided by Sid (which is not explicitly stored but obtained via iCSA), we can recreate the original text, as Sid indicates which Huffman tree to access when decoding words from CT.

To enable decoding from any random word position in the text we provide synchronism the codewords of CT, by using a vector B. Given a position i in Sid, $B[i] = p$ tells the offset in CT from which the corresponding variant of the canonical word $j = Sid[i]$ can be decoded (using the Huffman tree associated to the jth canonical word). After decoding one symbol from that point p in CT, we will find the beginning of the codeword of a separator, and after it the codeword of the variant of the canonical word in $Sid[i + 1]$, and so on. In our example, we can see that $B[5] = 16$ is the beginning in CT of the codeword '01' that corresponds to the third ($Sid[5] = 3$) canonical word ('01' $\rightarrow$ "COLOR"). Then, $CT[18, 19]$ ='01' is the codeword of the separator " of the ".

A second array, Z, is needed for *locate* and *display* operations. It maps any position i from vector Sid to its actual byte offset in the original text T: $Z[i] = j$ means that $T[j]$ is the first character of the word represented by $Sid[i]$.

To save space, both B and Z are sampled at regular positions $i \cdot k_b$ and $i \cdot k_z$, respectively, and only those positions are stored. A non-sampled value p from B $(ik_b < p < i(k_b + 1))$ is obtained by moving to position $B[i \cdot k_b]$ in CT and then decoding alternatively $p - (ik_b)$ words and separators. The number of decoded bits from CT added to the value $B[i \cdot k_b]$ tells us the value of $B[p]$. A non-sampled value p from Z is obtained similarly by adding to the previous sampled value $Z[ik_z]$ the number of characters decoded after processing $p - ik_z$ words and $p - ik_z$ separators. In this case decoding starts at position $B[ik_z]$ of CT.

For lack of space we omit the detailed structures of the presentation layer and the details of the search operations on FWCSA: *countWords*, *locateWords*, and *extractSnippet*.

5   Experimental Results

We used a large text collection with 1023MiB, obtained by aggregating several corpora from TREC-2: AP Newswire 1988 (AP) and Ziff Data 1989-1990 (ZIFF), as well as from TREC-4: Congressional Record 1993 (CR) and Financial Times 1991 to 1994, and finally Calgary corpus[6]. An isolated Intel®Pentium®-IV 3.00 GHz, with 4 GB RAM was used. It ran Debian (kernel 2.4.27), using gcc version 3.3.5 with -O9 optimizations. Time results measure CPU user time.

We compared our self-indexes WCSA and FWCSA against two in-memory block-addressing inverted indexes (II and FII) with similar features[7]. II is the same index from authors in [8] and FII is its *Flexible* counterpart. Therefore, the text is compressed with ETDC [9]; whereas postings are encoded differentially with ETDC and absolute samples are kept every k values to speed-up intersections. This approach differs only slightly from that in [10], and obtains similar results in practice. The normalization process of FWCSA and FII consisted of: (1) choosing as valid words maximal alphanumeric sequences, (2) skipping separators and stopwords, (3) folding to lowercase.

We measured *locateWords* time and also the time needed to extract a snippet containing 20-words around all the occurrences of a given pattern. We used 100 test patterns from 4 different groups of single-word patterns (with different frequency ranges) and also 4 groups of phrase-patterns composed of 2, 4, 6, and 8 words. Results for both *locateWords* and for *snippet* extraction refer to average time per occurrence (in msec/occurrence).

WCSA vs II. We consider two configurations varying the memory usage of the indexes. We used two setups of WCSA depending on the parameters of its iCSA layer; that is, on the sampling periods for its structures: $\{t_\psi, t_A, t_{A^{-1}}\}$. One, named WCSA$_1$, used $\{t_\psi, t_A, t_{A^{-1}}\} = \{16, 16, 64\}$; the other, WCSA$_2$, was set to $\{32, 32, 64\}$. For II, two parameters are needed, $\{k, b\}$, that refer to the sampling period to index its compressed postings lists, and the block size (in Kbytes). We call II$_1$ the setup $\{k, b\} = \{8, 16\}$, and II$_2$ to $\{k, b\} = \{32, 256\}$.

Table 1 shows that WCSA$_2$ overcomes II$_2$ in all aspects. In practice, when little memory is available the WCSA is clearly the best choice. Only when we use more memory, II$_1$ can compete with WCSA$_1$ in the extraction of snippets for either single-word patterns or short phrases. However, WCSA$_1$ is still faster than II$_1$ for locating. When we search for phrase patterns, the performance gaps between WCSA and II increase with the number of words in the phrase.

FWCSA vs FII. We used three setups of FWCSA using fixed values $B = 32$ and $Z = 512$ (presentation layer) and depending on the three sampling

---

[6] ftp://ftp.cpsc.ucalgary.ca/pub/projects/text.compression.corpus

[7] Some freely available inverted indexes were checked, but they either: *i)* used a different retrieval model than that of (F)WCSA, for example *Zettair* (retrieves passages, http://www.seg.rmit.edu.au/zettair/), *Wumpus* (needs the text separately, http://www.wumpus-search.org), and *Lemur* (ranked document retrieval, http://www.lemurproject.org); or *ii)* were not ready, or not public, or we could not install them, such as *Galago* (http://www.galagosearch.org/ [30]), those in [10] and [29], and *Terrier* (http://ir.dcs.gla.ac.uk/terrier/).

Table 1. Results comparing WCSA against II and FWCSA against FII

		WCSA$_i$		II$_i$		FWCSA$_i$			FII$_i$		
		i=1	i=2	i=1	i=2	i=1	i=2	i=3	i=1	i=2	i=3
	Ratio (%)	45.03	38.08	45.54	39.07	41.42	38.84	37.54	41.32	38.93	37.50
		Locate				*Locate*					
Words	1-100	0.009	0.018	0.018	0.246	0.030	0.058	0.070	0.042	0.161	0.503
(freq	101-1000	0.007	0.019	0.019	0.237	0.030	0.059	0.070	0.019	0.074	0.200
range)	1001-10000	0.006	0.019	0.023	0.163	0.030	0.058	0.069	0.021	0.089	0.171
	10000+	0.006	0.019	0.014	0.029	0.028	0.057	0.067	0.011	0.020	0.022
phrases	2	0.005	0.014	0.028	0.113	0.027	0.054	0.063	0.044	0.118	0.159
	4	0.005	0.009	1.128	3.737	0.030	0.058	0.069	0.026	0.064	0.089
#words	6	0.069	0.069	14.028	76.319	0.032	0.062	0.074	0.077	0.304	0.485
	8	0.059	0.059	7.396	50.118	0.044	0.059	0.074	3.086	15.551	27.795
		Snippet				*Snippet*					
Words	1-100	0.055	0.091	0.027	0.255	0.086	0.148	0.160	0.041	0.161	0.512
(freq	101-1000	0.053	0.083	0.021	0.238	0.087	0.151	0.161	0.022	0.078	0.204
range)	1001-10000	0.054	0.084	0.024	0.164	0.085	0.149	0.159	0.024	0.093	0.174
	10000+	0.054	0.084	0.015	0.030	0.083	0.145	0.155	0.014	0.023	0.025
phrases	2	0.046	0.070	0.028	0.114	0.078	0.139	0.148	0.047	0.121	0.163
	4	0.029	0.043	1.130	3.737	0.085	0.148	0.158	0.029	0.067	0.092
#words	6	0.069	0.139	14.028	76.389	0.092	0.159	0.170	0.080	0.307	0.486
	8	0.118	0.118	7.396	50.059	0.084	0.153	0.162	3.110	15.463	27.717

parameters of its iCSA. The first, named FWCSA$_1$, used the values $\{16, 16, 32\}$; FWCSA$_2$ was obtained by setting $\{32, 16, 64\}$; and FWCSA$_3$ used the values $\{32, 32, 64\}$. For FII, $\{k, b\}$ were set to $\{64,16\}$ to obtain FII$_1$; FII$_2$ was created with the values $\{64, 128\}$, and finally, II$_3$ used values $\{64, 1024\}$.

The results show that, when compression ratio is around 40% there is not a clear winner. FWCSA is better than FII for dealing with long phrases, but FII obtains the best results on high frequency words. However, as the amount of memory decreases, the results of FII worsen much faster than those of FWCSA.

Moreover, it is noticeable that II and FII versions are lower bounded in size by around 35%. However, with that amount of available memory (F)WCSA performs much better. Furthermore, we can always set (F)WCSA space to around 30%, yet with a clear loss in performance.

Non-II Alternatives. Other competitors to the inverted index, in a spirit similar to our WCSA, have recently appeared in the literature. We briefly compare with these in this section.

We first compare WCSA$_1$ and WCSA$_2$ with the wavelet-tree index on words (WT) [8]. We used two configurations of WT with different memory usage. WT$_1$ occupies 44.37% of the original text, whereas WT$_2$ uses 38.61%. Results in Table 2 show that, to search for single-word patterns, WT$_1$ is faster than WCSA$_1$. However, WCSA$_1$ overcomes WT$_1$ when locating phrases. Similar results are obtained for WT$_2$ versus WCSA$_2$.

We also briefly compared our WCSA against the approach called TH+AFFM in recent work [11] (word-based compression followed by character-wise self-indexing). We consider locating of phrases composed of 4 words (other choices give similar results). We adjust WCSA to work with the same memory of TH+AFFM, for different parameter combinations of both methods. It turns out that WCSA searches around 5 times faster in all cases.

Table 2. Results comparing WCSA against WT

		$WCSA_i$		WT_i		$WCSA_i$		WT_i	
		$i=1$	$i=2$	$i=1$	$i=2$	$i=1$	$i=2$	$i=1$	$i=2$
Ratio (%)		45.03	38.08	44.37	38.61	45.03	38.08	44.37	38.61
		Locate				*Snippet*			
Words	1-100	0.009	0.018	0.005	0.007	0.055	0.091	0.026	0.058
(freq	101-1000	0.007	0.019	0.004	0.044	0.053	0.083	0.025	0.093
range)	1001-10000	0.006	0.019	0.002	0.007	0.054	0.084	0.021	0.051
	10000+	0.006	0.019	0.002	0.003	0.054	0.084	0.019	0.045
phrases	2	0.005	0.014	0.010	0.021	0.046	0.070	0.025	0.058
	4	0.005	0.009	0.458	0.926	0.029	0.043	0.476	0.958
#words	6	0.069	0.069	9.028	21.181	0.069	0.139	9.028	21.250
	8	0.059	0.059	5.562	15.562	0.118	0.118	5.621	15.562

6   Conclusions and Future Work

We have shown that a self-index applied to natural language text, seen as a sequence of words rather than symbols, offers a very relevant alternative to the traditional inverted indexes. With sizes around 40% the inverted indexes can still compete with our (F)WCSA in some operations (extraction of snippets), but when we aim at using less space, our proposal performs much better.

In this work we have focused on one self-index, Sadakane's CSA. We plan to try out others that have mild dependence on the alphabet size. In particular, adapting the LZ-index [2, 23] should offer fast locating of occurrences.

Inverted indexes are also used for other purposes, as explained [4]. For example they are used. to implement the tf-idf model by recording the number of occurrences of each word in each document, in decreasing order of frequency. Only a short prefix of the posting list is fetched to solve queries. Can we provide similar functionalities with a self-index? Some initial advances have been made by Sadakane [27], but we are far from a definitive answer.

References

1. Apostolico, A.: The myriad virtues of subword trees. In: Combinatorial Algorithms on Words. NATO ISI Series, pp. 85–96. Springer, Heidelberg (1985)
2. Arroyuelo, D., Navarro, G., Sadakane, K.: Reducing the space requirement of LZ-index. In: Lewenstein, M., Valiente, G. (eds.) CPM 2006. LNCS, vol. 4009, pp. 319–330. Springer, Heidelberg (2006)
3. Baeza-Yates, R., Navarro, G.: Block-addressing indices for approximate text retrieval. J. of the American Society for Information Science 51(1), 69–82 (2000)
4. Baeza-Yates, R., Ribeiro-Neto, B.: Modern Information Retrieval. Addison-Wesley, Reading (1999)
5. Baeza-Yates, R., Salinger, A.: Experimental analysis of a fast intersection algorithm for sorted sequences. In: Proc. 12th SPIRE, pp. 13–24 (2005)
6. Barbay, J., López-Ortiz, A., Lu, T.: Faster adaptive set intersections for text searching. In: Proc. 5th WEA, pp. 146–157 (2006)
7. Bell, T., Cleary, J., Witten, I.: Text compression. Prentice Hall, Englewood Cliffs (1990)
8. Brisaboa, N., Fariña, A., Ladra, S., Navarro, G.: Reorganizing compressed text. In: Proc. 31st ACM SIGIR. ACM Press, New York (to appear, 2008)

9. Brisaboa, N., Fariña, A., Navarro, G., Paramá, J.: Lightweight natural language text compression. Information Retrieval 10, 1–33 (2007)
10. Culpepper, J., Moffat, A.: Compact set representation for information retrieval. In: Proc. 14th SPIRE, pp. 137–148 (2007)
11. Fariña, A., Navarro, G., Paramá, J.: Word-based statistical compressors as natural language compression boosters. In: Proc. 18th DCC, pp. 162–171 (2008)
12. Ferragina, P., Manzini, G., Mäkinen, V., Navarro, G.: Compressed representations of sequences and full-text indexes. ACM Transactions on Algorithms (TALG) 3(2) article 20 (2007)
13. Grossi, R., Gupta, A., Vitter, J.: High-order entropy-compressed text indexes. In: Proc. 14th ACM-SIAM SODA, pp. 841–850 (2003)
14. Heaps, H.: Information Retrieval - Computational and Theoretical Aspects. Academic Press, London (1978)
15. Huffman, D.: A method for the construction of minimum-redundancy codes. Proc. of the I.R.E. 40(9), 1090–1101 (1952)
16. Jacobson, G.: Space-efficient static trees and graphs. In: Proc. 30th FOCS, pp. 549–554 (1989)
17. Manber, U., Myers, G.: Suffix arrays: a new method for on-line string searches. SIAM Journal on Computing 22(5), 935–948 (1993)
18. Manzini, G.: An analysis of the Burrows-Wheeler transform. Journal of the ACM 48(3), 407–430 (2001)
19. Moffat, A.: Word-based text compression. Software Practice and Experience 19(2), 185–198 (1989)
20. Moffat, A., Katajainen, J.: In-place calculation of minimum-redundancy codes. In: Sack, J.-R., Akl, S.G., Dehne, F., Santoro, N. (eds.) WADS 1995, vol. 955, pp. 393–402. Springer, Heidelberg (1995)
21. Moura, E., Navarro, G., Ziviani, N., Baeza-Yates, R.: Fast and flexible word searching on compressed text. ACM Transactions on Information Systems (TOIS) 18(2), 113–139 (2000)
22. Munro, I.: Tables. In: Chandru, V., Vinay, V. (eds.) FSTTCS 1996, vol. 1180, pp. 37–42. Springer, Heidelberg (1996)
23. Navarro, G.: Indexing text using the Ziv-Lempel trie. Journal of Discrete Algorithms 2(1), 87–114 (2004)
24. Navarro, G., Mäkinen, V.: Compressed full-text indexes. ACM Computing Surveys 39(1) article 2 (2007)
25. Navarro, G., Moura, E., Neubert, M., Ziviani, N., Baeza-Yates, R.: Adding compression to block addressing inverted indexes. Information Retrieval 3(1), 49–77 (2000)
26. Sadakane, K.: New text indexing functionalities of the compressed suffix arrays. Journal of Algorithms 48(2), 294–313 (2003)
27. Sadakane, K.: Succinct data structures for flexible text retrieval systems. Journal of Discrete Algorithms (JDA) 5(1), 12–22 (2007)
28. Sanders, P., Transier, F.: Intersection in integer inverted indices. In: Proc. 9th ALENEX (2007)
29. Sanders, P., Transier, F.: Compressed inverted indexes for in-memory search engines. In: Proc. 10th ALENEX (2008)
30. Strohman, T., Croft, B.: Efficient document retrieval in main memory. In: Proc. 30th ACM SIGIR, pp. 175–182. ACM Press, New York (2007)
31. Weiner, P.: Linear pattern matching algorithm. In: Proc. 14th Annual IEEE Symposium on Switching and Automata Theory, pp. 1–11 (1973)
32. Witten, I., Moffat, A., Bell, T.: Managing Gigabytes, 2nd edn. Morgan Kaufmann Publishers, San Francisco (1999)

New Perspectives on the Prefix Array*

W.F. Smyth[1,2] and Shu Wang[1]

[1] Algorithms Research Group, Department of Computing & Software
McMaster University, Hamilton, Ontario, Canada L8S 4K1
{smyth,shuw}@mcmaster.ca
www.cas.mcmaster.ca/cas/research/algorithms.htm
[2] Digital Ecosystems & Business Intelligence Institute
Curtin University, GPO Box U1987, Perth WA 6845, Australia
W.Smyth@curtin.edu.au

Abstract. In this paper we consider the **prefix array** $\pi = \pi[1..n]$ of a string $x = x[1..n]$ in which $\pi[1] = 0$ and, for $i > 1$, $\pi[i] = k$ iff k is the largest integer such that $x[i..i+k-1] = x[1..k]$. The prefix array π is closely related to the **border array** β: an integer array $[1..n]$ such that $\beta[i] = k$ iff the length of the longest border of $x[1..i]$ is k. Border arrays or their variants are used in many string algorithms and prefix arrays can be used directly for pattern-matching. It is well known that for regular strings π provides all the information that β does; we show however that for indeterminate strings (those containing entries that match a subset of the alphabet) π actually provides more information, in fact still enabling all the borders of every prefix of x to be specified. Since a lot of the entries of π are expected to be zeros, it is natural to represent π in compressed form using integer arrays POS$[1..m]$ and LEN$[1..m]$, where m is the number of nonzero entries in π and $\pi[\text{POS}[j]] = \text{LEN}[j]$ iff the jth nonzero entry in π occurs in position POS$[j]$ and takes the value LEN$[j]$. The expected value of m is $n/\sigma - 1$, where σ is the alphabet size. The straightforward way of computing POS/LEN requires computing π first, therefore requires $O(n)$ extra space. We describe two $\Theta(n)$-time algorithms PL1 & PL2 to compute POS/LEN for regular strings using only $8m$ bytes of storage in addition to the n bytes required for x. PL1 requires about one-third the time of the standard border array algorithm MP on English-language strings; PL2 executes faster than MP on both English-language and highly periodic strings on $\{a, b\}$. For indeterminate strings, we describe an extension IPL of PL1 that computes POS/LEN in $O(n^2)$ worst-case time (though generally much faster), still using only $8m$ bytes of additional storage. For both regular and indeterminate strings, the compressed form of π can be used for efficient pattern-matching.

* This work was supported in part by the Natural Sciences & Engineering Research Council of Canada. The authors thank Maxime Crochemore for helpful discussions and anonymous referees for valuable suggestions.

A. Amir, A. Turpin, and A. Moffat (Eds.): SPIRE 2008, LNCS 5280, pp. 133–143, 2008.

1   Introduction

A **border** of a string $x = x[1..n]$ is a proper prefix of x that is also a suffix. The **border array** β of x is an integer array $[1..n]$ such that $\beta[i] = k$ iff the length of the longest border of $x[1..i]$ is k. Apparently first introduced in 1970 [MP70] as the **failure function**, the border array in various guises has numerous applications; for example, pattern-matching [AC75, KMP77, BM77, G79, CCGJ94], computing repetitions [ML84, M89], Lyndon decomposition [D83], determining quasiperiodicity [LS02]. In a field to which periodicity is so important computationally, this ubiquity is perhaps not surprising: to know the length ℓ of any border of a string $x = x[1..n]$ is to know that x has period $n-\ell$. Furthermore, since for every $i \in 1..n$ such that $\beta[i] > 0$, the longest border of $x[1..\beta[i]]$ is the second longest border of $x[1..i]$, it follows that the border array determines all the periods of every prefix of x. (These properties hold for strings on a regular alphabet Σ; later we introduce **indeterminate strings**, whose entries may match not only letters in Σ, but also specified subsets of Σ — for such strings, these properties do not necessarily hold.)

However, a property that holds for every border array, regardless of the nature of the underlying alphabet of x, is the following:

(P) If for some $i \in 2..n$, $\beta[i] > 0$, then $x[1..i-1]$ has a border of length $\beta[i]-1$.

Property (P) implies that all the borders of the prefixes of x occur in arithmetic sequences fully determined by maximum border lengths k; more precisely, if the maximum border length of $x[1..i]$ is $k > 0$, then

$$x[1..i-1], x[1..i-2], \ldots, x[1..i-k+1] \tag{1}$$

must have borders of (not necessarily maximum) lengths

$$k-1, k-2, \ldots, 1,$$

respectively. (Of course every nonempty prefix of x has a border of length zero.) Thus to describe all the borders of every nonempty prefix of x, it suffices to specify only the maximum border k at each position $i > 0$. Since in each of the sequences (1), $x[i-k+1] = x[1]$, it becomes clear that to describe all the borders of every prefix of x, it suffices to specify in every position $j \in 2..n$ the length $k > 0$ of the longest substring $x[j..j+k-1] = x[1..k]$ — that is, the longest substring that matches a prefix of x. Following [CHL01, CHL07], we define the **prefix array** $\pi = \pi[1..n]$ in which $\pi[1] = 0$ and, for $i > 1$, $\pi[i] = k$ iff k is the largest integer such that $x[i..i+k-1] = x[1..k]$. Based on the above discussion, we claim

Lemma 1. *The prefix array π describes all the borders of every prefix of x.* □

The standard border array algorithm [MP70], that we call Algorithm MP, is well known. [CHL01] outlines a basis for $\Theta(n)$-time algorithms to compute β from π and *vice versa*.

Since for every $i \in 2..n$, $\pi[i] = 0$ if and only if $x[i] \neq x[1]$, the number of nonzero elements in π is exactly m, where m is the number of occurrences of $x[1]$ in x other than at position 1. If we let $\sigma = |\Sigma|$, the expected value of m is $n/\sigma - 1$, a quantity less than $n/2$ for $\sigma \geq 2$. Thus it is natural to represent π in compressed form using integer arrays POS[1..m] and LEN[1..m], defined as follows: $\pi[\text{POS}[j]] = \text{LEN}[j]$ iff the jth nonzero entry in π occurs in position POS[j] and takes the value LEN[j]. If we allow q bytes for storage of an integer (generally $q = 4$), the expected storage requirement for this compressed form will be $2q(n/\sigma - 1)$ bytes, a substantial saving for larger alphabets over the qn bytes required for the border array (and no more even for $\sigma = 2$).

The straight forward approach to compute POS/LEN, is of course to compute π first and then re-write it in the compress form. However, this approach requires at least additional qn (temporary) space. In Section 2 we describe a new algorithm, executing on regular strings in worst-case $\Theta(n)$ time using (peak) $qm + O(1)$ bytes of additional storage, that computes π in compressed form POS/LEN. We show how to use POS/LEN for efficient pattern-matching. In Section 3 we introduce indeterminate strings and discuss the advantages and use of the prefix array in this context, again including in particular efficient pattern-matching. Section 4 gives the results of test runs of our algorithms on large regular strings of various kinds, and Section 5 outlines future related work.

2   The Compressed Prefix Array on Regular Strings

In [CHL07] an elegant algorithm is described for computing π in its usual uncompressed form. In Figure 1 we display a related algorithm PL1 that computes π in its compressed POS/LEN form.

PL1 executes in two main phases:

(1) In the first phase a scan of π identifies the positions that match $x[1]$, thus determining m. Then the POS array is formed with these positions in ascending order, while each position in the LEN array is initialized to 1 — the length of the shortest possible matching prefix.

(2) For every $j \in 1..m$, the second phase computes the length LEN[j] of the longest substring $x[\text{POS}[j]]$ that equals a prefix of x, making use of the knowledge that LEN[j] ≥ 1.

For every $j \in 1..m$, let $i = \text{POS}[j]$. Then the second phase depends on three main parameters:

- *pref*: initially the current length, then the maximum length, of the substring beginning at $x[i]$ that matches a prefix of x;
- *end*: the rightmost position in x such that $x[i..end]$ matches a prefix of x;
- *lim*: the largest value of *end* that has so far been computed over all the previous positions i in x.

The second phase of PL1 uses two routines, match and copy. Beginning with the current matching prefix *pref* at position i, the function match extends the

— *Initialize POS/LEN for m positions* $x[i] = x[1]$, $i \in 2..n$.
$\lambda \leftarrow x[1]$; $x[n+1] \leftarrow \$$; $m \leftarrow 0$
for $i \leftarrow 2$ **to** n **do**
    if $x[i] = \lambda$ **then**
        $m \leftarrow m+1$; POS$[m] \leftarrow i$; LEN$[m] \leftarrow 1$
— *For each* $j \in 1..m$, *determine longest match with prefix of* x.
$j \leftarrow 1$; $lim \leftarrow 1$
while $j \leq m$ **do**
    $i \leftarrow$ POS$[j]$; $pref \leftarrow$ LEN$[j]$
    if $i+pref > lim$ **then**
        $pref \leftarrow$ match$(1+pref, i+pref) - 1$
        LEN$[j] \leftarrow pref$
        $end \leftarrow i+pref-1$
        if $end > lim$ **then**
            $lim \leftarrow end$; copy$(j+1, lim)$
    $j \leftarrow j+1$
— *Return number of matching positions* $x[pos1 \cdots] : x[pos2 \cdots]$.
function match$(pos1, pos2)$
while $x[pos1] = x[pos2]$ **do**
    $pos1 \leftarrow pos1+1$; $pos2 \leftarrow pos2+1$
return $pos1$
— *Update LEN for every* λ *in* $x[i+1..lim-1] = x[2..pref-1]$.
procedure copy(J, lim)
$J' \leftarrow 1$; $I \leftarrow$ POS$[J]$
while $I < lim$ **do**
    LEN$[J] \leftarrow \min\{$LEN$[J'], lim-I+1\}$
    $J \leftarrow J+1$; $J' \leftarrow J'+1$; $I \leftarrow$ POS$[J]$

Fig. 1. Algorithm PL1: compute the POS/LEN arrays for $x = x[1..n]$

match as far to the right as possible, essentially adding to *pref* (if possible) and determining *end*. (Note that explicit end-of-string detection is avoided in match by adding a sentinel letter $\$$ at position $n+1$ of x that does not match any other letter in x.) If it turns out that $end > lim$, then it is possible that there may exist positions $I > i$ such that $x[I] = x[1]$ and $I < end$ – in such cases, there is a substring beginning at I that matches a substring beginning at position $I' < i$ (hence already computed) that in turn matches a prefix of x. For all such I and corresponding I', the procedure copy extends the match as far as *end* (now *lim*) by copying as much as possible of the already computed LEN value corresponding to I' into the LEN value corresponding to I. Note that copy depends critically on the transitivity of matching ($a = b$ and $b = c$ implies $a = c$), a property that always holds for regular strings. The use of copy avoids repeating letter comparisons and ensures that for each value of i considered, *pref* is as large as possible.

The first phase of PL1 requires $n-1$ letter comparisons and executes in $\Theta(n)$ time. The second phase requires at most $n-1$ letter comparisons and otherwise

$x[n+1] \leftarrow \$; \ pref \leftarrow 0; \ i \leftarrow 1; m \leftarrow 0$
while $i < n$ **do**
    if $pref = 0$ **then** $i \leftarrow i+1$
    $pref \leftarrow \mathtt{match}(i, pref)$
    if $(pref \neq 0)$
        $m \leftarrow m+1; \ POS[m] \leftarrow i; \ LEN[m] \leftarrow pref$
        $\mathtt{copy}(i, pref, m)$

Fig. 2. Algorithm PL2: compute the prefix array π for $x = x[1..n]$

    — *Return number of matching positions* $x[pos1 \cdots] : x[pos2 \cdots]$.
function $\mathtt{match}(i, pref)$
while $x[i+pref] = x[1+pref]$ **do**
    $pref \leftarrow pref+1;$
return $pref$

procedure $\mathtt{copy}(i, pref, m)$
$J \leftarrow 1$
while $POS[J] \leq pref$ **do**
    $max \leftarrow pref-POS[J]+1$
    if $LEN[J] < max$ **then**
        $m \leftarrow m+1; \ POS[m] \leftarrow i+POS[J]-1; \ LEN[m] \leftarrow LEN[J]$
    elsif $i+pref \geq n$ **then**
        $m \leftarrow m+1; \ POS[m] \leftarrow i+POS[J]-1; \ LEN[m] \leftarrow max$
    else
        $pref \leftarrow -1; J \leftarrow J-1$
    $J \leftarrow J+1$
if $pref = -1$ **then**
    $i \leftarrow i+POS[J]-1; \ pref \leftarrow max$
else
    $i \leftarrow i+pref-1; pref \leftarrow 0$

Fig. 3. Routines $\mathtt{match}$ and $\mathtt{copy}$ for Algorithm PL2

requires only constant time for each position i in x that needs to be considered. Thus PL1 is a $\Theta(n)$ time algorithm. As noted earlier, additional storage required for POS/LEN is $2qm$ bytes, where q is the number of bytes needed for integer storage.

Because it deals efficiently with positions in x that do not match $x[1]$, Algorithm PL1 is more efficient on strings with larger alphabets that occur frequently in practice. In Figure 2 we describe an alternative algorithm PL2 that is fast (faster than MP) on highly periodic strings on $\{a, b\}$ as well as on strings on larger alphabets.

The new routines $\mathtt{match}$ and $\mathtt{copy}$ are shown in Figure 3. PL2 differs from PL1 primarily in its approach to $\mathtt{copy}$: rather than copying as much as possible up to the current value of lim, PL2 copies only if the match to be copied can be copied in its entirety (or if end-of-string is reached). If the copy cannot be completed, a nonzero value of $pref$ is returned, indicating the minimum length of the match at the next occurrence of $x[1]$.

Given a pattern $p = p[1..n_1]$ and a text $y = y[1..n_2]$, the positions of all occurrences of p in y can be identified by first forming

$$x = p\#y,$$

where $\# \neq \$$ is a sentinel that does not appear in p or x and $n = n_1 + n_2 + 1$, then executing PL1 or PL2. The positions in y at which p occurs are identified by the values of $j \in 1..m$ such that $\text{LEN}[j] = n_1$. The time requirement is $\Theta(n)$ with a maximum of $2n - 2$ letter comparisons, a minimum of $n - 1$.

3   Prefix Array on Indeterminate Strings

Let λ_i, $|\lambda_i| \geq 2$, $1 \leq i \leq s$, be pairwise distinct subsets of Σ. We form a new alphabet $\Sigma' = \Sigma \cup \{\lambda_1, \lambda_2, .., \lambda_s\}$ and define a new relation **match** ($\approx$) on Σ' as follows:

- for every $\mu_1, \mu_2 \in \Sigma$, $\mu_1 \approx \mu_2$ if and only if $\mu_1 = \mu_2$;
- for every $\mu \in \Sigma$ and every $\lambda \in \Sigma' - \Sigma$, $\mu \approx \lambda$ and $\lambda \approx \mu$ if and only if $\mu \in \lambda$;
- for every $\lambda_i, \lambda_j \in \Sigma' - \Sigma$, $\lambda_i \approx \lambda_j$ if and only if $\lambda_i \cap \lambda_j \neq \emptyset$.

Observe that $match$ is reflexive and symmetric but not necessarily transitive; for example, if $\lambda = \{a, b\}$, then $a \approx \lambda$ and $b \approx \lambda$ does not imply $a \approx b$. This idea seems to have first been mentioned in [FP74] and existed in varies guises such as generalized strings [A87], subset matching [CHI99, CH03], partial words [BSH02], and degenerate strings [IRVV07].

Here we define the letters in $\Sigma' - \Sigma$ to be **indeterminate**, and a string containing indeterminate letters is called an **indeterminate string**. A position in x at which an indeterminate letter appears is called a **hole**. An important special case of an indeterminate letter is $\lambda = \{\Sigma\}$, usually denoted by $*$ and called a **don't-care**. In [IMMP03] an average-case $\Theta(n)$-time (worst-case $O(n^2)$-time) algorithm was described for the calculation of all the borders of every prefix of a string $x = x[1..n]$ containing don't-cares. In [HS03] this algorithm was extended to strings with arbitary holes.

Because of the nontransitivity of the match operation on indeterminate letters, it is no longer necessarily true, as remarked in the Introduction, that $\beta[\beta[i]]$ necessarily gives the length of the second-longest border of $\beta[1..i]$. For example, if $x = aba * b$, a border array $\beta = 00122$ would give correctly the longest borders of $x[1..i]$, $i \in 1..5$, but because $\beta[\beta[4]] = 0$, it would not report the fact that actually $\beta[1..4]$ has a border of length 1 in addition to its longest border of length 2. Thus the algorithms given in [IMMP03, HS03] all need to compute a linked list

— Initialize POS/LEN/EQUAL for m positions $x[i] \approx x[1]$, $i \in 2..n$.
$\lambda \leftarrow x[1]$; $x[n+1] \leftarrow \$$; $m \leftarrow 0$
for $i \leftarrow 2$ to n do
  if $x[i] \approx \lambda$ then
    $m \leftarrow m+1$; POS$[m] \leftarrow i$; LEN$[m] \leftarrow 1$
    if $x[i] = \lambda$ then EQUAL$[m] \leftarrow$ TRUE
    else EQUAL$[m] \leftarrow$ FALSE
— For each $j \in 1..m$, determine longest match with prefix of x.
$j \leftarrow 1$; $lim \leftarrow 1$
while $j \leq m$ do
  $i \leftarrow$ POS$[j]$; $pref \leftarrow$ LEN$[j]$
  if $i+pref > lim$ then
    $pref \leftarrow$ match$(1+pref, i+pref, j)-1$
    LEN$[j] \leftarrow pref$
    $end \leftarrow i+pref-1$
    if $end > lim$ then
      $lim \leftarrow end$; copy$(j+1, lim)$
  $j \leftarrow j+1$
— Return number of matching positions $x[pos1 \cdots] : x[pos2 \cdots]$.
function match$(pos1, pos2, j)$
while $x[pos1] \approx x[pos2]$ do
  $pos1 \leftarrow pos1+1$; $pos2 \leftarrow pos2+1$
  if $x[pos1] \neq x[pos2]$ then EQUAL$[j] \leftarrow$ FALSE
return $pos1$
— Update LEN for every λ in $x[i+1..lim-1] = x[2..pref-1]$.
procedure copy(J, lim)
$J' \leftarrow 1$; $I \leftarrow$ POS$[J]$
while $I < lim$ do
  if EQUAL$[J']$ then LEN$[J] \leftarrow \min\{$LEN$[J'], lim-I+1\}$
  $J \leftarrow J+1$; $J' \leftarrow J'+1$; $I \leftarrow$ POS$[J]$

Fig. 4. Algorithm IPL: compute the POS/LEN arrays for indeterminate $x = x[1..n]$

at each position i of β in order to give specifically all the borders of the prefixes of the indeterminate string x. In the worst case, this requires $O(n^2)$ storage. However, due to the validity of Lemma 1 for both regular and indeterminate strings, the computation of π rather than β eliminates the requirement for lists: for an indeterminate string, the nonzero positions $i > 1$ in π will be those for which $x[i] \approx x[1]$, and the values of these $\pi[i]$ will be sufficient to specify all the borders of all the prefixes of x, just as in the indeterminate case, with no additional storage required. In the above example, it would suffice to return $\pi = 00220$.

Another consequence of the intransitivity of indeterminate matching is that the relationship between border and period, mentioned in the Introduction, is modified. Following [BSH02] we say that an indeterminate string x has **strong** *period* p if and only if for every $i_1, i_2 \in 1..n$ such that $i_1 \equiv i_2 \pmod{p}$, $x[i_1] \approx x[i_2]$; x has **weak period** p if and only if for every $i_1, i_2 \in 1..n$ such that $i_2 = i_1 + p$, $x[i_1] \approx x[i_2]$. If an indeterminate string x has a border of length ℓ, we can say only that x has weak period $n - \ell$, not necessarily a strong period. Consider, for example, $x = abcab * abd$ of length $n = 9$ with a border of length $\ell = 6$.

Figure 4 gives pseudocode of Algorithm IPL, the indeterminate version of Algorithm PL1 (Figure 1). The main change is the introduction of a bit array EQUAL[$1..m$]: EQUAL[j] = TRUE iff the substring starting at position POS[j] of length LEN[j] is equal to (rather than merely matching) a prefix of x. If equality holds, then the `copy` procedure can be performed as usual; if not, then it must be skipped. EQUAL is maintained in function `match`: if at any position there is a match ($\approx$) with a prefix of x but not equality ($=$), EQUAL[j] is permanently set to FALSE. An alternative, brute-force (and perhaps in practice faster) approach would have been to simply eliminate `copy` altogether and to perform all matches directly with $x[2 \cdots]$. In any case IPL has worst-case time $O(n^2)$, attained for example by $x = *\lambda^{n-1}$, but only requires $9(n/\sigma - 1)$ bytes of storage. The implementation of letter-matching ($\approx$) in indeterminate strings can be nontrivial; various alternatives are discussed in [HSW06, HSW08].

By computing π rather than borders, Algorithm IPL makes available all the border information in a generally much more compact form, without the need to store linked lists of possible borders for each position of x.

4   Experimental Results

We have tested Algorithms PL1 and PL2 against the standard border array algorithm MP and the algorithm PI for computation of the uncompressed π array of [CHL07], using two kinds of test data: highly periodic strings on $\{a, b\}$ and English-language texts. The highly periodic strings are concatenations of long strings in the family of strings with many runs identified in [FSS03]; the English files are concatenations of various novels (for example, *The Mysterious Affair at Styles*, *War and Peace*). The results are presented in Figures 5 and 6. Experiments were conducted using a 2.6GHz Opteron 885 processor with 2GB main memory available, under GNU Linux (kernel release 2.6.18-92.1.1.el5). The compiler was g++ with the -O3 option. The run times used were the minima over 10 runs, not including input/output.

On periodic strings Algorithms PL2 and PI are comparable, slightly faster than MP, that in turn requires only two-thirds the time of PL1. On English-language strings, however, PL1 is the fastest algorithm, requiring about one-third the time of MP, and slightly faster than PL2.

5   Summary and Future Work

In this paper we have presented new efficient algorithms PL1 and PL2 to compute the prefix array in compressed form. More comprehensive testing is required

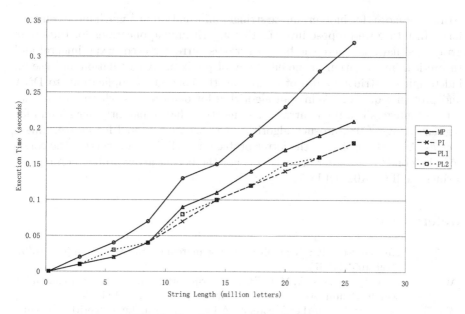

Fig. 5. Test Result on Periodic Strings

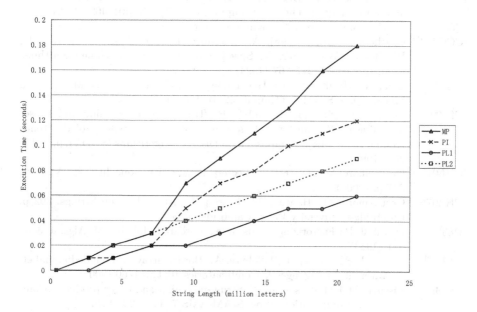

Fig. 6. Test Result on English Strings

to determine the kinds of strings on which they execute most quickly. It appears
that, on strings arising in practice, these algorithms generally execute considerably faster than the border array algorithm while providing the same (for regular

strings) or more (for indeterminate strings) information in a much more compact form. In future we propose investigating algorithmic applications for the prefix array in situations where the border array is currently used. Experimental and theoretical investigation of the efficiency of prefix array calculation and use on indeterminate strings is also of interest in the context of applications to DNA and protein sequences. While an algorithm for determining whether or not an array of integers is a prefix array for some string has apparently been found for regular strings [C08], no such algorithm has been discovered for indeterminate strings. As we have shown, the latter problem is well defined, whereas the corresponding one for a border array is not. For the border array problem on regular strings see [FGLR02, DLL05].

References

[A87]   Abrahamson, K.: Generalized string matching. SIAM J. Comput. 16(6), 1039–1051 (1987)

[AC75]   Aho, A.V., Corasick, M.J.: Efficient string matching: an aid to bibliographic search. Comm. Assoc. Comput. Mach. 18(6), 333–340 (1975)

[BM77]   Boyer, R.S., Strother Moore, J.: A fast string searching algorithm. Comm. Assoc. Comput. Mach. 20(10), 762–772 (1977)

[BSH02]   Blanchet-Sadri, F., Hegstrom, R.A.: Partial words and a theorem of Fine and Wilf revisited. Theoret. Comput. Sci. 270(1/2), 401–409 (2002)

[C08]   Crochemore, M.: private communication

[CCGJ94]   Crochemore, M., Czumaj, A., Gąsieniec, L., Jarominek, S., Lecroq, T., Plandowski, W., Rytter, W.: Speeding up two string-matching algorithms. Algorithmica 12, 247–267 (1994)

[CH03]   Cole, R., Hariharan, R.: Tree Pattern Matching to Subset Matching in Linear Time. SIAM J. Comput. 32(4), 1056–1066 (2003)

[CHI99]   Cole, R., Hariharan, R., Indyk, P.: Tree pattern matching and subset matching in deterministic $O(nlog^3m)$ time. In: Proceedings of the Tenth Annual ACM-SIAM Symposium on Discrete Algorithms (SODA), pp. 245–254 (1999)

[CHL01]   Crochemore, M., Hancart, C., Lecroq, T.: Algorithmique du Texte, Vuibert, 347 p. (2001)

[CHL07]   Crochemore, M., Hancart, C., Lecroq, T.: Algorithms on Strings, 392 p. Cambridge University Press, Cambridge (2007)

[D83]   Duval, J.-P.: Factorizing words over an ordered alphabet. J. Algs. 4, 363–381 (1983)

[DLL05]   Duval, J.-P., Lecroq, T., Lefebvre, A.: Border array on bounded alphabet. J. Automata, Languages & Combinatorics 10(1), 51–60 (2005)

[FP74]   Fischer, M.J., Paterson, M.S.: String-matching and other products. Complexity of Computation, Proc. SIAM-AMS 7, 113–125 (1974)

[FGLR02]   Franek, F., Gao, S., Lu, W., Ryan, P.J., Smyth, W.F., Sun, Y., Yang, L.: Verifying a border array in linear time. J. Combinatorial Maths. & Combinatorial Comput. 42, 223–236 (2002)

[FSS03]   Franek, F., Simpson, R.J., Smyth, W.F.: The maximum number of runs in a string. In: Miller, M., Park, K. (eds.) Proc. 14th Australasian Workshop on Combinatorial Algs., pp. 36–45 (2003)

[G79]   Galil, Z.: On improving the worst case running time of the Boyer- Moore
           string matching algorithm. Comm. Assoc. Comput. Mach. 22(9), 505–508
           (1979)
[HS03]   Holub, J., Smyth, W.F.: Algorithms on indeterminate strings. In: Proc.
           14th Australasian Workshop on Combinatorial Algs., pp. 36–45 (2003)
[HSW06]   Holub, J., Smyth, W.F., Wang, S.: Hybrid pattern-matching algorithms on
           indeterminate strings. In: Daykin, J., Mohamed, M., Steinhoefel, K. (eds.)
           Texts in Algorithmics. King's College London Series, pp. 115–133 (2006)
[HSW08]   Holub, J., Smyth, W.F., Wang, S.: Fast pattern-matching on indeterminate
           strings. J. Discrete Algs. 6(1), 37–50 (2008)
[IMMP03]   Iliopoulos, C.S., Mohamed, M., Mouchard, L., Perdikuri, K.G., Smyth,
           W.F., Tsakalidis, A.K.: String regularities with don't cares. Nordic J. Com-
           put. 10(1), 40–51 (2003)
[IRVV07]   Iliopoulos, C.S., Sohel Rahman, M., Voráček, M., Vagner, L.: The con-
           strained longest common subsequence problem for degenerate strings. Im-
           plementation and Application of Automata, 309–311 (2007)
[KMP77]   Knuth, D.E., Morris, J.H., Pratt, V.R.: Fast pattern matching in strings.
           SIAM J. Comput. 6(2), 323–350 (1977)
[LS02]   Li, Y., Smyth, W.F.: Computing the cover array in linear time. Algorith-
           mica 32(1), 95–106 (2002)
[M89]   Main, M.G.: Detecting leftmost maximal periodicities. Discrete Applied
           Maths. 25, 145–153 (1989)
[ML84]   Main, M.G., Lorentz, R.J.: An $O(n \log n)$ algorithm for finding all repeti-
           tions in a string. J. Algs. 5, 422–432 (1984)
[MP70]   Morris, J.H., Pratt, V.R.: A Linear Pattern-Matching Algorithm, Tech.
           Rep. 40, University of California, Berkeley (1970)

Indexed Hierarchical Approximate String Matching

Luís M.S. Russo[1,3,*], Gonzalo Navarro[2,**], and Arlindo L. Oliveira[1,4]

[1] INESC-ID, R. Alves Redol 9, 1000 Lisboa, Portugal
aml@algos.inesc-id.pt
[2] Dept. of Computer Science, University of Chile
gnavarro@dcc.uchile.cl
[3] CITI, Departamento de Informática, Faculdade de Ciências e Tecnologia,
Universidade Nova de Lisboa, Portugal
lsr@di.fct.unl.pt
[4] Instituto Superior Técnico, Universidade Técnica de Lisboa, Portugal

Abstract. We present a new search procedure for approximate string matching over suffix trees. We show that hierarchical verification, which is a well-established technique for on-line searching, can also be used with an indexed approach. For this, we need that the index supports bidirectionality, meaning that the search for a pattern can be updated by adding a letter at the right or at the left. This turns out to be easily supported by most compressed text self-indexes, which represent the index and the text essentially in the same space of the compressed text alone. To complete the symbiotic exchange, our hierarchical verification largely reduces the need to access the text, which is expensive in compressed text self-indexes. The resulting algorithm can, in particular, run over an existing fully compressed suffix tree, which makes it very appealing for applications in computational biology. We compare our algorithm with related approaches, showing that our method offers an interesting space/time tradeoff, and in particular does not need of any parameterization, which is necessary in the most successful competing approaches.

1 Introduction and Related Work

Approximate string matching (ASM) is an important problem that arises in applications related to text searching, pattern recognition, signal processing, and computational biology, to name a few. The problem consists in locating all the occurrences O of a given pattern string P, of size m, in a larger text string T, of size n, where the distance between P and O is less than a given threshold k. We focus on the edit distance, that is, the minimum number of character insertions, deletions, and substitutions of single characters to convert one string into the other.

---

* Partially funded by the Portuguese Science and Technology Foundation by project ARN, PTDC/EIA/67722/2006.
** Partially funded by Millennium Institute for Cell Dynamics and Biotechnology, Grant ICM P05-001-F, Mideplan, Chile.

A. Amir, A. Turpin, and A. Moffat (Eds.): SPIRE 2008, LNCS 5280, pp. 144–154, 2008.

The most successful indexed approach to this problem, in practice, is so-called "hybrid" indexing. It starts with a *filtration* phase that determines the positions of potential occurrences. Those positions are then sequentially verified in the text. The pattern pieces searched for in the filtration phase are short enough to control the exponential cost of this search, and long enough so that the number of occurrences to verify in the text is also controlled. By carefully optimizing this partitioning, hybrid indexes achieve $O(mn^\lambda)$ average time, for some $0 < \lambda < 1$, and work well for reasonably high error levels. Hybrid methods have been implemented over q-gram indexes [1], suffix arrays [2], and q-sample indexes [3]. Yet, many of those linear-space indexes are very large anyway. For example, suffix arrays require 4 times the text size and suffix trees require at least 10 times [4]. Compressed indexes, based on succinct and compressed data structures, provide less space-demanding indexes [5]. Their space requirements are measured in terms of the empirical text entropy, H_k, which gives a lower bound for the number of bits per symbol achievable over that text by a k-th order compressor.

There have been several approaches to ASM over compressed indexes. The most successful one in practice is that of Russo *et al.* [6], which builds over a Ziv-Lempel-based compressed index, and approaches hybrid performance in practice. This is faster than our new index, still ours is significantly smaller, in theory and in practice. In addition, our algorithm can run over most compressed text indexes, in particular over *fully-compressed suffix trees* [7] (FCSTs), which offer complete suffix-tree functionality. Hence, our algorithm can be used as a subroutine in other suffix-tree-based algorithms.

2   Our Contribution

In this work we explore the impact of *hierarchical verification* on hybrid search. Hierarchical verification means that an area that needs to be verified is not immediately checked with the maximum number of errors; instead the error threshold is raised gradually. Curiously enough, this technique was originally proposed by Myers [1] in his hybrid index and later extended and used by Navarro *et al.* [8] for an on-line algorithm. However, these approaches used hierarchical verification directly over the text T, meaning that none of the repeated computation was factorized. We investigate precisely how to do this computation over the index, thus allowing us to avoid repeated computation.

Simultaneously, our result achieves compressed space, because we use FCSTs, which are functional representations of suffix trees and in particular are bidirectional. Typical indexes, classical suffix trees in particular, are unidirectional, meaning that they can search only by using the letters at the end of the pattern. Due to the RANK/SELECT duality [5], bidirectionality arises naturally in a class of compressed indexes, which we will refer to as *bidirectional compressed indexes*.

Bidirectional indexes are one important ingredient of our approach. Another crucial piece is computing the edit distance. Algorithms for this purpose are typically unidirectional, computed from left to right, because they are based on dynamic programing or automata. Interestingly this computation was made bidirectional, more than 10 years ago, by Landau *et al.* [9]. They showed how to obtain the edit distance for strings A and cB by extending that for for strings A and B, where c is a letter.

Combining these bidirectional algorithms we can use hierarchical verification directly over the index, instead of over T. Thus, we fill an important gap in indexed ASM. Moreover, while hybrid methods need careful tuning (where a small error can be disastrous), ours achieve close performance without need of tuning (and can be improved by tuning as well).

In addition, our work addresses a very important practical issue. Compressed indexes are usually self-indexes, meaning that they do not store the text T but even so they are able to consult it. Even when in theory reading ℓ consecutive letters takes $O(\ell)$ time, experimental results show [10] that this is still two orders of magnitude slower than storing T. This can easily be explained as the penalty of missing cache in modern computer architectures. Efficient algorithms for ASM over compressed indexes must therefore minimize their accesses to T. Hence hierarchical verification directly over the index is a very important technique in this context, both in theory and in practice.

3   Basic Concepts

We denote by T a **string**; by Σ the **alphabet** of size σ; by $T[i]$ the symbol at position $(i \bmod n)$; by $T.T'$ **concatenation**; by $T = T[..i - 1].T[i..j].T[j + 1..]$ respectively a **prefix**, a **substring** and a **suffix**; by $S \sqsubseteq S'$ that S is a substring of S'. We refer indifferently to nodes and to their path-labels, also denoted by v. The **suffix tree** of T is the deterministic compact labeled tree for which the path-labels of the leaves are the suffixes of $T\$$, where $\$$ is a terminator symbol not belonging to Σ. We will assume n is the length of $T\$$. For a detailed explanation see Gusfield's book [11]. The **suffix array** $A[0, n - 1]$ stores the suffix indexes of the leaves in lexicographical order.

3.1   Bidirectional Compressed Indexes

Our algorithm can be implemented over any bidirectional index. This means that, from the index point corresponding to a text substring $T[i..j]$ we can efficiently move to that of $T[i..j + 1]$ but also to that of $T[i - 1..j]$.

Although classical text indexes are not usually bidirectional, most compressed indexes are. For example, FM-indexes [12] offer a so-called LF mapping operation, which moves from the suffix array position k such that $A[k] = i$, to position k' such that $A[k'] = i - 1$. Compressed suffix arrays [13], instead, offer function ψ, moving to a k' such that $A[k'] = i + 1$, thus the inverse of ψ serves as an LF mapping as well.

FCSTs [7] build complete suffix tree functionality on top of a compressed bidirectional index, in particular an FM-index fits best. The LF mapping allows FCSTs implement Weiner links [14]: WEINERLINK(v, a), for node v and letter a,

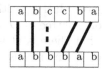

Fig. 1. Schematic representation of the edit distance between *abccba* and *abbbab*

Fig. 2 (left):

	a	b	c	c	b	a	j:
	0	1	2	3	4	5	6	0
a	1	**0**	1	2	3	4	5	1
b	2	1	**0**	1	2	3	4	2
b	3	2	1	**1**	**1**	2	3	3
b	4	3	2	2	**2**	**2**	3	4
a	5	4	3	3	3	3	**2**	5
b	6	5	4	4	4	**3**	**3**	6
i:	0	1	2	3	4	5	6	

Fig. 2 (right):

	a	b	c	c	b	a	j:
	0							0
a	0							1
b	1	0	1					2
b	2	1	1	1	2			3
b		2	2	2	2			4
a						2		5
b								6
i:	0	1	2	3	4	5	6	

Fig. 2. *D* table computation for strings *abccba* and *abbbab*. (left) The numbers in bold refer to the alignment shown in Fig. 1. (right) Computation with increasing error bound.

gives the suffix tree node v' with path-label $a.v[0..]$, and it is the key to move from a v representing $T[i..j]$ to a v' representing $T[i-1..j]$, that is, to birectionality. The other direction, that is, from $T[i..j]$ to $T[i..j+1]$, is supported just by moving to a child of v. FCSTs support all of the usual suffix tree navigation operations, including suffix links (via ψ) and lowest common ancestors (LCA(v, v')).

3.2 Approximate String Matching

The *edit* or *Levenshtein* distance between two strings, $ed(A, B)$, is the smallest number of edit operations that transform A into B. We consider as operations insertions, deletions, and substitutions. There is a well-known dynamic programming (DP) algorithm that computes the D matrix, where $D[i, j]$ is the edit distance, $ed(A[..i-1], B[..j-1])$, between the prefixes $A[..i-1]$ and $B[..j-1]$ of A and B. Fig. 2(left) shows an example of the D matrix for $A = abccba$ and $B = abbbab$. Therefore by looking at cell $D[6, 6] = 3$ we can conclude that $ed(abccba, abbbab) = 3$. Let the size of A and B be m and m' respectively. This matrix can be computed, in $O(mm')$ time, by setting $D[0, 0] = 0$ and

$$D[i, j] = \min \left\{ \begin{array}{ll} D[i-1, j] + 1 & \text{if } i > 0 \\ D[i, j-1] + 1 & \text{if } j > 0 \\ D[i-1, j-1] + \delta_{A[i-1] = B[j-1]} & \text{if } i, j > 0 \end{array} \right\},$$

where $\delta_{x=y}$ is 0 if $x = y$ and 1 otherwise. Ukkonen [15] noted that in order to find cells in D whose value is k there is no need to compute cells with value larger than k; those can be replaced by $+\infty$. The remaining cells are referred to as active cells. With this method, extending the computation of $ed(A, B)$ to $ed(Ac, B)$ or $ed(A, Bc)$ requires only $O(k)$ time.

Assuming we have a text T, previously pre-processed into a FCST, the problem we are interested in solving in this paper is: given a pattern P and error limit k, determine all the substrings O of T for which $ed(P, O) \leq k$. As our running example consider that $P = abccba$, $k = 2$ and $T = abbbab$. The only substring O of T is *abbba*. A way to find this string, not always the most efficient one, is to perform a depth-first search over the suffix tree of T, moving one letter at a time, simultaneously computing the D table, for P and O', where O' is the path-label of the node we are visiting. This table can be used to control the search. When we reach a point O' and $ed(P, O') \leq k$, which can be checked as $D[|P|, |O'|] \leq k$, this string is reported as an occurrence. Usually we also report all the positions in T at which O' occurs, which means traversing the whole subtree of O' and reporting all its leaf positions. Otherwise if $ed(P, O') > k$ but there is at least one active cell in the last row, $i.e.$ $D[i, |O'|] \leq k$ for some i, this means that $ed(P[..i-1], O') \leq k$ and, therefore O' can potentially be extended into an occurrence and the search is allowed to proceed. If, on the other hand, there are no active cells in the last row of D, the search can be abandoned, not proceeding to deeper points. For example by looking at Fig. 2 we can conclude that the search should not proceed further after $abbbab$ because there are no active cells in the last row of the table. Also, since all the other rows contain active cells, this point is indeed reached by the search. It helps to think of D as a stack of rows that is growing downwards. Note that it is a convenient coincidence that the difference between the D tables of $ed(P, O')$ and $ed(P, O'c)$ is only the last row. This means that we can move between these two tables simply by adding or removing a row. At each step the DFS algorithm either pushes a new element into the stack, $i.e.$ moves from $ed(P, O')$ to $ed(P, O'c)$, or it removes a row from the stack, $i.e.$ moves from $ed(P, O'c)$ to $ed(P, O')$. This process is known as neighborhood generation and it will be a key ingredient in our algorithm. The problem with this process is that it might have a very low success rate, $i.e.$ only a small percentage of the nodes visited by the process turn out to be occurrences of P.

4   Bidirectional Traversal

Our algorithm will proceed in a slightly more sophisticated fashion. Instead of extending O' only in one direction, to the right, we will use a bidirectional search. Landau et $al.$ [9] obtained the surprising result that it is possible to compute $ed(A, cB)$ from $ed(A, B)$, also in time $O(k)$. The resulting algorithm is very sophisticated and the reader should consult the original paper. For our purposes all we need are the following observations. The extension is not restricted to B, $i.e.$ we can also extend $ed(A, B)$ to $ed(cA, B)$. The number of errors does not have to be fixed, $i.e.$ we can extend a computation with k errors to a computation on $k+1$ errors in $O(k+1)$ time. Finally, the data structure they use in their algorithm are two doubly linked lists organized in a grid. This means that if we compute $ed(A, cB)$ from $ed(A, B)$ we can revert back to the $ed(A, B)$ state by simply keeping a rollback log of which pointers to revert, which requires $O(k)$ computer words[1]. For our algorithm this idea suffices since, as in the previous paragraph, the states we need to visit are always organized in a stack. Therefore

---

[1] It seems to us that it is possible to extend their algorithm to support this directly, but if that is not the case we can still use the rollback log idea.

we never need to compute a sequence such as $ed(A, B)$ to $ed(A, cB)$ to $ed(dA, cB)$ to $ed(dA, B)$.

To improve the success rate of the process described above we should start our search from an area of P that is well preserved. To limit the number of errors we divide the pattern into smaller pieces. We will use the following filtration lemma.

Lemma 1 ([10]). *Let A and B be strings, let $A = A_0 A_1 \ldots A_j$, for strings A_i and some $j \geq 1$. Let $k_i \in \mathbb{R}$ such that $ed(A, B) < \sum_{i=0}^{j} k_i$. Then there is a substring B' of B and an i such that $ed(A_i, B') < k_i$.*

In our algorithm we will use $A = P$ and $B = O$ and divide the errors in a homogeneous fashion, *i.e.* choose $k_i = \alpha |A_i| + \epsilon$, where $\alpha = k/m$ and $\epsilon > 0$ is a number that can be as small as we want and it is only used to guarantee that $ed(A, B) < \sum_{i=0}^{j} k_i$. Recall our running example with $O = abbba$ and $P = abc.cba$, assuming this is the partition of A. Therefore we should have $k_0 = k_1 = (2/6) \times 3 + \epsilon$. Hence the lemma says that in any O there is at least one substring O' such that $ed(O', abc) < 1 + \epsilon$ or $ed(O', cba) < 1 + \epsilon$. In our example there are in fact two substrings O' that satisfy this property, $ed(abb, abc) \leq 1$ and $ed(bba, cba) \leq 1$. On one hand this is good because it validates the lemma. On the other hand it is excessive because the same string will be found in more than one way. To solve this redundancy notice that we do not need to add ϵ to both k_i's, *i.e.* we can choose k_0 as before and $k_1 = 1$. This means that the conclusion of the lemma now states that there should be an O' such that $ed(O', abc) \leq 1$ or $ed(O', cba) < 1 \Rightarrow ed(O', cba) \leq 0$, and hence the redundancy is eliminated.

Note that the condition on O' is no guarantee that there exists an occurrence O of P, since it is a one-way implication. Hence the area around O' must be verified to determine whether there is an occurrence or not. Note that in previous work the usual verification procedure is computed in T, not taking advantage of the index. Therefore, verifying those occurrences can cost $O(k(m + k))$ operations. The problem with dividing P too much, such as when $j = k + 1$, is that the number of positions to verify can become excessively large and again we get a low success rate, *i.e.* only a small percentage of the O's verified by the process turn out to be occurrences of P.

The hybrid approach tries to maximize the overall success rate by finding an optimal balance between filtration and neighborhood generation. It was shown [2] that the optimal point occurs for $j = \Theta(m/\log_\sigma n)$, with a complicated constant. Our approach can have a slightly different optimal point, but if we use their j the resulting algorithm is never worse than theirs. Moreover we also attempt to automatically determine the hybrid point and hence eliminate the need for parameterization.

5   Indexed Hierarchical Verification

We modify the verification phase, after filtration, in two ways. (1) We will perform it over the FCSTs instead of over T, to factor our possibly repeated computations. (2) We use hierarchical instead of direct verification, which also provides a strategy to approximate the optimal point.

The idea of hierarchical verification is to gradually extend the error level instead of jumping directly to k. This is obtained by iterating Lemma 1.This technique was shown to be extremely efficient for the on-line approach [2]. We use the following lemma (proof omitted).

Lemma 2. *Let A and B be strings, let $A = A_0 A_1 \ldots A_j$, for strings A_i and some $j + 1 = 2^h \geq 1$. Let $k_i \in \mathbb{R}$ such that $ed(A, B) < \sum_{i=1}^{j} k_i$. For some fixed $0 \leq i \leq j$, define $A'_{i'} = A_{2^{i'} \lfloor i/2^{i'} \rfloor} \ldots A_{2^{i'}(1+\lfloor i/2^{i'} \rfloor)-1}$, for any $0 \leq i' \leq h$, as the hierarchical upward path from A_i to A, and define accordingly $k'_{i'} = \sum_{i''=2^{i'} \lfloor i/2^{i'} \rfloor}^{2^{i'}(1+\lfloor i/2^{i'} \rfloor)-1} k_{i''}$ as the error level corresponding to each $A'_{i'}$. Then there are strings $B_0 \sqsubseteq \ldots \sqsubseteq B_h = B$ and an i such that for any $0 \leq i' \leq h$ we have $ed(A'_{i'}, B_{i'}) < k'_{i'}$. Moreover, for each i', if $A'_{i'}$ is a prefix(suffix) of $A'_{i'+1}$ then $B_{i'}$ is a prefix(suffix) of $B_{i'+1}$.*

Consider our running example with $k = 2$ and $P = abccba$. Instead of applying Lemma 2 we will instead iterate Lemma 1, which is actually the way we compute the partition in practice. We divide $P = A = abc.cba$ into pieces of size 3 and therefore we have $k'_0 = k'_1 = 3 \times (2/6) + \epsilon = 1 + \epsilon$, which in practice means 1 error per piece. Now we divide these pieces as $ab.c.cb.a$ and we have $k_0 = k_2 = 2 \times (2/6) + \epsilon$ and $k_1 = k_3 = 1 \times (2/6) + \epsilon$, this means 0 errors for all the pieces. Notice that we can refine our method by adding ϵ to only one k_i, as we did in Section 3.2. Hence we can choose $k_0 = k_2 = 2/3$ and $k_1 = 1/3 + \epsilon$ and $k_3 = 1/3$. Notice that in our example the occurrence $abbba$ verifies this lemma because $ed(ab, ab) < 2/3$ and $ed(abb, abc) < (2/3) + (1/3) + \epsilon$, where ab and abc are substrings of P.

This lemma is used to reduce the cost of verifying an occurrence. Instead of directly verifying the space around a B_0 when $ed(A_i, B_0) < k_i$ for a string B such that $ed(A, B) < k$, we extend the error level gradually. Assuming i is even, this means checking for $ed(A_i.A_{i+1}, B_1) < k_i + k_{i+1}$ first, for some B_1. Fig. 2(right) shows an example of this process, computed with table D.Whenever a row reaches a certain level in the hierarchy and contains active cells, the computation on that row is extended to activate the cells that are $< k_i + k_{i+1}$. For example since $D[2, 2] = 0$ the cells in row 2 that can be $< 1 + \epsilon$ are activated, *i.e.* cells $D[1, 2]$ and $D[3, 2]$, that correspond to $ed(a, ab)$ and $ed(abc, ab)$. A similar process happens at row 3. In theory we can compute all the cells that are $\leq k$ all the time. Still, we can also start to compute them at a given row, especially since it is not necessary to fill upwards the missing cells in the table. That is, we can compute the missing cells, up to $< k_i + k_{i+1}$, from the ones already in the table. There is no problem if the value of the new cells is larger than their value on the complete D table. In fact it is desirable. This will only make the algorithm skip occurrences that, because of Lemma 2, will be found in another case.

To determine that $ed(A_i, B_0) < k_i$ we must compute the DP table for these two strings. Extending this computation to $ed(A_i.A_{i+1}, B_1) < k_i + k_{i+1}$ is simple because table D only needs to be updated in its natural directions (to the right and downwards). From the suffix tree point of view this situation is also natural because it involves descending in the tree.

When i is odd the situation is a bit trickier. This time we must check for $ed(A_{i-1}.A_i, B_1) < k_{i-1} + k_i$. This is much more difficult because we need to move in the FCST by prepending letters to the current point. This is possible with the WEINERLINK operation, recall Section 3.1. Moreover we need to extend the DP in unnatural directions (to the left and upwards). For this we use the result [9] mentioned in Section 3.2. Hence computing each new row requires only $O(k)$ operations. Note that the underlying operation on which their algorithm relies is the longest common prefix of any two suffixes of A and B. To solve this we build a FCST for P, in $O(m)$ time, in uncompressed format so that the LCA operation takes $O(1)$ time. Note that this FCST is built only once at the beginning of the algorithm and adds $O(m \log m)$ bits to the space requirements of the algorithm. We determine the positions of $O'[i..]$ in that suffix tree, in $O(m')$ time, with the PARENT and WEINERLINK operations. Together with the LCA operation we can compute the size of the necessary longest common prefixes. Note that whenever O' is extended to/contracted from cO', this information must be updated, by recomputing in $O(m')$ time.

Our algorithm consists in neighborhood generation, where the error bound is gradually increased. Depending on the position of current P's substring in the hierarchical verification the string O' is extended either to the left or to the right. Hence, as mentioned before, the $ed(P, O')$ states are stored in a stack, whereas the O' string being generated is stored in a double stack structure that can be pushed/popped at both ends.

6   Practical Issues and Testing

We implemented a prototype, BiFMI, to test our algorithm. Lacking a FCST implementation, we simulated it with a bidirectional FM-Index over one wavelet tree [5]. We reverse the search so that the most common search (forwards) is done using LF (where the FM-index is faster) instead of ψ. We use efficient sequential algorithms as a baseline (namely *BPM*, the bit-parallel DP matrix of Myers [16], and *EXP*, the exact pattern partitioning by Navarro and Baeza-Yates [17]). We also included in the comparison authors' implementation of several competing indexes: *Hybrid* is the classical hybrid technique over plain suffix arrays [2]; *LZI* and *DLZI* are basic and improved algorithms based on the LZ-index [18], which partition into $j = k + 1$ exact searches for pattern pieces and decompress the candidate text areas for (non-hierarchical) verification [19]; *FMIndex* is the same strategy applied over Navarro's fast and large FM-index implementation (which is much faster than our own FM-index); and finally *ILZI* is a recent ASM algorithm [6] over the ILZI compressed index [10].

The machine was a Pentium 4, 3.2 GHz, 1 MB L2 cache, 1GB RAM, running Fedora Core 3, and compiling with `gcc-3.4 -09`. We used the texts from the Pizza&Chili corpus[2], with 50 MB of English and DNA and 64 MB of proteins.

---

[2] `http://pizzachili.dcc.uchile.cl`

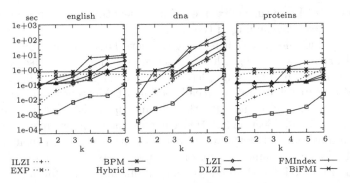

Fig. 3. Average user time for finding the occurrences of patterns of size 30 with k errors. The y axis units are in seconds and common to the three plots.

Table 1. Memory peaks, in Megabytes, for the different approaches when $k = 6$

	ILZI	Hybrid	LZI	DLZI	FMIndex	BiFMI
English	55	257	145	178	131	54
DNA	45	252	125	158	127	40
Proteins	105	366	217	228	165	63

The pattern strings were sampled randomly from the text and each character was distorted with 10% of probability into an insertion, deletion, or substitution. All the patterns had length $m = 30$. Every configuration was tested during at least 60 seconds using at least 5 repetitions. Hence the number of repetitions varied between 5 and 130,000. To parameterize the hybrid index we tested all the j values from 1 to $k + 1$ and reported the best time. We did a similar process on the ILZI index. We tested our algorithm, BiFMI, in automatic mode, *i.e.* not using any parameterization.

The average query time, in seconds, is shown in Fig. 3 and the respective memory heap peaks for indexed approaches are shown in Table 1. The hybrid index provides the fastest approach to the problem. However it also requires the most space. Our BiFMI index, on the other hand, achieves the smallest space (and it can still be reduced). We maintain a sparse sampling for our prototype, to show that even within little space we can achieve competitive performance. The *FMIndex*, on the other hand, needs a much denser sampling to be competitive. Thus our hierarchical and bidirectional verification method was faster than the basic one, even if run on a much slower index (our versus Navarro's FM-Index).

Aside from the hybrid index, the fastest approach in reduced space is the ILZI-based one. The performance of our prototype closely follows that of ILZI, except for the DNA file. This indicates that we were able to approach hybrid performance. We were also, mostly, able to reduce the gap caused by cache misses. Notice that the ILZI index is consistently at most one order of magnitude slower than Hybrid, for $k \leq 3$. Our algorithm was not so effective in the DNA file but was still able to avoid two orders of magnitude slowdown for proteins and English. Notice that this is important, since aside from the ILZI, the other compressed approaches seem to saturate at a given performance for low error levels: in English $k = 1$ to 3, in DNA $k = 1$ to 2, and in proteins $k = 1$ to 5. This is particularly troublesome since indexed approaches are the best alternative only for low error levels. In fact the sequential approaches outperform the compressed indexed approaches for higher error levels. In DNA this occurs at $k = 4$ and in English at $k = 5$.

We did not implement the algorithm of Landau *et al.* [9]; instead we used the bit-parallel NFA of Wu *et al.* [20] and recomputed the D table whenever it was necessary to change the computing direction. Note this requires $O(m)$ time when we switch from right to left or vice versa, but after the change it will require only $O(k)$ time for each new row. Although in theory this process could slow down our algorithm by a factor of $O(\log k)$, in practice this factor was negligible.

7   Conclusions and Future Work

In this paper we studied the impact of hierarchical verification in ASM. We obtained an automatic hybrid index that uses fully-compressed suffix trees. This a very important result because it is the first algorithm that approximates the performance of the hybrid index automatically and effectively in practice. Our result is also very important because FCSTs require only compressed space, *i.e.* $nH_k + O(n \log \sigma)$ bits. Compared to other compressed indexes, our approach was more efficient for low error levels. Although it was less efficient than the ILZI-based algorithm, it requires less space in theory and in practice. In theory, the ILZI requires $5nH_k + o(n \log \sigma)$ bits, but, in practice that is closer to $3nH_k$, including the sublinear term. On the other hand, a FCST requires $nH_k + o(n \log \sigma)$ bits in theory, but this becomes a bit higher in practice if we consider the sublinear term. Moreover our algorithm can be used as a subroutine in a suffix tree algorithm whereas the ILZI-based algorithm cannot.

References

1. Myers, E.W.: A sublinear algorithm for approximate keyword searching. Algorithmica 12(4/5), 345–374 (1994)
2. Navarro, G., Baeza-Yates, R.: A hybrid indexing method for approximate string matching. Journal of Discrete Algorithms 1(1), 205–239 (2000)
3. Navarro, G., Sutinen, E., Tarhio, J.: Indexing text with approximate q-grams. J. Discrete Algorithms 3(2-4), 157–175 (2005)
4. Kurtz, S.: Reducing the space requirement of suffix trees. Softw., Pract. Exper. 29(13), 1149–1171 (1999)
5. Navarro, G., Mäkinen, V.: Compressed full-text indexes. ACM Comp. Surv. 39(1) article 2 (2007)
6. Russo, L.M.S., Navarro, G., Oliveira, A.L.: Approximate string matching with Lempel-Ziv compressed indexes. In: Ziviani, N., Baeza-Yates, R. (eds.) SPIRE 2007. LNCS, vol. 4726, pp. 264–275. Springer, Heidelberg (2007)

7. Russo, L., Navarro, G., Oliveira, A.: Fully-Compressed Suffix Trees. In: Laber, E.S., Bornstein, C., Nogueira, L.T., Faria, L. (eds.) LATIN 2008. LNCS, vol. 4957, pp. 362–373. Springer, Heidelberg (2008)
8. Navarro, G., Baeza-Yates, R.: Improving an algorithm for approximate pattern matching. Algorithmica 30(4), 473–502 (2001)
9. Landau, G.M., Myers, E.W., Schmidt, J.P.: Incremental string comparison. SIAM J. Comput. 27(2), 557–582 (1998)
10. Russo, L.M.S., Oliveira, A.L.: A compressed self-index using a Ziv-Lempel dictionary. In: 13th SPIRE. LNCS, vol. 4029, pp. 163–180. Springer, Heidelberg (2006)
11. Gusfield, D.: Algorithms on Strings, Trees and Sequences. Cambridge University Press, Cambridge (1997)
12. Ferragina, P., Manzini, G.: Indexing compressed text. Journal of the ACM 52(4), 552–581 (2005)
13. Sadakane, K.: New text indexing functionalities of the compressed suffix arrays. J. of Algorithms 48(2), 294–313 (2003)
14. Weiner, P.: Linear pattern matching algorithms. In: IEEE Symp. on Switching and Automata Theory, pp. 1–11 (1973)
15. Ukkonen, E.: Finding approximate patterns in strings. Journal of Algorithms, 132–137 (1985)
16. Myers, G.: A fast bit-vector algorithm for approximate string matching based on dynamic programming. Journal of the ACM 46(3), 395–415 (1999)
17. Navarro, G., Baeza-Yates, R.: Very fast and simple approximate string matching. Information Processing Letters 72, 65–70 (1999)
18. Navarro, G.: Indexing text using the Ziv-Lempel trie. J. Discrete Algorithms 2(1), 87–114 (2004)
19. Morales, P.: Solving complex queries over a compressed text index. Undergraduate thesis, Dept. Comp. Sci., Univ. Chile (2005) Spanish. G. Navarro, advisor
20. Wu, S., Manber, U.: Fast text searching allowing errors. Commun. ACM 35(10), 83–91 (1992)

An Efficient Linear Space Algorithm for Consecutive Suffix Alignment under Edit Distance (*Short Preliminary Paper*)

Heikki Hyyrö

Department of Computer Sciences, University of Tampere, Finland
heikki.hyyro@cs.uta.fi

Abstract. We discuss the following variant of incremental edit distance computation: Given strings A and B with lengths m and n, respectively, the task is to compute, in n successive iterations $j = n \dots 1$, an encoding of the edit distances between A and all prefixes of $B_{j..n}$. Here $B_{j..n}$ is the suffix of B that begins at its jth character. This type of *consecutive suffix alignment* [3] is powerful e.g. in solving the cyclic string comparison problem [3]. There are two previous efficient algorithms that are capable of consecutive suffix alignment under edit distance: the algorithm of Landau et al. [2] that runs in $O(kn)$ time and uses $O(m+n+k^2)$ space, and the algorithm of Kim and Park [1] that runs in $O((m + n)n)$ time and uses $O(mn)$ space. Here k is a user-defined upper limit for the computed distances ($0 \leq k \leq \max\{m, n\}$). In this paper we propose the first efficient linear space algorithm for consecutive suffix alignment under edit distance. Our algorithm uses $O((m + n)n)$ time and $O(m + n)$ space.

1   Introduction and Preliminaries

Computing edit distance between two given strings A and B is a classic string processing problem that has applications e.g. in spelling correction and computational biology. In this paper we concentrate on the classic form of edit distance, which is defined as the minimum number of single-character insertions, deletions and/or substitutions that are needed in order to transform A into B or vice versa. We denote such an edit distance between A and B by $ed(A, B)$. For example $ed(\text{"wrong"}, \text{"string"}) = 3$, which reflects how we may transform "wrong" into "string" by substituting 'w' $\to$ 't' and 'o' $\to$ 'i' and inserting 's' to the front.

For a string S, the notation S_i refers to its ith character and $S_{i..j}$ to its substring whose beginning and ending positions are i and j, respectively. Throughout this paper A will be a string of length m and B a string of length n, and so $A = A_{1..m}$ and $B = B_{1..n}$.

The fundamental solution for computing $ed(A, B)$ is based on filling an $(m + 1) \times (n+1)$ dynamic programming matrix D, where eventually $D[i, j] = ed(A_{1..i}, B_{1..j})$. The initially known values are $D[i, 0] = i$ for $i = 0 \dots m$ and $D[0, j] = j$ for $j = 0 \dots n$, and the remaining cells can be filled in $O(mn)$ time using the recurrence $D[i, j] = \min\{1 + D[i-1, j], 1 + D[i, j-1], \delta(i, j) + D[i-1, j-1]\}$. Here

A. Amir, A. Turpin, and A. Moffat (Eds.): SPIRE 2008, LNCS 5280, pp. 155–163, 2008.
© Springer-Verlag Berlin Heidelberg 2008

$\delta(i,j)$ is a substitution penalty function: $\delta(i,j) = 0$ if $A_i = B_j$, and $\delta(i,j) = 1$ if $A_i \neq B_j$.

Landau et al. [2] stated the problem of *incremental* string comparison in the following manner. Given that we have previously computed a solution for comparing A and B, the task is to compute efficiently a solution for comparing A and B', where either $B' = Bb$ or $B' = bB$ for some character b, and B' has length $n' = n + 1$. That is, B' is formed from B by inserting some character b either to the front or rear of B. Within the context of this paper, comparing A and B refers to computing $ed(A, B)$, but generally also other measures of string distance (or similarity) could be used. The character insertions to B can be done to the front or rear in any order. We might first be asked to compute $ed(A, bB)$, then $ed(A, bBc)$, then $ed(A, abBc)$, and so on, where a, b and c are some characters. We note that a solution refers to an appropriate encoding of information that not only gives the currently asked value $ed(A, B)$, but also enables efficient computation of the possibly next asked value $ed(A, B')$.

Later Landau et al. [3] defined *consecutive suffix alignment* as follows: Given the strings A and B, the task is to compare A against each suffix $B_{h..n}$ of B in the order $h = n \ldots 1$, ie. from the shortest suffix to the longest. This task is essentially a restricted form of incremental string comparison where each new character must always be inserted to the front (ie. of form $B' = bB$), and the string compared with A evolves from B_n to the full string $B_{1..n}$.

In similar fashion we may define consecutive prefix alignment as the task of comparing A against each prefix $B_{1..j}$ of B in the order $j = 1 \ldots n$.

We define these alignment tasks under edit distance more formally as follows.

Definition 1. *The task of consecutive suffix alignment under edit distance: Compute a representation of the distances $ed(A, B_{h..j})$ for $j = h \ldots n$ in the order $h = n \ldots 1$.*

In other words, when we compare A and a suffix $B_{h..n}$, we wish to compute the edit distance between A and each prefix $B_{h..j}$ of $B_{h..n}$, where $h \leq j \leq n$. By "a representation of the distances" we mean that the distances may be stored in some other form than their actual values. As will be seen later, we will use a representation that allows enumerating the values $ed(A, B_{h..j})$ $j = h \ldots n$ at iteration h in $O(1)$ time per value.

Definition 2. *The task of consecutive prefix alignment under edit distance: Compute a representation of the distances $ed(A, B_{1..j})$ for $j = 1 \ldots h$ in the order $h = 1 \ldots n$.*

When comparing A and a prefix $B_{1..n}$, we again wish to compute the edit distance between A and each prefix $B_{h..j}$ of $B_{h..n}$, where $h \leq j \leq n$.

The basic recurrence for filling D inherently performs consecutive prefix alignment. If we have previously filled D when computing $ed(A, B_{1..h-1})$, the values $ed(A, B_{1..j})$ are already known for $j = 1 \ldots h - 1$. The missing value $ed(A, B_{1..h})$ may be computed simply by adding column h into D and filling it according to the recurrence and the character B_h. Hence for each value of h, only one new value of form $ed(A, B_{1..j})$ needs to be computed.

As noted by Landau et al. [2], the case of consecutive suffix alignment (inserting to the front of B) is much more difficult. On the other hand, consecutive suffix alignment has many advantages [2,3]. One example is the task of *approximate string matching*, in which we are given a text string T, a pattern string P and an error threshold k, and the task is to find each j for which $ed(P, T_{h..j}) \leq k$ for some $h \leq j$. This problem can be solved by using the basic dynamic programming recurrence with the modified initial values $D[0, j] = 0$ for $j = 0 \ldots n$. One limitation of this solution is that it is not able to report the lengths (or starting positions) of the approximate occurrences, but only their end positions within T. We might wish to know at least some h, or perhaps the largest h, for which $ed(P, T_{h..j}) \leq k$ at a given position j. This type of questions are easy to answer during consecutive suffix alignment. We initially set $A = P$ and $B = T$. When the suffix alignment is comparing A and $B_{h..n}$, we will find out the values $ed(A, B_{h..j}) = ed(P, T_{h..j})$ for $j = h \ldots n$. This makes it simple to gather information regarding both starting and ending positions of each found occurrence.

Landau et al. [2,3] list also several other problems, e.g. the cyclic string comparison problem, that can be solved efficiently with consecutive suffix alignment.

There are currently two algorithms that can perform efficient consecutive suffix alignment under edit distance. By an efficient algorithm we mean one that takes at most linear time $O(m+n)$ when computing the values $ed(A, B_{h..j})$ after the values $ed(A, B_{h+1..j})$ have been computed before, ie. the total time is at most $O(n(m + n))$. For example using the basic dynamic programming algorithm for D would require $O(mn^2)$ time.

The first efficient algorithm is the algorithm of Landau et al. [2], which performs the computation in $O(kn)$ time while using $O(n+k^2)$ space. This algorithm receives an error threshold k, where $0 \leq k \leq \max\{m, n\}$, as a parameter and does not compute values $ed(A, B_{1..j}) > k$. Note that if we wish to compute all values $ed(A, B_{1..j})$, the running time of this algorithm can be stated in the form $O((m + n)n)$. The second efficient algorithm is the algorithm of Kim and Park that runs in $O((m + n)n)$ time and uses $O(mn)$ space. Both of these afore mentioned algorithms are capable of general incremental edit distance computation.

In terms of related work, there exists an efficient algorithm by Landau et al. [3] for consecutive suffix alignment under the longest common subsequence (LCS) similarity function. After a single preprocessing phase that takes $O(n \log n)$ time, the algorithm runs in $O(Ln)$ time and uses $O(m+n)$ space, where L is the length of the LCS between A and B.

In what follows we will propose the first linear space algorithm for consecutive suffix alignment under edit distance. Our algorithm is inspired by the algorithm of Landau et al. [3] for consecutive suffix alignment under LCS similarity, but the details are very different. The running time of our algorithm is $O((m+n)n)$.

2   A Suitable Linear Space Representation

We use the notation D_h^g to denote a dynamic programming matrix D that corresponds to $ed(A_{g..m}, B_{h..n})$. The matrix D_h^g exists for $g = 1 \ldots m$ and $h = 1 \ldots n$.

The values in D_h^g are defined to be $D_h^g[i,j] = ed(A_{g..i}, B_{h..j})$ for $i = g \ldots m$ and $j = h \ldots n$. In addition, $D_h^g[g-1,j] = j - h + 1$ for $j = h - 1 \ldots n$ and $D_h^g[i, h-1] = i - g + 1$ for $i = g - 1 \ldots m$. These latter values correspond to row 0 and column 0 of D. Note that D_h^g is addressed according to the indices of the corresponding character positions in the full strings A and B: the matrix has rows $g - 1 \ldots m$ and columns $h - 1 \ldots n$.

Choosing a suitable representation for the distance information is the most crucial step in designing a linear space algorithm for consecutive suffix alignment. We have chosen to represent row m of D_h^g in incremental fashion in a vector R_h^g, which is defined as follows:

1. The vector R_h^g contains $n - h + 1$ positions, whose indices are $h \ldots n$.
2. $R_h^g[h] = D_h^g[m, h]$.
3. $R_h^g[j] = D_h^g[m, j] - D_h^g[m, j-1]$, for $j = h + 1 \ldots n$ when $h < n$.

The first value $R_h^g[h]$ is the base value of the vector, and the remaining values tell the differences between adjacent values on row m of D_h^g. Under this representation $D_h^g[m, j] = ed(A_{g..m}, B_{h..j}) = R_h^g[h] + \Sigma_{q=h+1}^{j} R_h^g[q]$ for $h \le j \le n$.

Our goal is to compute the vector R_h^1 in iteration h of consecutive suffix alignment. The vector R_h^1 represents the distances $ed(A, B_{h..j})$ in such a manner that they can be enumerated in $O(1)$ time per value in the order $j = h \ldots n$.

It is useful to also define special vectors R_h^{m+1} for $h = 1 \ldots n$. These correspond to $A = \varepsilon$, where ε is the empty string. Because R_h^{m+1} represents the values $ed(\varepsilon, B_{h..j}) = h - j + 1$ for $j = h \ldots n$, we define $R_h^{m+1}[j] = 1$ for $j = h \ldots n$.

The motivation for selecting this particular representation comes from the fact that the vectors R_h^g have the following three useful properties. We omit proofs of Lemmas in this short paper (see Appendix for proof-sketches).

Lemma 1. *When $1 \le g \le m$, the nonequality $R_h^g[j] \ne R_h^{g+1}[j]$ holds in at most three positions j, where $h \le j \le n$.*

Lemma 2. *When $1 \le h < n$, the nonequality $R_h^g[j] \ne R_{h+1}^g[j]$ holds in at most three positions j, where $h + 1 \le j \le n$.*

Lemma 3. *When $1 \le g \le m$ and $1 \le h < n$, the nonequality $R_h^g[j] \ne R_{h+1}^{g+1}[j]$ holds in at most three positions j, where $h + 1 \le j \le n$.*

Given some vectors R_t^s and R_v^u, we define $\Delta(R_t^s, R_v^u)$ to be a difference list that contains in the order of increasing j the pairs $(j, R_t^s[j])$ for those positions j where $R_t^s[j] \ne R_v^u[j]$. We allow the case where $t \ne v$, that is, the vectors may have different size. We interpret the nonequality $R_t^s[j] \ne R_v^u[j]$ to hold whenever exactly one of the two values does not exist, and insert the special value $(j, -1)$ in $\Delta(R_t^s, R_v^u)$ if $s \le j < t$. Here -1 signals a non-existing value.

Lemma 1 enables us to represent the m vectors R_h^g for $g = 1 \ldots m$ using only linear space. At each iteration h, we use the trivially known vector R_h^{m+1} as the base vector. Then for $g = m \ldots 1$, each vector R_h^g is represented by the difference list $\Delta(R_h^g, R_h^{g+1})$. Lemma 1 ensures that each list $\Delta(R_h^g, R_h^{g+1})$ consists of at most three pairs $(j, R_h^g[j])$ where $R_h^g[j] \neq R_h^{g+1}[j]$. Hence the base vector R_h^{m+1} and the lists $\Delta(R_h^g, R_h^{g+1})$ take overall $O(m + n)$ space to represent all vectors R_h^g at a given h.

3   Computing Vectors R_h^g from Vectors R_{h+1}^g in Linear Time

Now we have a linear space representation of the distances. The next question is how to compute the vectors R_h^g efficiently when the vectors R_{h+1}^g are known.

To get the computation started, we note that the vectors R_n^g and the lists $\Delta(R_n^g, R_n^{g+1})$ are easy to compute for $g = m \ldots 1$ in $O(m)$ time and space: we know that $R_n^g[n] = D_n^g[m, n] = ed(A_{g..m}, B_n) = m - g$ if the character B_n matches a character within $A_{g..m}$, and otherwise $R_n^g[n] = ed(A_{g..m}, B_n) = m - g + 1$. The lists $\Delta(R_n^g, R_n^{g+1})$ can then be created for $g = m \ldots 1$ during a single-pass traversal over the values $R_n^g[n]$. Therefore we may assume that the lists $\Delta(R_{h+1}^g, R_{h+1}^{g+1})$ are available when we wish to compute the vectors R_h^g.

For a given $h < n$, the vectors R_h^g are computed in the order $g = m \ldots 1$. The special vectors R_h^{m+1} and R_{h+1}^{m+1} are simple to build in $O(n)$ time and space and we know that $\Delta(R_h^{m+1}, R_{h+1}^{m+1}) = \{(h, 1)\}$. So we will assume that the difference list $\Delta(R_{h+1}^g, R_{h+1}^{g+1})$ is available. Moreover, the vector R_{h+1}^g can be computed in $O(1)$ time from R_{h+1}^{g+1} when $\Delta(R_{h+1}^g, R_{h+1}^{g+1})$ is known. Hence we also assume that the vectors R_h^{g+1} and R_{h+1}^{g+1} are explicitly known when we next compute R_h^g.

Let us introduce some further notation and useful properties.

Lemma 4. *Let R_t^s, R_v^u and R_y^x be some vectors. Assume that we have the difference lists $\Delta(R_t^s, R_v^u)$ and $\Delta(R_v^u, R_y^x)$ whose sizes are ℓ_1 and ℓ_2, respectively. Then the difference list $\Delta(R_t^s, R_y^x)$ can be constructed in $O(\ell_1 + \ell_2)$ time/space.*

Lemma 5. *Let R_t^s, R_v^u and R_y^x be some vectors. Assume that we have the difference lists $\Delta(R_v^u, R_t^s)$ and $\Delta(R_y^x, R_t^s)$ whose sizes are ℓ_1 and ℓ_2, respectively. Then the difference list $\Delta(R_v^u, R_y^x)$ can be constructed in $O(\ell_1 + \ell_2)$ time/space.*

We use the notation $\hat{R}_{h+1}^g$ to denote an extended version of the vector R_{h+1}^g that contains also position h. This new position corresponds to an initial value in column 0 of D and is set to contain the value $\hat{R}_{h+1}^g[h] = D_{h+1}^g[m, h] = m - g + 1$. After this, the representation of $\hat{R}_{h+1}^g$ is corrected to include only a single base value by setting $\hat{R}_{h+1}^g[h + 1] = R_{h+1}^g[h + 1] - \hat{R}_{h+1}^g[h]$. Note that if $R_{h+1}^g[h+1] - (m-g+1) \neq R_{h+1}^g[h+1]$, then $\Delta(\hat{R}_{h+1}^g, R_{h+1}^g) = \{(h, m-g+1), (h+1, R_{h+1}^g[h+1] - (m-g+1))\}$, and otherwise $\Delta(\hat{R}_{h+1}^g, R_{h+1}^g) = \{(h, m-g+1)\}$.

We will represent the vectors R_h^{g+1}, $\hat{R}_{h+1}^{g+1}$ and $\hat{R}_{h+1}^g$ implicitly as a combination of a base vector R_{h+1}^{g+1} and the difference lists $\Delta(R_h^{g+1}, R_{h+1}^{g+1})$, $\Delta(\hat{R}_{h+1}^{g+1}, R_{h+1}^{g+1})$ and $\Delta(\hat{R}_{h+1}^g, R_{h+1}^{g+1})$. The lists $\Delta(R_h^{g+1}, R_{h+1}^{g+1})$ and $\Delta(\hat{R}_{h+1}^{g+1}, R_{h+1}^{g+1})$ are known, and $\Delta(\hat{R}_{h+1}^g, R_{h+1}^{g+1})$ may be computed in $O(1)$ time and space from the known lists $\Delta(\hat{R}_{h+1}^g, R_{h+1}^g)$ and $\Delta(R_{h+1}^g, R_{h+1}^{g+1})$ (Lemma 4). After this, the remaining process of computing R_h^g has the five steps:

Step 1. Construct the difference list $\Delta(R_h^g, R_{h+1}^{g+1})$.

Step 2. Construct the difference list $\Delta(R_h^g, R_h^{g+1})$.

Step 3. Construct the difference list $\Delta(R_h^g, R_{h+1}^g)$.

Step 4. Transform the vector R_{h+1}^{g+1} into R_{h+1}^g according to $\Delta(R_{h+1}^g, R_{h+1}^{g+1})$.

Step 5. Transform the vector R_h^{g+1} into R_h^g according to $\Delta(R_h^g, R_h^{g+1})$.

Once $\Delta(R_h^g, R_{h+1}^{g+1})$ has been constructed in step 1, steps 2-3 can be done in $O(1)$ time and space (Lemma 5). Also steps 4-5 are easy to do in $O(1)$ time during a scan of the lists $\Delta(R_{h+1}^g, R_{h+1}^{g+1})$ and $\Delta(R_h^g, R_h^{g+1})$. So let us concentrate on step 1. The construction of R_h^g is based on the recurrence given in Lemma 6. The notation $\langle R_h^g[j] \rangle$ refers to the actual value $D_h^g[m,j] = R_h^g[h] + \Sigma_{q=h+1}^j R_h^g[q]$ that $R_h^g[j]$ represents incrementally. We use this notation instead of $D_h^g[m,j]$ to emphasize that the actual values $D_h^g[m,j]$ are not stored explicitly.

Lemma 6. $\langle R_h^g[j] \rangle = \min\{1 + \langle R_h^{g+1}[j] \rangle, \delta(g,h) + \langle \hat{R}_{h+1}^{g+1}[j] \rangle, 1 + \langle \hat{R}_{h+1}^g[j] \rangle\}$, for $j = h \ldots n$.

The recurrence allows us to construct the list $\Delta(R_h^g, R_{h+1}^{g+1})$ during a $O(1)$ time and space "merging" of the lists $\Delta(R_h^{g+1}, R_{h+1}^{g+1})$, $\Delta(\hat{R}_{h+1}^{g+1}, R_{h+1}^{g+1})$ and $\Delta(\hat{R}_{h+1}^g, R_{h+1}^{g+1})$ in relation to their common base vector R_{h+1}^{g+1}. First we (temporarily) add 1 to $R_h^{g+1}[h]$ and $\hat{R}_{h+1}^g[j]$, and $\delta(g,h)$ to $\hat{R}_{h+1}^{g+1}[h]$. Adding to the base value takes care of these additions for all j. Assume that we are currently at position j in R_{h+1}^{g+1}, and let X, Y and Z be such references to the vectors R_h^{g+1}, $\hat{R}_{h+1}^{g+1}$ and $\hat{R}_{h+1}^g$ that $\langle X[j] \rangle \leq \langle Y[j] \rangle \leq \langle Z[j] \rangle$. Initially at $j = h$ the references can be set appropriately by an $O(1)$ explicit comparison of the base values at position h. If $X[h] \neq R_{h+1}^{g+1}[h]$, the pair $(h, X[h])$ is inserted to $\Delta(R_h^g, R_{h+1}^{g+1})$. The ordering of X, Y and Z cannot change, and $\Delta(R_h^g, R_{h+1}^{g+1})$ contain a difference, before the next position $j' > j$ that contains a difference in one of the lists $\Delta(R_h^{g+1}, R_{h+1}^{g+1})$, $\Delta(\hat{R}_{h+1}^{g+1}, R_{h+1}^{g+1})$ and $\Delta(\hat{R}_{h+1}^g, R_{h+1}^{g+1})$. Based on the difference(s) at such j', we check the ordering $\langle X[j'] \rangle \leq \langle Y[j'] \rangle \leq \langle Z[j'] \rangle$ and possibly reassign the references accordingly. If X is not reassigned and $\Delta(X, R_{h+1}^{g+1})$ contains a difference at j', it is inserted also to $\Delta(R_h^g, R_{h+1}^{g+1})$. If Y (or same way Z) takes over X, we know that $R_h^g[j'] = \langle Y[j'] \rangle - \langle X[j'-1] \rangle = Y[j'] + \langle Y[j] \rangle - \langle X[j] \rangle$. This is because $\langle Y[j'-1] \rangle - \langle X[j'-1] \rangle = \langle Y[j] \rangle - \langle X[j] \rangle$. If $R_{h+1}^{g+1}[j'] \neq R_h^g[j']$, we insert $(j', R_h^g[j'])$ to $\Delta(R_h^g, R_{h+1}^{g+1})$. Then we move to the next position with a difference and repeat. As there are $O(1)$ positions with differences, the total time for this process of constructing $\Delta(R_h^g, R_{h+1}^{g+1})$ is $O(1)$.

Initializing R_h^{m+1} and R_{h+1}^{m+1} takes $O(n)$ time, and each of the remaining m steps takes $O(1)$ time. Hence iteration h takes $O(m+n)$ time, and the total time over $h = n \ldots 1$ is $O((m+n)n)$. The used space is $O(m+n)$ when we store the size-$O(n)$ vector R_h^g and the $O(m)$ size-$O(1)$ difference lists only for h and $h+1$.

References

1. Kim, S.R., Park, K.: A dynamic edit distance table. Journal of Discrete Algorithms 2(2), 303–312 (2004)
2. Landau, G.M., Myers, E.W., Schmidt, J.: Incremental string comparison. SIAM Journal of Computing 27(2), 557–582 (1998)
3. Landau, G.M., Myers, E.W., Ziv-Ukelson, M.: Two algorithms for LCS consecutive suffix alignment. Journal of Computer and System Sciences 73(7), 1095–1117 (2007)

Appendix

Here we present some sketches for the proofs of the Lemmas in order to help the reviewers evaluate their correctness.

Proof of Lemma 1

D_h^g and D_h^{g+1} determine R_h^g and R_h^{g+1}. Since the difference between the corresponding compared strings is only the one character A_g, we know that the values $D_h^g[m, j] = ed(A_{g..m}, B_{h..j})$ and $D_h^{g+1}[m, j] = ed(A_{g+1..m}, B_{h..j})$ can differ by at most one. That is, $D_h^g[m, j] = D_h^{g+1}[m, j] - 1$, $D_h^g[m, j] = D_h^{g+1}[m, j]$, or $D_h^g[m, j] = D_h^{g+1}[m, j] + 1$. We show later that if there are indices u and w such that $D_h^g[m, u] = D_h^{g+1}[m, u] - c$ and $D_h^g[m, w] = D_h^{g+1}[m, w] - c$, where $c \in \{-1, 0, 1\}$, then $D_h^g[m, v] = D_h^{g+1}[m, v] - c$ for all $v = u \ldots w$. This means that all indices j where $D_h^g[m, j] = D_h^{g+1}[m, j] - c$ form a consecutive block. And since c has three possibilities, there can be at most three such blocks. When $j - 1$ and j belong to the same such block, then $D_h^g[m, j - 1] = D_h^{g+1}[m, j - 1] - c$ and $D_h^g[m, j] = D_h^{g+1}[m, j] - c$. This leads to $D_h^{g+1}[m, j - 1] - D_h^g[m, j - 1] = c = D_h^{g+1}[m, j] - D_h^g[m, j]$. From this we have that $D_h^g[m, j] - D_h^g[m, j - 1] = D_h^{g+1}[m, j] - D_h^{g+1}[m, j - 1]$, which implies $R_h^g[j] = R_h^{g+1}[j]$ when $j > h$. Hence the nonequelity $R_h^g[j] \neq R_h^{g+1}[j]$ is possible only if $j = h$ or $j - 1$ and j belong to different blocks. There are at most two borders between three blocks, so the total number of differing positions j is at most three.

Now we show that if there are indices u and v such that $D_h^g[m, u] = D_h^{g+1}[m, u] - c$ and $D_h^g[m, v] = D_h^{g+1}[m, v] - c$, where $c \in \{-1, 0, 1\}$, then $D_h^g[m, w] = D_h^{g+1}[m, w] - c$ for all $w = u \ldots v$

It is well-known that edit distance corresponds to a shortest path in D and that such paths are piecewise optimal. Consider Fig. 1. It is related to D_h^g and D_h^{g+1}. We show paths that correspond to distances, and each path is labeled with its length (corresponding distance). Path a corresponds to $ed(A_{g..m}, B_{h..w}) = a$. Path $b_1 b_2$ corresponds to $ed(A_{g..m}, B_{h..v}) = b_1 + b_2$. Path $c_1 c_2$ corresponds to $ed(A_{g..m}, B_{h..u}) = c_1 + c_2$. Path $d_1 d_2$ corresponds to $ed(A_{g+1..m}, B_{h..w}) = d_1 + d_2$. Path $e_1 e_2$ corresponds to $ed(A_{g+1..m}, B_{h..v}) = e_1 + e_2$. Finally, path f corresponds to $ed(A_{g+1..m}, B_{h..u}) = f$. The piece-ends are connected at the crossings marked with black points.

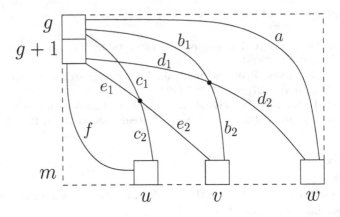

Fig. 1.

If $u = w$ or $u = w - 1$ there is nothing to prove. So assume $u < w - 1$ and consider v for which $u < v < w$. There are three cases.

Case 1: $c = -1$, ie. $D_h^g[m, u] = D_h^{g+1}[m, u] + 1$ and $D_h^g[m, w] = D_h^{g+1}[m, w] + 1$. In terms of Fig. 1, in this setting it is known that $a > d_1 + d_2$. We use proof by contradiction. Assume that $D_h^g[m, v] \leq D_h^{g+1}[m, v]$. In Fig. 1 this means that $b_1 + b_2 \leq e_1 + e_2$. Since a is a shortest path, $b_1 + d_2 \geq a > d_1 + d_2$. This implies $b_1 > d_1$. In similar fashion, $d_1 + b_2 \geq e_1 + e_2$. Putting these together we have $b_1 + b_2 > d_1 + b_2 \geq e_1 + e_2$, which contradicts our assumption that $b_1 + b_2 \leq e_1 + e_2$.

Case 2: $c = 1$, ie. $D_h^g[m, u] = D_h^{g+1}[m, u] - 1$ and $D_h^g[m, w] = D_h^{g+1}[m, w] - 1$. In terms of Fig. 1, in this setting it is known that $f > c_1 + c_2$. The proof is symmetric with case 1. Assume that $D_h^g[m, v] \geq D_h^{g+1}[m, v]$. In Fig. 1 this means that $b_1 + b_2 \geq e_1 + e_2$. Since f is a shortest path, $e_1 + c_2 \geq f > c_1 + c_2$. This implies $e_1 > c_1$. In similar fashion, $c_1 + e_2 \geq b_1 + b_2$. Putting these together we have $e_1 + e_2 > c_1 + e_2 \geq b_1 + b_2$, which contradicts our assumption that $b_1 + b_2 \geq e_1 + e_2$.

Case 3: $c = 0$, ie. $D_h^g[m, u] = D_h^{g+1}[m, u]$ and $D_h^g[m, w] = D_h^{g+1}[m, w]$. In terms of Fig. 1, in this setting it is known that $a = d_1 + d_2$ and $f = c_1 + c_2$. We prove two subcases by contradiction. Assume first that $D_h^g[m, v] < D_h^{g+1}[m, v]$. In Fig. 1 this means that $b_1 + b_2 < e_1 + e_2$. Since a is a shortest path, $b_1 + d_2 \geq a = d_1 + d_2$. This implies $b_1 \geq d_1$. In similar fashion, $d_1 + b_2 \geq e_1 + e_2$. Putting these together we have $b_1 + b_2 \geq d_1 + b_2 \geq e_1 + e_2$, which contradicts our assumption that $b_1 + b_2 < e_1 + e_2$. Assume then that $D_h^g[m, v] > D_h^{g+1}[m, v]$. In Fig. 1 this means that $b_1 + b_2 > e_1 + e_2$. Since f is a shortest path, $e_1 + c_2 \geq f = c_1 + c_2$. This implies $e_1 \geq c_1$. In similar fashion, $c_1 + e_2 \geq b_1 + b_2$. Putting these together we have $e_1 + e_2 \geq c_1 + e_2 \geq b_1 + b_2$, which contradicts our assumption that $b_1 + b_2 > e_1 + e_2$.

This completes the proof.

Proofs of Lemma 2 and Lemma 3

We do not give separate proofs here, but mention just that the proofs are done in similar fashion as the proof of Lemma 1. Especially Lemma 2 is quite symmetric with Lemma 1.

Proofs of Lemma 4 and Lemma 5

These should be quite obvious. The process is in some sense similar to merging two sorted lists into a single sorted list. Traversing the difference lists takes time proportional to their combined length. The traversal can be done in parallel in both lists so that if one of the two lists has a difference at position j, we can check in $O(1)$ time whether also the other has one. In both Lemma 4 and Lemma 5, it is not difficult to combine the values of the two lists in a suitable manner to get the desired merged result.

Proof of Lemma 6

The situation is actually the same, although for reverse strings, as the situation in the traditional recurrence for filling D. Let $\overline{A}$ denote the reverse string of A. Then $\overline{A}_i = A_{m+1-i}$ for $i = 1 \ldots m$. Define $\overline{B'}$ as the reverse of the string $B_{1..j}$, so that $\overline{B'}_i = B_{j+1-i}$ for $i = 1 \ldots j$.

Now $\langle R_h^g[j]\rangle$ corresponds to $ed(A_{g..m}, B_{h..j}) = ed(\overline{A}_{1..m-g+1}, \overline{B'}_{1..j-h+1})$, $\langle R_h^{g+1}[j]\rangle$ to $ed(A_{g+1..m}, B_{h..j}) = ed(\overline{A}_{1..m-g}, \overline{B'}_{1..j-h+1})$, $\langle \hat{R}_{h+1}^{g+1}[j]\rangle$ to $ed(A_{g+1..m}, B_{h+1..j}) = ed(\overline{A}_{1..m-g}, \overline{B'}_{1..j-h})$, and $\langle \hat{R}_{h+1}^g[j]\rangle$ to $ed(A_{g..m}, B_{h+1..j}) = ed(\overline{A}_{1..m-g+1}, \overline{B'}_{1..j-h})$.

From the basic recurrence for D we get that $ed(\overline{A}_{1..m-g+1}, \overline{B'}_{1..j-h+1})$ is the minimum over the choices

- $1 + ed(\overline{A}_{1..m-g}, \overline{B'}_{1..j-h+1})$
- $\delta'(m-g+1, j-h+1) + ed(\overline{A}_{1..m-g}, \overline{B'}_{1..j-h})$, where $\delta'(i,j) = 0$, if $\overline{A}_i = \overline{B'}_j$, and otherwise $\delta'(i,j) = 1$.
- $1 + ed(\overline{A}_{1..m-g+1}, \overline{B'}_{1..j-h})$

This leads to $\langle R_h^g[j]\rangle = \min\{1 + \langle R_h^{g+1}[j]\rangle, \delta(g,h) + \langle \hat{R}_{h+1}^{g+1}[j]\rangle, 1 + \langle \hat{R}_{h+1}^g[j]\rangle\}$, for $j = h \ldots n$, as in Lemma 6.

Run-Length Compressed Indexes Are Superior for Highly Repetitive Sequence Collections

Jouni Sirén[1,*], Niko Välimäki[1,**], Veli Mäkinen[1,**], and Gonzalo Navarro[2,***]

[1] Dept. of Computer Science, Univ. of Helsinki, Finland
{jltsiren,nvalimak,vmakinen}@cs.helsinki.fi
[2] Dept. of Computer Science, Univ. of Chile
gnavarro@dcc.uchile.cl

Abstract. A repetitive sequence collection is one where portions of a *base sequence* of length n are repeated many times with small variations, forming a collection of total length N. Examples of such collections are version control data and genome sequences of individuals, where the differences can be expressed by lists of basic edit operations. This paper is devoted to studying ways to store massive sets of highly repetitive sequence collections in space-efficient manner so that retrieval of the content as well as queries on the content of the sequences can be provided time-efficiently. We show that the state-of-the-art entropy-bound full-text *self-indexes* do not yet provide satisfactory space bounds for this specific task. We engineer some new structures that use run-length encoding and give empirical evidence that these structures are superior to the current structures.

1 Introduction

Self-indexing [5, 9, 20, 24] is a new algorithmic approach to storing and retrieving sequential data. The idea is to represent the text (a.k.a. sequence or string) compressed so that random access to the content of the text is maintained, and pattern retrieval queries on the content of the text are supported as well.

The self-indexing approach becomes especially interesting when applied to collections of texts. A special case of a text collection is one which contains several *versions* of one or more *base sequences*. Such collections are not uncommon. For example, a *version control system* needs to store several versions of the same file with only small edit differences between the consecutive entries. If the entries are stored independently of each others, the version control system will end up spending unnecessarily large amounts of memory. If the system stores only the edits, queries on the content of one specific version becomes non-trivial.

An analogy to the storage and retrieval of version control data is soon becoming reality in the field of molecular biology. Once the DNA sequencing technologies become faster and more cost-effective, it may be that in the near future the

---

* Funded by the Research Foundation of the University of Helsinki.
** Funded by the Academy of Finland under grant 119815.
*** Partially funded by Millennium Institute for Cell Dynamics and Biotechnology, Grant ICM P05-001-F, Mideplan, Chile.

A. Amir, A. Turpin, and A. Moffat (Eds.): SPIRE 2008, LNCS 5280, pp. 164–175, 2008.

sequencing of individual genomes becomes a feasible task [3, 12, 21]. With such data in hand, many fundamental issues such as storing and analyzing thousands of individual genomes become a top concern. For the analysis of such collections of biological sequences, one would need to use some variant of a *generalized suffix tree* [11] as that provides a variety of algorithmic tools to do analysis in linear or near-linear time. The memory requirement of such solution is unimaginable with current random access memories and also challenging in permanent storage.

Self-indexes should, in principle, cope well with the two applications above as both data types contain high amounts of repetitive structure. In particular, as the main building blocks of *compressed suffix trees* [7, 22, 23, 25] they enable compressing the collections to close to their *high-order entropy* and enabling flexible analysis tasks to be executed. However, there is a fundamental problem with the fact that the high-order entropies are defined by the frequencies of symbols in their fixed-length contexts. These contexts do not change *at all* when more *identical* sequences are added to the collection. Hence, these self-indexes are unable of exploiting the fact that the texts in the collection are highly similar.

In this paper, we propose new self-indexes based on run-length compression, that are suitable for storing highly repetitive collections of texts. We implemented the new structures and compared them experimentally to existing structures. The experiments show that our new structures achieve superior compression both on DNA collections and on version control data. The superiority can be explained in theory as well; the theoretical analysis together with related extended results (see Sect. 7) is part of subsequent work [16].

The paper is structured as follows. Section 2 introduces the basic concepts and goes through the related literature. Sections 3, 4, and 5 derive the new run-length compressed indexes. Section 6 gives the experimental results and Sect. 7 discusses the subsequent work.

2   Basic Concepts

A *string* $S = S_{1,n} = s_1 s_2 \cdots s_n$ is a sequence of *symbols* (a.k.a. character or letter). Each symbol is an element of a *alphabet* $\Sigma = \{1, 2, \ldots, \sigma\}$. A *substring* of S is written $S_{i,j} = s_i s_{i+1} \ldots s_j$. A *prefix* of S is a substring of the form $S_{1,j}$, and a *suffix* is a substring of the form $S_{i,n}$. If $i > j$ then $S_{i,j} = \varepsilon$, the empty string of length $|\varepsilon| = 0$. A *text* string $T = T_{1,n}$ is a special string with $t_n = \$$. The *lexicographical order* "$<$" among strings is defined in the obvious way.

We assume the reader is familiar with the *empirical k-th order entropy* $H_k(T)$ for which holds $0 \leq H_k(T) \leq H_{k-1}(T) \leq \cdots \leq H_0(T) \leq \log \sigma$ [18].

The compressors to be discussed are derivatives of the *Burrows-Wheeler transform (BWT)* [2]. The transform produces a permutation of T, denoted by T^{bwt}, as follows: (i) Build *suffix array* [17] $\mathsf{SA}[1, n]$ of T, that is an array of pointers to all the suffixes of T in the lexicographic order; (ii) The transformed text is $T^{bwt} = L$, where $L[i] = T[\mathsf{SA}[i] - 1]$, taking $T[0] = T[n]$.

The BWT is reversible, that is, given $T^{bwt} = L$ we can obtain T as follows: (a) Compute the array $C[1, \sigma]$ storing in $C[c]$ the number of occurrences of characters $\{\$, 1, \ldots, c-1\}$ in the text T; (b) Define the *LF mapping* as follows: $LF(i) = C[L[i]] + rank_{L[i]}(L, i)$, where $rank_c(L, i)$ is the number of occurrences of character c in the prefix $L[1, i]$; (c) Reconstruct T backwards as follows: set $s = 1$, for each $n - 1, \ldots, 1$ do $t_i \leftarrow L[s]$ and $s \leftarrow LF[s]$. Finally, append the end marker $t_n \leftarrow \$$. We study the following problem.

Definition 1. *Given a collection $\mathcal{C}$ of r sequences $T^k \in \mathcal{C}$ such that $|T^1| = n$ and $\sum_{k=1}^{r} |T^k| = N$, where $T^2, T^3, \ldots, T^r$ contain overall s mutations (i.e., symbol substitutions) from the base sequence T^1, the* repetitive collection indexing problem *is to store $\mathcal{C}$ in as small space as possible such that the following operations are supported as efficiently as possible:* count(P) *(How many times P appears as a substring of the texts in $\mathcal{C}$?);* locate(P) *(List the occurrence positions of P in $\mathcal{C}$); and* display(k, i, j) *(Return $T_{i,j}^k$).*

The above is an extension of the well-known *basic indexing problem*, where the collection has only one sequence T. We call a data structure a *self-index* if it does not need T to solve the three queries above.

A comprehensive solution to the basic indexing problem uses the suffix array $SA[1, n]$. Two binary searches are enough to find the interval $SA[sp, ep]$ such that count and locate are immediately solved [17]. The solution is not as space-efficient as possible, since array SA requires $n \log n$ bits, and the solution is not yet a self-index, since T is needed.

The *FM-index* [5] is a self-index based on the BWT. It solves counting queries by finding the interval $SA[sp, ep]$ that contains the occurrences of pattern P. The FM-index uses the array C and function $rank_c(L, i)$ in the so-called *backward search* algorithm, calling function $rank_c(L, i)$ $O(m)$ times. The two other basic indexing problem queries are solved e.g. using sampling of SA and its inverse SA^{-1}, and LF-mapping to derive the unsampled values from the sampled ones. Many variants of the FM-index have been derived that differ mainly in the way the $rank_c(L, i)$-queries are solved [20]. For example, on small alphabet sizes, one can achieve $nH_k(1 + o(1))$ space with constant time support for $rank_c(L, i)$ [6].

Now, the (repetitive) collection indexing problem can be solved using the normal self-index for the concatenation $T^1 \# T^2 \# \cdots T^r \$$, where $\# \notin \Sigma$ is a special symbol. However, the space requirement achieved even with a high-entropy compressed index is not attractive for the case of repetitive collections. For example, the solution by Ferragina et al. [6] requires $NH_k(\mathcal{C}) + o(N \log \sigma)$ bits. Notice that even with $s = 0$, $H_k(\mathcal{C}) \approx H_k(T^1)$, and hence the space is about r times more than what the same solution uses for the basic indexing problem.

In this paper, we derive solutions whose space requirements depend on the number of *runs in the Burrows-Wheeler transform*. We will introduce some notations to talk about runs. A *self-repetition* is a maximal interval $SA[i, i+l]$ of suffix array SA having a *target interval* $SA[j, j+l]$ such that $SA[j+r] = SA[i+r]+1$ for all $0 \leq r \leq l$. Let $\Psi(i) = SA^{-1}[SA[i]+1]$ [9, 24]. The intervals of Ψ corresponding to a self-repetition in the suffix array are called *runs*. We have $\Psi(i+1) = \Psi(i)+1$ when both $\Psi(i)$ and $\Psi(i + 1)$ are contained in the same run.

Let $R_\Psi(T)$ be the number of runs in Ψ of text T and $R(T) = R_{bwt}(T)$ the number of equal letter runs in T_{bwt}. Both are tightly connected, R_Ψ and R_{bwt}, namely $R_\Psi \leq R_{bwt} \leq R_\Psi + \sigma$ [14], allowing one to use them interchangeably under most circumstances. We will denote both with R when clear from context.

Now, it is easy to see that quantities $R_{bwt}(T)$ and $R_{bwt}(\mathcal{C})$ are the same when $s = 0$. Mutations make $R_{bwt}(\mathcal{C})$ grow. It is possible to derive expected case bounds on how these terms are related; these analyses are omitted here. Instead, we introduce structures whose space depends on $R_{bwt}(\mathcal{C})$ and study empirically the growth of $R_{bwt}(\mathcal{C})$ on varying s. We limit our attention to self-indexes providing query count(P).

3 RLCSA: Run-Length Compressed Suffix Array

The *Run-Length Compressed Suffix Array* is based on the Compressed Suffix Array by Mäkinen, Navarro and Sadakane [15]. We use run-length encoding of the differences $\Psi(i) - \Psi(i-1)$ to store the array. Absolute $\Psi(i)$ values are sampled at regular intervals of the *compressed* array. The resulting structure supports counting queries with backward searching.

Differential encoding of Ψ transforms a run $\Psi(i)\Psi(i+1) \cdots \Psi(i+l)$ into $\Psi(i) - \Psi(i-1)$ followed by l 1s, where $\Psi(i) - \Psi(i-1) > 1$. We say that the run is *trivial* if $l = 0$. If we use run-length encoding on the 1s, we encode the trivial runs simply as $\Psi(i) - \Psi(i-1)$. A nontrivial run, instead, is encoded as three numbers, $\Psi(i) - \Psi(i-1), 1, l$. That is, each time we encode a difference equal to 1, the length of the run of 1s follows. This way, run-length compression pays nothing for trivial runs, only for nontrivial runs where it has a potential benefit.

Let N be the total size of the collection and R' the number of nontrivial runs. The sum of all the differences $\Psi(i) - \Psi(i-1)$ is at most σN [15], and the total length of the runs of 1s is $N - R$. Hence by using Elias delta coding to encode the integers, we need at most

$$|\Psi| \leq \left(R \log \frac{\sigma N}{R} + R' \left(1 + \log \frac{N-R}{R'} \right) \right) (1 + o(1))$$

bits for the array Ψ. By using sampling step of B bits, we need $O((\frac{|\Psi|}{B} + \sigma) \log N)$ bits for the sampled $\Psi(i)$ values, effectively making the total size of RLCSA $|\Psi|(1 + \varepsilon)$ for any $\varepsilon > 0$.

To retrieve $\Psi(i)$, we first binary search the samples and then sum up the differences in the corresponding part of the Ψ array until we reach position i. This gives us count(P) queries in $O(|P|(\log \frac{|\Psi|}{B} + B))$ time by using bacward searching [15].

4 RLWT: Run-Length Encoded Wavelet Tree

Next we will describe a new data structure that we call *Run-Length encoded Wavelet Tree*. We exploit well-known bit-vector operations: For a bit vector B of length u, $rank_b(B,i)$ gives the number of b-bits in $B[1,i]$ for all $1 \leq i \leq u$

and $b \in \{0,1\}$. The inverse function $select_b(B, x)$ gives the position of the x'th b-bit in the bit vector B.

Wavelet tree [8] is a binary tree structure whose leaves represent the symbols in the alphabet. The root is associated with the sequence $T = T_{1,N}$. In a *balanced* wavelet tree, the left (right) child of the root is a wavelet tree of the sequence $T_<$ ($T_\geq$) obtained by concatenating all positions i having $t_i < \sigma/2$ ($t_i \geq \sigma/2$). This subdivision is represented by a bit vector of length n that marks which positions go to the left subtree (by 0) and which go right (by 1). Recursion is continued until the concatenated sequence contains a repeat of one symbol. One can reveal t_i, compute $rank_c(T, i)$, and $select_c(T, j)$ with $O(\log \sigma)$ *rank/select* queries on the bit-vectors on the path to the leaf (or back) containing c [8].

The space required by a balanced wavelet tree depends on how we encode the bit vectors. Let R be the number of runs in a text $T_{1,N}$. Let B^{all} be the level-wise concatenation of all the bit vectors in the balanced wavelet tree for the sequence T. In the worst case, each run in T equals one 0/1-bit run on each of the $\log \sigma$ levels of the wavelet tree, so that the upper-bound for the number of 0/1-bit runs in B^{all} is $R \log \sigma$ (the best case is $1 \cdot \log \sigma$). Let $b \leq \lceil \frac{1}{2} R \log \sigma \rceil$ be the number of 1-bit runs in B^{all}. The RLWT data structure encodes B^{all} into two separate bit vectors B^1 and B^{rl} such that the number of 1-bits in both bit vectors is exactly b: bit vector B^1 marks all the starting positions of 1-bit runs in B^{all}, and bit vector B^{rl} encodes the run-lengths of these runs in *unary coding*. More precisely, $B^1[i] = 1$ only if $B^{all}[i] = 1$ and $B^{all}[i-1] = 0$, for all $1 < i \leq N \log \sigma$, and $B^1[1] = 1$ if $B^{all}[1] = 1$. Unary code for a bit run of length j contains $j - 1$ zero bits concatenated with one 1-bit. The length of B^{rl} is the sum of the lengths of 1-bit runs in B^{all}, which is always at most $N \log \sigma$ bits.

Query $rank_1(B^{all}, i)$ can be solved using only the bit vectors B^1 and B^{rl} by calculating the number of 1-bits in two closed intervals $[0, j-1]$ and $[j, i]$, where j is the starting position of the 1-bit run that precedes position i in B^{all}. For the first interval, let r be the number of 1-bit runs in B^{all} that start before or at the position i, i.e. $r = rank_1(B^1, i)$. From the definition of B^{rl} follows that $rank_1(B^{all}, j - 1)$ equals $select_1(B^{rl}, r - 1)$. Now it remains to calculate the number of 1-bits in the closed interval $[j, i]$ of the bit vector B^{all}: Let k be the length of the rth run, that is to say $k \leftarrow select_1(B^{rl}, r) - rank_1(B^{all}, j-1)$. The number of 1-bits in the closed interval is

$$rank_1(B^{all}, i) - rank_1(B^{all}, j - 1) = \begin{cases} k & \text{if } i - j \geq k, \\ i - j + 1 & \text{otherwise.} \end{cases}$$

Finally, the answer to the original $rank_1(B^{all}, i)$ query is just the sum of the above values $rank_1(B^{all}, j - 1)$ and $rank_1(B^{all}, i) - rank_1(B^{all}, j - 1)$.

Gupta et al. [10] have shown that a *binary searchable dictionary representation (BSD)* of a bit-vector B of u bits containing b 1-bits, requires $|gap(B)| + O(|gap(B)|/\log b) = |gap(B)|(1 + o(1))$ bits of space and supports *rank* queries in $t_{\mathsf{AT}} = AT(u, b)$ time, where $AT(u, b) = o((\log \log u)^2)$, and and *select* in $O(\log \log b)$ time. In the worst case, length of the *gap encoded sequence* $|gap(B)|$ is $b \log(u/b) + O(b \log \log(u/b))$ bits.

For the bit vectors B^1 and B^{rl}, we have strict upper-bounds of $u \leq N \log \sigma$ and $b \leq \lceil \frac{1}{2} R \log \sigma \rceil$. Using the BSD, the bit vectors can be represented in at most $R \log \sigma \log \frac{2N}{R}(1 + o(1)) + O\left(R \log \sigma \log \log \frac{2N}{R}\right)$ bits. All the wavelet tree queries can be supported without storing the bit vector B^{all} itself.

Using the RLWT structure with *backward searching* [5], we can count the number of occurrences of a pattern of length m in $O(m \log(\sigma t_{AT}))$ time. Table C adds $\sigma \log N$ bits to the space requirement.

5   RLFM+: Improved Run-Length FM-Index

The RLWT structure can be improved in the case the input text is T^{bwt}: The *Run-Length FM-Index (RLFM)* of [14] uses a reduction such that the equal letter runs of T^{bwt} are marked into two bit-vectors, and the sequence of run heads of length R is encoded using a normal wavelet tree. We can represent the two bit-vectors using BSD, giving immediately the following result: The RLFM data structure for the sequence T^{bwt} takes $(R \log \sigma + 2R \log \frac{N}{R})(1 + o(1)) + O\left(R \log \log \frac{N}{R}\right) + \sigma \log N$ bits of space. The structure supports count(P) in time $O(|P|(\log(\sigma) + t_{AT}))$.

6   Experimental Results

We implemented the three proposed structures RLCSA, RLWT, and RLFM+, each supporting count()-queries. Standard strategies to support display() and locate() are trivial to add. (Almost all space/time tradeoffs are possible for those queries, so the base structure for supporting count() is the crucial one.)

For comparison, we selected several well-engineered implementations of self-indexes from the *Pizza&Chili* site[1]. Unless otherwise noted, we used no extra space for display() and locate(), and left the default options for the rest. We also compared our indexes to several compressors and a version control system.

We performed experiments on two data sets. The synthetic DNA sequence collections were based on the DNA sequences from *Pizza&Chili*. We took a 1, 4 or 16 MB prefix and repeated it 25, 50 or 100 times. Each character in the repetitions was individually mutated into another character in $\{A, C, G, T\}$ with ten different probabilities ranging from 0 to 0.05. This was intended to simulate the case of one base sequence and $r - 1$ mutated sequences.

Our other data set is based on the source code for portable versions of OpenSSH[2]. We used a 4.44 MB tar archive containing the source code for version 4.7p1, as well as on another 176.55 MB archive containing the source code for all 75 versions up to version 4.7p1. The latter contained multiple copies of the same files as well as many highly similar files, making it highly compressible.

The experiments were performed on a 3 GHz Intel Pentium 4 Northwood machine with 3 GB RAM running Fedora Core 7 based Linux.

---

[1] http://pizzachili.dcc.uchile.cl/ or http://pizzachili.di.unipi.it/.
[2] http://www.openssh.com/

6.1   Implementations and Parameters

The implementations of Succinct Suffix Array (SSA, version 2) [6, 14], Run-Length FM-index (RLFM) [14], Alphabet-Friendly FM-index (AFFM, version 2) [6] were taken from the *Pizza&Chili* site. All of them use a Huffman-shaped wavelet tree to achieve compression. SSA achieves zero-order compression by building the wavelet tree directly on the BWT, and is the fastest. RLFM builds it on the run heads of the BWT, and thus its space is related to the number of runs in the collection, yet the two extra bit-vectors it uses are not compressed. AFFM achieves high-order compression, $NH_k + o(N \log \sigma)$ bits, by partitioning the BWT into suitable chunks and building a wavelet tree per chunk. Its space is not related to the runs in the BWT.

Sadakane's Compressed Suffix Array (CSA) [24] implementation was also taken from *Pizza&Chili*. It achieves high-order compression related to the runs in Ψ, yet also includes less compressed bit vectors. We used sample rates 128 (default; CSA-128) and 1024 (CSA-1024 or CSA) for the Ψ values. The total size of the samples for a 400 MB collection is 3.1 MB for CSA-1024 and 25 MB for CSA-128. Suffix array sample rate was set to 65536 to make the size of these unused samples negligible (not to confuse with the sampling to access Ψ).

Also included in the comparison was a self-index based on Lempel-Ziv parsing (LZ-index, *Pizza&Chili* version 4) [1]. We selected $1/\epsilon = 15$ as a reasonable space/time tradeoff and subtracted the space (41 MB for a 400 MB collection) used for `display()` and `locate()` queries, for fairness with the other structures (although the implementation does not let one discard it).

Our indexes RLCSA, RLWT, and RLFM+ can be seen as versions of CSA, SSA, and RLFM, respectively, enhanced to profit from highly repetitive collections. The implementation of RLCSA is optimized for secondary memory. Hence we have used 32 kilobyte sampling step for Ψ (RLCSA-32k or RLCSA) in addition to the more reasonable 128 bytes (RLCSA-128). In practice, RLCSA-128 is at most 20% larger than RLCSA-32k. The difference can be reduced by changing the size of the samples from 24 bytes to $3\lceil \log N \rceil$ bits per sample. In RLWT and RLFM+, we used simpler encoding for the bit vectors than the original BSD. The implemented structure solves *rank* in $O(\log b)$ time.

In addition to the existing self-indexes, we compared our new indexes to several plain compressors. The well-known `gzip` and `bzip2` compressors were used with parameter `-9` to achieve maximum compression. Due to their small block sizes, they cannot profit from the large-scale repetitiveness in our data sets. We have also used the highly efficient LZ77-based compressor `p7zip`[3] with options `-mx=9 -md=30` to see how much we pay for the retrieval functionality. With a window of length up to 1 GB, `p7zip` can compress texts with long repeats much better than standard Lempel-Ziv based compressors.

Finally we have used the Subversion (SVN)[4] version control system for the OpenSSH source code data set. The source codes were inserted into a repository

---

[3] http://p7zip.sourceforge.net/

[4] http://subversion.tigris.org/

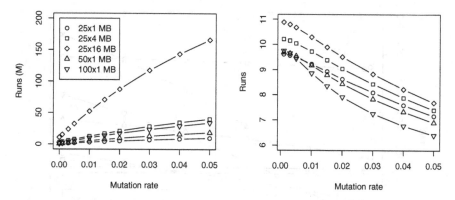

Fig. 1. The number of runs in Ψ (left) and the average number of new runs per mutation (right) on repeated DNA sequences

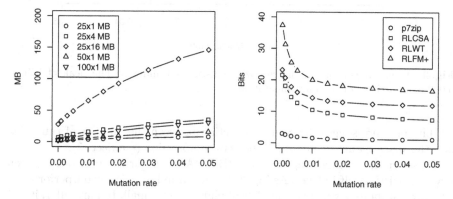

Fig. 2. The size of RLCSA on repeated DNA sequences (left) and the average number of bits required to encode a run on 25x16 MB DNA (right)

using FSFS file system in a chronological order one version at a time. We measured the sizes of subdirectory db/ of the repository, using utility du.

6.2 Results

Fig. 1 shows the number of runs in Ψ of repeated DNA sequences. The number of runs grows somewhat sublinearly in the number of mutations. New runs are created when mutations move suffixes to new positions in the suffix array. However, as the mutations accumulate, it becomes more likely that a similar mutation has already happened before, reducing the number of new runs created.

Fig. 2 shows the sensitivity of the sizes of our new self-indexes to the number of runs in the collection. RLCSA is clearly smaller than the two other indexes. It is interesting to see that with high mutation rates, p7zip requires only about

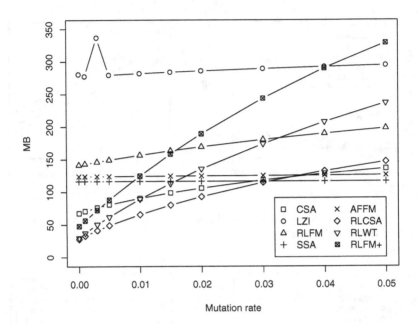

Fig. 3. A comparison of our indexes with existing self-indexes. The peak of LZI at 0.003 is an artifact of the implementation.

1.2 bits per run, suggesting some connection between the number of runs in Ψ and the space requirements of Lempel-Ziv compression (see Sect. 7).

We select the 25 times repeated 16 MB DNA prefix for comparisons between new and existing self-indexes. As Fig. 3 shows, our indexes clearly outperform the existing self-indexes when the number of mutations is small. In particular, it can be seen that our indexes are the most sensitive to high repetitiveness, followed by CSA and RLFM, and then LZ-index. SSA and AFFM are completely insensitive.

As predicted by the theoretical space bounds, RLCSA outperforms RLWT. Surprisingly, RLWT outperforms RLFM+. This is explained by the fact that RLFM+ *always* uses two bit-vectors with R bits set, and a separate wavelet tree taking close to $R \log \sigma$ bits (or slightly less in practice due to the Huffman shape). RLWT instead uses a wavelet tree formed by $\log \sigma$ levels of bit vectors each with *at most* R bits set. This worst case does not happen in practice. On random text the expected number of bits set is $\frac{\sigma/2}{\sigma-1} R \log \sigma$, and this decreases on non-random text due to the BWT effect. For example on DNA ($\log \sigma = 2$) there are only $1.25R$ bits set in RLWT. Assuming a δ-encoding of the run lengths, we get a pretty good approximation of 14.34 bits for RLWT, and 18.58 for RLFM+.

The size difference between RLCSA and RLFM+ is also surprising, given the similar high-order terms in the space bounds. This is partially explained by the ratio of non-trivial runs to total runs R'/R decreasing from 0.80 at mutation rate 0.001 to 0.37 at 0.05. Additionally, the size bound for RLFM+ has a significant low-order term. Also note the size difference of the similar CSA and RLFM.

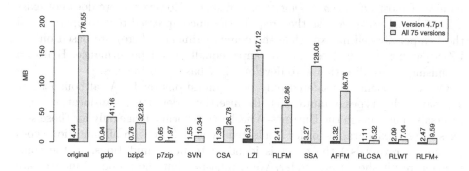

Fig. 4. Compression results for OpenSSH sources

Table 1. Time for counting on the different indexes. We remind that the LZ-index is not designed for counting.

Structure	Time (μs)	Size (MB)
CSA-128	103.0	112.29
CSA-1024	347.0	90.41
LZ-index	198596.8	281.92
RLFM	29.5	156.50
SSA	13.0	116.37
AFFM	19.4	124.15

Structure	Time (μs)	Size (MB)
RLCSA-128	72.7	77.54
RLCSA-32k	11130.0	65.52
RLWT	1050.0	89.30
RLFM+	189.7	124.48

Next we compare our indexes with existing self-indexes as well as plain compressors on OpenSSH sources. As seen in Fig. 4, our indexes clearly outperform the existing self-indexes. Again RLWT outperforms RLFM+ even with this larger alphabet size, indicating that the average RLWT space requirement is better than the worst case (see also [4] for a more rigorous analysis of runs in wavelet tree). It is interesting to note that despite the search functionality, our indexes remain smaller than the SVN repository.

The increased space efficiency of our indexes has been paid in time efficiency. To test this, we extract 1000 random substrings of length 10 from the 16 MB DNA prefix. We then repeat the prefix 25 times with mutation rate 0.01 and measure counting query times. Table 1 gives average query times and structure sizes, showing the competitiveness of RLCSA-128.

7   Discussion

In this study, we have mainly considered self-indexes based on the Burrows-Wheeler framework. There is also a family of (self-)indexes which is based on the Lempel-Ziv parsing, see [13, 19, 20]. It is easy to see that the LZ77 parsing of a repetitive text collection consists of at most of $P(T^1) + s + 1$ phrases, where $P(T^1)$ gives the number of phrases in T^1. It follows that LZ77 based indexes require at most $O(n \log \sigma + s \log n)$ bits of space. However, there does not exist a LZ77 based *self-index*, as they require the uncompressed text to operate. All the Lempel-Ziv self-indexes (like the one experimented here) are based on the LZ78 parsing, which does not guarantee equally good performance. Hence, a promising future direction is to develop LZ77 based self-indexes.

Our experiments considered only point mutations on DNA, although there are many other types of mutations, like insertions, deletions, translocations, and reversals. The runs in the Burrows-Wheeler transform change only for those suffixes whose lexicographic order is affected by a mutation. In all mutation types (except in reversals[5]) the effect is identical to point mutations, so the compression result should be similar. We emphasize that the proposed indexes are completely universal, as they do not need to know what and where the mutations are. This is also illustrated by the experiment on version control data, where the changes are cumulative, and there is no base sequence, but rather a "founder sequence". The founder model also characterizes genome collections, but again the index does not need to know the phylogeny to succeed in compression.

In subsequent (still theoretical) work [16], we have derived dynamic versions of all the proposed self-indexes, where sequences can be deleted from and inserted to the collection at any time. These indexes take basically the same space as the static ones discussed here. We have also considered new structures for `display` and `locate`, where the number of suffix array samples depend on s as well. One can use both the static and the dynamic versions of these indexes as building blocks of recent compressed suffix trees [7, 22, 23].

References

1. Arroyuelo, D., Navarro, G., Sadakane, K.: Reducing the space requirement of LZ-index. In: Lewenstein, M., Valiente, G. (eds.) CPM 2006. LNCS, vol. 4009, pp. 318–329. Springer, Heidelberg (2006)
2. Burrows, M., Wheeler, D.: A block sorting lossless data compression algorithm. Technical Report Technical Report 124, Digital Equipment Corporation (1994)
3. Church, G.M.: Genomes for all. Scientific American 294(1), 47–54 (2006)
4. Ferragina, P., Giancarlo, R., Manzini, G.: The myriad virtues of wavelet trees. In: Bugliesi, M., Preneel, B., Sassone, V., Wegener, I. (eds.) ICALP 2006. LNCS, vol. 4051, pp. 560–571. Springer, Heidelberg (2006)
5. Ferragina, P., Manzini, G.: Indexing compressed texts. J. of the ACM 52(4), 552–581 (2005)
6. Ferragina, P., Manzini, G., Mäkinen, V., Navarro, G.: Compressed representations of sequences and full-text indexes. ACM TALG 3(2) article 20 (2007)
7. Fischer, J., Mäkinen, V., Navarro, G.: An(other) entropy-bounded compressed suffix tree. In: Ferragina, P., Landau, G.M. (eds.) CPM 2008. LNCS, vol. 5029, pp. 152–165. Springer, Heidelberg (2008)
8. Grossi, R., Gupta, A., Vitter, J.: High-order entropy-compressed text indexes. In: Proc. 14th SODA, pp. 841–850 (2003)
9. Grossi, R., Vitter, J.: Compressed suffix arrays and suffix trees with applications to text indexing and string matching. SIAM J. on Computing 35(2), 378–407 (2006)

---

[5] Adding the reverse complement of the base sequence to the collection solves this.

10. Gupta, A., Hon, W.-K., Shah, R., Vitter, J.S.: Compressed data structures: Dictionaries and data-aware measures. In: Proc. 16th DCC, pp. 213–222 (2006)
11. Gusfield, D.: Algorithms on Strings, Trees and Sequences: Computer Science and Computational Biology. Cambridge University Press, Cambridge (1997)
12. Hall, N.: Advanced sequencing technologies and their wider impact in microbiology. The Journal of Experimental Biology 209, 1518–1525 (2007)
13. Kärkkäinen, J.: Repetition-based text indexes. Technical Report A-1999-4, Department of Computer Science, University of Helsinki, Finland (1999)
14. Mäkinen, V., Navarro, G.: Succinct suffix arrays based on run-length encoding. Nordic Journal of Computing 12(1), 40–66 (2005)
15. Mäkinen, V., Navarro, G., Sadakane, K.: Advantages of backward searching — efficient secondary memory and distributed implementation of compressed suffix arrays. In: Fleischer, R., Trippen, G. (eds.) ISAAC 2004, vol. 3341, pp. 681–692. Springer, Heidelberg (2004)
16. Mäkinen, V., Navarro, G., Sirén, J., Välimäki, N.: Run-length compressed indexes for repetitive sequence collections. Technical Report C-2008-42, Department of Computer Science, University of Helsinki, Finland (2008)
17. Manber, U., Myers, G.: Suffix arrays: a new method for on-line string searches. SIAM J. on Computing 22(5), 935–948 (1993)
18. Manzini, G.: An analysis of the Burrows-Wheeler transform. J. of the ACM 48(3), 407–430 (2001)
19. Navarro, G.: Indexing text using the ziv-lempel trie. J. of Discrete Algorithms (JDA) 2(1), 87–114 (2004)
20. Navarro, G., Mäkinen, V.: Compressed full-text indexes. ACM Computing Surveys 39(1) article 2 (2007)
21. Pennisi, E.: Breakthrough of the year: Human genetic variation. Science 21, 1842–1843 (2007)
22. Russo, L., Navarro, G., Oliveira, A.: Dynamic fully-compressed suffix trees. In: Ferragina, P., Landau, G.M. (eds.) CPM 2008. LNCS, vol. 5029, pp. 191–203. Springer, Heidelberg (2008)
23. Russo, L., Navarro, G., Oliveira, A.: Fully-compressed suffix trees. In: Laber, E.S., Bornstein, C., Nogueira, L.T., Faria, L. (eds.) LATIN 2008. LNCS, vol. 4957, pp. 362–373. Springer, Heidelberg (2008)
24. Sadakane, K.: New text indexing functionalities of the compressed suffix arrays. J. of Algorithms 48(2), 294–313 (2003)
25. Sadakane, K.: Compressed suffix trees with full functionality. Theory of Computing Systems 41(4), 589–607 (2007)

Practical Rank/Select Queries over Arbitrary Sequences*

Francisco Claude and Gonzalo Navarro

Department of Computer Science, Universidad de Chile
{fclaude,gnavarro}@dcc.uchile.cl

Abstract. We present a practical study on the compact representation of sequences supporting *rank*, *select*, and *access* queries. While there are several theoretical solutions to the problem, only a few have been tried out, and there is little idea on how the others would perform, especially in the case of sequences with very large alphabets. We first present a new practical implementation of the compressed representation for bit sequences proposed by Raman, Raman, and Rao [SODA 2002], that is competitive with the existing ones when the sequences are not too compressible. It also has nice local compression properties, and we show that this makes it an excellent tool for compressed text indexing in combination with the Burrows-Wheeler transform. This shows the practicality of a recent theoretical proposal [Mäkinen and Navarro, SPIRE 2007], achieving spaces never seen before. Second, for general sequences, we tune wavelet trees for the case of very large alphabets, by removing their pointer information. We show that this gives an excellent solution for representing a sequence within zero-order entropy space, in cases where the large alphabet poses a serious challenge to typical encoding methods. We also present the first implementation of Golynski et al.'s representation [SODA 2006], which offers another interesting time/space trade-off.

1 Introduction

During the past years, there has been an increasing interest in compressed data structures, since they allow one to manipulate more data in main memory, waiving the painful overcost of accessing the disk. Apart from saving space, this largely improves the execution time of an algorithm even when the compressed version makes several times more operations than the uncompressed counterpart. This applies, albeit less sharply, to all levels of memory hierarchy.

Probably the most basic tool, used in virtually all compressed data structures, is the sequence of symbols supporting *rank*, *select* and *access*. $Rank(a,i)$ counts the number of as until position i. $Select(a,i)$ finds the position of the i-th occurrence of a in the sequence. $Access(i)$ returns the symbol at position i in the sequence. The most basic case is when the sequence is drawn from a binary alphabet. Theoretically and practically appealing solutions have been proposed

---

* Partially funded by Fondecyt Grant 1-080019 (Chile).

A. Amir, A. Turpin, and A. Moffat (Eds.): SPIRE 2008, LNCS 5280, pp. 176–187, 2008.

for this case, achieving space close to the zero-order entropy of the sequence and good time performance.

The general case, when the alphabet has size $\sigma > 2$, has many applications to compressed representation of texts [8, 9, 14], trees [1, 18], graphs [5], binary relations [1], etc. For example, it has been shown [7, 14] that a compressed representation of a sequence that supports *rank* and *access* suffices to build a compressed full-text index if combined with the Burrows-Wheeler transform (BWT) [3]. Many solutions for general sequences have been proposed [8, 9], but as far as we know only some implementations of wavelet trees [6, 11] have been tried out.

In this paper we propose and study practical implementations of sequences. Our first contribution is a compressed representation of binary sequences based on Raman, Raman, and Rao's (RRR) [18] theoretical proposal. We combine faithful implementation of the theory with commonsense decisions. The result is compared, on uniformly distributed bitmaps, with a number of very well-engineered implementations for compressible binary sequences [16], and found to be competitive when the sequence is not too compressible, that is, when the fraction of 1s raises over 10%.

Still this result does not serve to illustrate the local compressibility property of RRR data structure, that is, it adapts well to local variations in the sequence. Mäkinen and Navarro [12] showed that the theoretical properties of RRR structure makes it an excellent alternative for full-text indexing: By combining it with the BWT, a high-order compressed self-index is immediately obtained, without all the extra sophistications used up to then [14]. In this paper we show experimentally that the proposed combination does work well in practice, achieving (sometimes significantly) better space than any other existing self-index, with moderate or no slowdown. The other compressed bitmap representations do not achieve this result: the bitmaps are globally balanced, but they exhibit long runs of 0s or 1s that only the RRR technique exploits so efficiently.

We then turn our attention to representing sequences over larger alphabets. Huffman-shaped wavelet trees have been used to approach zero-order compression of sequences [6, 11]. This requires $O(\sigma \log n)$ bits for the symbol table and the tree pointers, where n is the sequence length. On large alphabets, this factor can be prohibitive in space and ruin the compression ratios. We propose an alternative representation that uses no (or just $\log \sigma$) pointers, and concatenates all the bitmaps of the wavelet tree levelwise. As far as we know, no previous direct solution to *select* over this representation existed. Combined with our compressed bitmap representation, the result is an extremely useful tool to represent a sequence up to its zero-order entropy, disregarding any problem related to alphabet size. We illustrate this point by improving an existing result on graph compression [5], in a case where no other considered technique succeeds.

Finally, we present the (as far as we know) first implementation of Golynski et al.'s data structure for sequences [9], again combining faithful implementation of the theory with common sense. The result is a representation that does not compress the sequence, yet it answers queries very fast without using too much extra space. In particular, its performance over a sequence of word identifiers provides a sequence representation that uses about 70% of the original space of the text (in character form) and gives the same functionality of an inverted index. It might become an interesting alternative to recent wavelet-tree-based proposals for representing text collections [2], and to inverted indexes in general.

2   Related Work

We divide the related work into two subsections, the first one covering *rank*, *select* and *access* for binary sequences, and the second covering the case of larger alphabets. We omit the base of logarithms when it is 2. We make heavy use of the definition of zero-order empirical entropy for a sequence S of length n drawn from an alphabet Σ of size σ: $H_0(S) = \sum_{a \in \Sigma} \frac{n_a}{n} \log \frac{n}{n_a}$, where n_a is the number of occurrences of symbol a in S. In the case where $\Sigma = \{0, 1\}$ and $n_1 = m << n$, it is interesting to write $H_0 = m \log \frac{n}{m} + O(m)$.

2.1   Binary Sequences

Many solutions have been proposed for the case of binary sequences. Consider a bitmap $B[1, n]$ with m ones. The first compact solution to this problem is capable of answering the queries in constant time and uses $n + o(n)$ bits [4] (i.e., B itself plus $o(n)$ extra space); the solution is straightforward to implement [10]. This was later improved by Raman, Raman and Rao (RRR) [18] achieving $nH_0(B) + o(n)$ bits while answering the queries in constant time, but the technique is not anymore simple to implement. Several practical alternatives achieving very close results have been proposed by Sadakane and Okanohara [16], tailored to the case of small m: esp, recrank, vcode, sdarray, and darray. Most of them are very good for *select* queries, yet *rank* queries are slower. The variant esp is indeed a practical implementation of RRR structure that saves space by replacing some pointers by estimations based on entropy.

   In this work we implement the RRR data structure [18]. It divides the sequence into blocks of length $u = \frac{\log n}{2}$ and every block is represented as a tuple (c_i, o_i). The first component, c_i, represents the *class* of the block, which corresponds to its number of 1s. The second, o_i, represents the *offset* of that block inside a list of all the possible blocks in class c_i. Three tables are defined: E, R and S. Table E stores every possible combination of u bits, sorted by class, and by offset within each class. It also stores all answers for *rank* at every position of each combination. Table R corresponds to the concatenation of all the c_i's, using $\lceil \log(u+1) \rceil$ bits per field. Table S stores the concatenation of the o_i's using $\left\lceil \log \binom{u}{c_i} \right\rceil$ bits per field. This structure also needs two partial sum structures [17], one for R and the other for the length of the o_i's in S, $posS$. For answering *rank* until position i we first compute $sum(R, \lfloor i/u \rfloor) = \sum_{j=0}^{\lfloor i/u \rfloor} R_j$, the number of 1s before the beginning of i's block, and then *rank* inside the block until position i using table E. For this we need to find o_i: using $sum(posS, \lfloor i/u \rfloor)$ we determine the starting position of o_i in S, and with c_i and u we know how many bits we need to read. For *select* queries, they store the same extra information as Clark

[4], but no practical implementation for this extra structure has been shown. *Access* can be answered with two *ranks*, $access(i) = rank(1, i) - rank(1, i - 1)$.

2.2 Arbitrary Sequences

Rank, *select* and *access* operations can be extended to arbitrary sequences drawn from an alphabet Σ of size σ. The two most prominent data structures that solve this problem are reviewed next.

Wavelet Trees. [8, 11, 15] are perfectly balanced trees that store a bitmap of length n in the root; every position in the bitmap is either 0 or 1 depending on the value of the most significant bit of the symbol in that position in the sequence.[1] A symbol with a 0 goes to the left subtree and a symbol with a 1 goes to the right subtree. This decomposition continues recursively with the next highest bit, and so on. The tree has σ leaves and requires $n\lceil \log \sigma \rceil$ bits, n bits per level. Every bitmap in the tree answers *access*, *rank* and *select* queries.

The *access* query for position i can be answered by following the path described for position i. At the root, if the bitmap at position i has a 0/1, we descend to the left/right child, switching to the bitmap position $rank(0/1, i)$ in the left/right subtree. This continues recursively until reaching the last level, when we finish forming the binary representation of the symbol.

Query *rank* for symbol a until position i can be answered in a similar way as *access*, the difference being that instead of considering the bit at position i in the first level, we consider the most significant bit of a; for the second level we consider the second highest bit, and so on. We update the position for the next subtree with $rank(b, i)$, where b is the bit of a considered at this level. At the leaves, the final bitmap position corresponds to the answer to $rank(a, i)$ in S.

The *select* query does a similar process as *rank*, but upwards. To select the i-th occurrence of character a, we start at the leaf where a is represented and do $select(b, i)$ where, as before, b is the bit of a corresponding to this level. Using the position obtained by the binary *select* query we move to the parent, querying for this new position. At the root, the position is the final result.

The cost of the operations is $O(\log \sigma)$ assuming constant-time *rank*, *select* and *access* over bitmaps. A practical variant to achieve $n(H_0(S) + 1)$ bits of space is to give the wavelet tree the shape of the Huffman tree of S [11, 15].

Golynski et al. [9] proposed a data structure capable of answering *rank*, *select* and *access* in time $O(\log \log \sigma)$ using $n \log \sigma + n\, o(\log \sigma)$ bits of space. The main idea is to reduce the problem over one sequence to n/σ *chunks* of length σ. For each symbol they concatenate, in unary, the number of its occurrences in each *chunk*, and then concatenate those sequences for all the symbols in a bitmap B. Armed with *rank* and *select* structures, B's total length is $2n + o(n)$ bits.

---

[1] In general wavelet trees are described as dividing alphabet segments into halves. The description we give here, based on the binary decomposition of alphabet symbols, is more convenient for the solutions shown in this paper.

With B it is possible to answer *rank* and *select* queries up to *chunk* granularity. Every *chunk* stores σ text symbols using a bitmap X and a permutation π. X stores the cardinality of every symbol of the alphabet in the *chunk* using the same encoding as B. π stores the permutation obtained by stably sorting the sequence represented by the *chunk*, and uses a data structure that allows computation of π^{-1} in $O(\log \log \sigma)$ time [13]. This adds up to $2n + n \log \sigma + n\,o(\log \sigma)$ bits, and permits completing all queries in constant time for *select*, and $O(\log \log \sigma)$ time for *rank* and *access*. The latter complexity needs a Y-Fast trie within each chunk to search π for the position of interest, among those corresponding to the occurrences of a single symbol within the chunk.

We note that the $n\,o(\log \sigma)$ extra term does not vanish asymptotically with n but with σ. This suggests, as we verify experimentally later, that the structure performs well only on large alphabets.

3   Practical Implementations

3.1   Raman, Raman and Rao's Structure

We fix $u = 15$ so that the c_i's need 4 bits to represent the class $(0 - 15)$. We store table E using 16-bit integers for the bitstring contents, and for the pointers to the beginning of each class in E. The answers to *rank* are not stored but computed on the fly from the bitstrings, so E uses just 64 KB. Table R is represented by a compact array using 4 bits per field, achieving fast extraction. Table S stores each offset using $\left\lceil \log \binom{u}{c_i} \right\rceil$ bits.

The partial sums are represented by a one-level sampling. For table R we sample the sum every k values, and store these values in a new table $sumR$ using $\lceil \log m \rceil$ bits per field, where m is the number of ones. To obtain the partial sum until position i we compute $sumR[j] + \sum_{p=jk}^{i} c_p$ where $j = \lfloor i/k \rfloor$, and the summation of the c_p's is done sequentially over the R entries. The positions in S are represented the same way: We store the sampled sums in a new table called $posS$ using $\lceil \log(\sum_{i=1}^{n/u} \lceil \log \binom{u}{c_i} \rceil) \rceil$ bits per field. We compute the position for block i as $posS[j] + \sum_{p=jk}^{i} \lceil \log \binom{u}{c_p} \rceil$. We precompute the 16 possible $\lceil \log \binom{u}{c_p} \rceil$ values in order to speed up this last sequential summation.

With this support, we answer *rank* queries by using the same RRR procedure. Yet, *select*$(1, i)$ queries are implemented in a simpler and more practical way. We use a binary search over $sumR$, finding the rightmost sampled block for which $sumR[k] \leq i$. Then we traverse table R looking for the block in which we expect to find the i-th bit set (i.e., adding up c_p's until we exceed i). Finally we access this block in table E and traverse it bit by bit until finding the i-th 1. *Select*$(0, i)$ can be implemented analogously.

3.2   Wavelet Trees without Pointers

There exist already Huffman-shaped wavelet tree implementations that achieve close to zero-order entropy space. Yet, those solutions are not efficient when the alphabet is very large: The overhead of storing the Huffman symbol assignment and the wavelet tree pointers, $O(\sigma \log n)$, ruins the compression if σ is large. In this section we present an alternative implementation that achieves zero-order entropy with a very mild dependence on σ (i.e. $O(\log \sigma \log n)$ bits of space), thus extending the existing results to the case of very large alphabets. We use two bitmaps: DA and Occ. $DA[1, \sigma]$ stores which symbols appear in the sequence, $DA[i] = 1$ if symbol i appears in S. This allows us to remap the sequence in order to get a contigous alphabet; using $rank$ and $select$ over DA we can map in both directions (we only use it if the alphabet is not contiguous). $Occ[1, n]$ records the number of occurrences of symbols $i \leq k$ by placing a one at $\sum_{i=1}^{k} n_i$. For example, the sequence 113213323 would generate $Occ = 001010001$.

Our implementation of the wavelet tree stores $\lceil \log \sigma \rceil$ bitmaps of length n. The tree is mapped to these bitmaps levelwise: the first bitmap corresponds to the root, the next one corresponds to the concatenation of left and right children of the root, and so on. In this set of bitmaps we must be able to calculate the interval $[s, e]$ corresponding to the bitmap of a node, and to obtain the new interval $[s', e']$ upon a child or parent operation. Assume the current node is at level l ($l = 1$ at the leaves) on a tree of h levels. Further, assume that a is the symbol related to the query, that $\Sigma = \{0, \ldots, \sigma-1\}$, and that $select_{Occ}(1, 0) = 0$.

We compute the left child as $s' = s$ and $e' = e - rank(1, e) + rank(1, s - 1)$, and the right child as $s' = e + 1 - rank(1, e) + rank(1, s - 1)$ and $e' = e$. Let us explain the left child formula. In the next level, the current bitmap segment is partitioned into a left child and right child parts. The left child starts at the same position of the current segment in this level, so $s' = s$. To know the end of its part, we must add the number of 0s in the current segment, $e' = s+rank(0, e)-rank(0, s-1)-1 = s+(e-rank(1, e))-((s-1)-rank(1, s-1))-1$.

The formula to compute the parent is $s' = select_{Occ}(1, \lfloor a/2^l \rfloor \cdot 2^l) + 1$ and $e' = select_{Occ}(1, (\lfloor a/2^l \rfloor + 1) \cdot 2^l)$. The idea is to consider the binary representation of a, as this is the way our wavelet tree is structured. A node at level l should contain all the combinations of the l lowest bits of a. For example, if $l = 1$ and $a = 5 = (101)_2$, its parent is the node at level $l = 2$ comprising the symbols $4 = (100)_2$ to $5 = (101)_2$. The parent of this node, at level $l = 3$, comprises symbols $4 = (100)_2$ to $7 = (111)_2$. We blur the last l bits of a and use $select_{Occ}$ to find the right segments at any level corresponding to the symbol intervals.

To achieve compression we represent the bitmaps of each level (as well as Occ) using RRR, whose sampling yields a time/space trade-off for the structure.

3.3 Golynski's Structure

We implement Golynski et al.'s proposal rather faithfully, except that we replace the Y-Fast trie by a binary search over the positions for the $rank$ query. In practice, this yields a gain in space and time except for large σ values and biased symbol distribution within the *chunk* (remind that we must search within the range of occurrences of a symbol of Σ in a *chunk* of size σ, i.e. the range is $O(1)$ size on average). Hence the time for $rank$ is $O(\log \sigma)$ worst case, and $O(1)$ on average. The version used for permutations [13] requires $(1 + \epsilon)n\lceil \log n \rceil$ bits for

n elements and computes π^{-1} in $O(1/\epsilon)$ worst-case time (code by D. Arroyuelo). This gives us a space/time tradeoff parameter for this data structure.

In the case when $n \approx \sigma$ we also experiment with using only one *chunk* to represent the structure. This speeds up all the operations since we do not need to compute which *chunk* should we query, and all the operations over B become unnecessary, as well as storing B itself.

4   Experimental Results

We first test the data structures for binary sequences, on random data and on the BWT of real texts, showing that RRR is an attractive option. Second, we compare the data structures for general sequences on various types of large-alphabet texts, obtaining several interesting results. Finally, we apply our machinery to obtain the best results so far on compressed text indexing.

The machine is a Pentium IV 3.0 GHz with 4GB of RAM using Gentoo GNU/Linux with kernel 2.6.13 and g++ with -O9 and -DNDEBUG options.

4.1   Binary Sequences

We generated three random uniformly and independently distributed bitmaps of length $n = 10^8$, with densities (fraction of 1s) of 5%, 10% and 20%. Fig. 1 compares our RRR implementation against the best practical ones in previous work [16], considering operations *rank* and *select* (*access* can be implemented as the difference of two *rank*'s, and in some cases it can be done slightly better, yet only some of the structures in [16] support it). As control data we include a fast uncompressed implementation [10], which is insensitive to the bitmap density.

Our RRR implementation is far from competitive for very low densities (5%), where it is totally dominated by sdarray, for example. For 10% density it is already competitive with esp, its equivalent implementation [16], while offering more space/time tradeoffs and achieving the best space. For 20% density, RRR is unparalleled in space usage, and alternative implementations need significantly

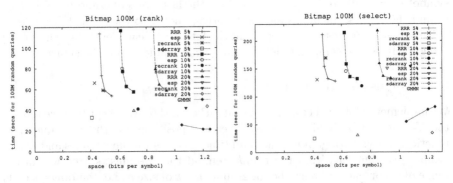

Fig. 1. Space in bits per symbol and time in seconds for answering 10^8 random queries over a bitmap of 10^8 bits

Table 1. Space (as a fraction of bitmap size) and *rank* time (in μsec/query) achieved by various data structures on the wavelet tree bitmaps of a BWT-transformed text

Variant	Size	*Rank* time
sdarray	2.05	> uncompressed
recrank	1.25	> uncompressed
esp	0.50	0.594
RRR (ours)	0.48	0.494
uncompressed	1.05	0.254

more space to beat its time performance. We remark that RRR implements $select(0, i)$ in the same time of $select(1, i)$ with no extra space, where competing structures would have to double the space they devote to *select*.

A property of RRR that is not apparent over uniformly distributed bitmaps, but it becomes very relevant for implementing the wavelet tree of a BWT-transformed text, is its ability to exploit local regularities in the bit sequence. To test this, we extracted the 50MB English text from *Pizza&Chili* (http://pizzachili.dcc.uchile.cl), computed the balanced wavelet tree of its BWT, and concatenated all the bitmaps of the wavelet tree levelwise. Table 1 shows the compression achieved by different methods. Global methods like sdarray and recrank fail, taking more space than an uncompressed implementation. RRR stands out as the clear choice for this problem, followed by esp (which is based on the same principle). The bitmap density is around 40%, yet RRR achieves space similar to 5% uniformly distributed density.

4.2 General Sequences

We compared our implementations of Golynski et al.'s and different variants of wavelet trees. We consider three alphabet sizes. The smaller one is byte-size: We consider our plain text sequences English and DNA, seeing them as character sequences. Next, we consider a large alphabet, yet not large enough to compete with the sequence size: We take the 200MB English text from *Pizza&Chili* and regard this as a sequence of *words* The result is a sequence of $46,582,195$ words over an alphabet of size $270,096$. Providing *access* and *select* over this sequence mimics a word-addressing inverted index functionality [2]. Finally, we consider a case where the alphabet is nearly as large as the text. The sequence corresponds to graph Indochina after applying *Re-Pair* compression on its adjacency list [5]. The result is a sequence of length $15,568,253$ over an alphabet of size $13,502,874$. The result of *Re-Pair* can still be compressed with a zero-order compressor, but the size of the alphabet challenges all of the traditional methods. Our techniques can achieve such compression and in addition provide *rank/select* functionality, which permits implementing backward traversal on the graph for free.

We consider full and 1-chunk variants of Golynski. On wavelet trees, variant *DA* maps the alphabet to a contiguous range, *RRR* compresses the bitmaps with RRR, *Ptrs* uses the standard version with pointers, and *Huff* gives Huffman tree shape to the pointer-based wavelet tree. We show several combinations of these.

Fig. 2 shows the results for byte alphabets. Golynski's structure is not competitive here. This shows that their $o(\log \sigma)$ term is not yet negligible for $\sigma = 256$. On wavelet trees, the Ptrs+Huff variant excells in space and in time. Adding RRR reduces the space very slightly in some cases; in others keeping balanced shape gives better time in exchange for much more space. We also included Naive, an implementation for byte sequences that stores the plain sequence plus regularly sampled *rank* values for all symbols [2]. The results show that we could improve their wavelet trees on words by replacing their Naive method by ours.

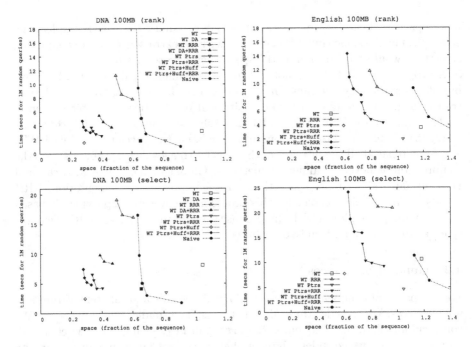

Fig. 2. Results for byte alphabets. Space is measured as a fraction of the sequence size (assuming one byte per symbol).

Fig. 3 shows experiments on larger alphabets. Here Golynski et al.'s structure becomes relevant. On the sequence of words it adds to the previous scenario a third space/time tradeoff point, offering much faster operations (especially *select*) in exchange for significantly more space. Yet, this extra space is acceptable for the sequence of word identifiers, as overall it requires 70% *of the original text*. As such it competes with a word-addressing inverted index, following a recent trend of replacing inverted indexes by a succinct data structure for representing sequences of word identifiers [2], which can retrieve the text using *access* but also find the consecutive positions of a word using *select*.

Again, adding RRR reduces the space of Ptrs+Huff, yet this time the reduction is more interesting, and might have to do with some locality in the usage of words across a text collection.

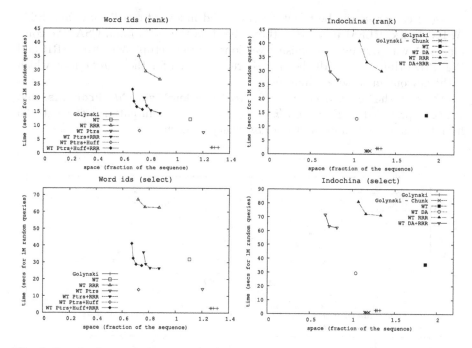

Fig. 3. Results for word identifiers (left) and a graph compressed with Re-Pair (right). Space is measured as a fraction of the sequence (using $\lceil \log \sigma \rceil$ bits per symbol).

For the case of graphs, where the alphabet size is close to the text length, the option of using just one chunk in Golynski et al.'s structure becomes extremely relevant. It does not help to further compress the Re-Pair output, but for 20% extra space it provides very efficient backward graph traversal. On the other hand, the wavelet trees with DA+RRR offer further compression up to 70% of the Re-Pair output size, which is remarkable for a technique that already achieved excellent compression results [5]. The price is much higher access time. An interesting point here is that the versions with pointers are not applicable here, as the alphabet overhead drives their space over 3 times the sequence size. Hence exploring versions that do not use pointers pays off.

4.3   Compressed Full-Text Self-indexes

It was recently proved [12] that the wavelet tree of the Burrows-Wheeler transform (BWT) of a text (the key ingredient of the successful FM-index family of text self-indexes [15]), can achieve high-entropy space without any further sophistication, provided the bitmaps of the wavelet tree are represented using RRR structure [18]. Hence a simple and efficient self-index emerges, at least in theory. In Section 4.1 we showed that RRR indeed takes unique advantage from the varying densities along the bitmap typical of the BWT transform. We can now show that this proposal [12] has much practical value.

Fig. 4 compares the best suffix-array based indexes from *Pizza&Chili*: SSA, AFFM-Index, RLFM-Index (the three based on the BWT) and CSA. We combine SSA with our most promising versions for this setup, WT Ptrs+RRR and WT Ptrs+Huff+RRR. All the spaces are optimized for the *count* query, which is the key one in these self-indexes

We built the index over the 100 MB texts English, DNA, Proteins, and Sources provided in *Pizza&Chili*. We chose 10^5 random patterns of length 20 from the same texts and ran the *count* query on each.

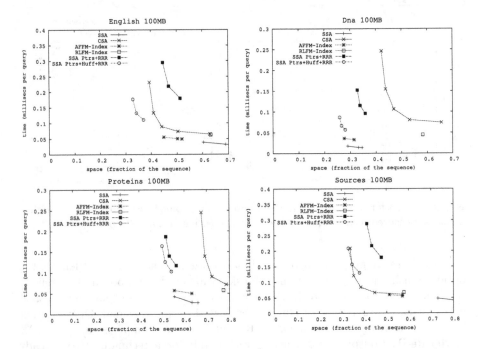

Fig. 4. Times for counting, averaged over 10^5 repetitions, for patterns of length 20

As can be seen, our new implementation is extremely space-efficient, achieving a space performance *never seen before* in compressed full-text indexing, and in some cases without a time penalty.

The reason why combining RRR with Huffman shape is better than RRR alone, when RRR by itself should in principle exploit all of the compressibility captured by Huffman, is in the c component of the (c,o) pairs of RRR. These pose a fixed overhead per symbol which is not captured by the entropy. Indeed, we measured the length of the o components (table S) in both cases (for English) and the difference was 0.02%. The Huffman shape helps reduce the total number of symbols to be indexed, and hence it reduces the overhead due to the c components.

Several lines of work are open to future work, in particular implementing an efficient version of RRR combined with run-length encoding, which should give even better spaces. Another line is to pursue on the idea of an inverted-index-like capability by encoding the sequence of words of a natural language text, or a word-based self-index.

References

1. Barbay, J., He, M., Munro, I., Srinivasa Rao, S.: Succinct indexes for strings, binary relations and multi-labeled trees. In: 18th SODA, pp. 680–689 (2007)
2. Brisaboa, N., Fariña, A., Ladra, S., Navarro, G.: Reorganizing compressed text. In: SIGIR (to appear, 2008)
3. Burrows, M., Wheeler, D.: A block sorting lossless data compression algorithm. Tech.Rep. 124, December (1994)
4. Clark, D.: Compact Pat Trees. Ph.D thesis, University of Waterloo (1996)
5. Claude, F., Navarro, G.: A fast and compact Web graph representation. In: Ziviani, N., Baeza-Yates, R. (eds.) SPIRE 2007. LNCS, vol. 4726, pp. 118–129. Springer, Heidelberg (2007)
6. Ferragina, P., González, R., Navarro, G., Venturini, R.: Compressed text indexes: From theory to practice (manuscript, 2007), `http://pizzachili.dcc.uchile.cl`
7. Ferragina, P., Manzini, G.: Indexing compressed texts. J. ACM 52(4), 552–581 (2005)
8. Ferragina, P., Manzini, G., Mäkinen, V., Navarro, G.: Compressed representations of sequences and full-text indexes. ACM TALG 3(2) article 20 (2007)
9. Golynski, A., Munro, I., Rao, S.: Rank/select operations on large alphabets: a tool for text indexing. In: SODA, pp. 368–373 (2006)
10. González, R., Grabowski, S., Mäkinen, V., Navarro, G.: Practical implementation of rank and select queries. Posters WEA, pp. 27–38 (2005)
11. Grossi, R., Gupta, A., Vitter, J.: High-order entropy-compressed text indexes. In: SODA, pp. 841–850 (2003)
12. Mäkinen, V., Navarro, G.: Implicit compression boosting with applications to self-indexing. In: SPIRE, pp. 214–226 (2007)
13. Munro, I., Raman, R., Raman, V., Srinivasa Rao, S.: Succinct representations of permutations. In: Baeten, J.C.M., Lenstra, J.K., Parrow, J., Woeginger, G.J. (eds.) ICALP 2003. LNCS, vol. 2719, pp. 345–356. Springer, Heidelberg (2003)
14. Navarro, G., Mäkinen, V.: Compressed full-text indexes. ACM Comp. Surv. 39(1) article 2 (2007)
15. Navarro, G., Mäkinen, V.: Compressed full-text indexes. ACM Comp. Surv. 39(1) article 2 (2007)
16. Okanohara, D., Sadakane, K.: Practical entropy-compressed rank/select dictionary. In: ALENEX (2007)
17. Raman, R., Raman, V., Srinivasa Rao, S.: Succinct dynamic data structures. In: Dehne, F., Sack, J.-R., Tamassia, R. (eds.) WADS 2001, vol. 2125, pp. 426–437. Springer, Heidelberg (2001)
18. Raman, R., Raman, V., Srinivasa Rao, S.: Succinct indexable dictionaries with applications to encoding k-ary trees and multisets. In: SODA, pp. 233–242 (2002)

Clique Analysis of Query Log Graphs

Alexandre P. Francisco[1,*], Ricardo Baeza-Yates[2], and Arlindo L. Oliveira[1]

[1] INESC-ID/IST, Technical University of Lisbon, Portugal
[2] Yahoo! Research Barcelona, Spain & Santiago, Chile

Abstract. In this paper we propose a method for the analysis of very large graphs obtained from query logs, using query coverage inspection. The goal is to extract semantic relations between queries and their terms. We take a new approach to successfully and efficiently cluster these large graphs by analyzing clique overlap and *a priori* induced cliques. The clustering quality is evaluated with an extension of the modularity score. Results obtained with real data show that the identified clusters can be used to infer properties of the queries and interesting semantic relations between them and their terms. The quality of the semantic relations is evaluated both using a tf-idf based score and data from the Open Directory Project. The proposed approach is also able to identify and filter out multitopical URLs, a feature that is interesting in itself.

1   Introduction

Knowledge discovery is one of the main problems in data mining and information retrieval. Human interaction through the Web generated implicit knowledge -or the wisdom of crowds [1]- represents an important path towards improving the discovery of interesting knowledge. Nowadays the Web is the biggest representation of human knowledge, where people contribute with content either explicitly or implicitly. An example of an implicit contribution is searching, as people contribute with their knowledge by clicking on retrieved documents. Thus, queries submitted to search engines carry implicit knowledge and they can be seen as equivalent to tags associated to clicked documents. An important and interesting challenge is then to extract relevant relations from query logs, namely semantic relations between queries and their terms.

Graphs are a natural way to view relations between queries and URLs. As Baeza-Yates and Tiberi [2], we start with the bipartite graph of queries and URLs, where a query q and a URL u are connected if a user clicked in the URL u that was an answer for the query q. Then, we generate a query graph by analyzing common URLs among queries. A more frequent approach is to define a similarity measure among queries. However it is more difficult to understand why queries are similar and it can add noise to data already noisy.

This paper proposes two different contributions. First, we propose a method to efficiently cluster very large graphs using clique percolation [3] and *a priori*

---

* Work done while visiting Yahoo! Research Barcelona.

A. Amir, A. Turpin, and A. Moffat (Eds.): SPIRE 2008, LNCS 5280, pp. 188–199, 2008.

induced cliques. The quality of the clustering is evaluated by computing an extension of the modularity score [4,5] for overlapping clusters. Second, we analyze the obtained clusters and extract semantic relations, inside and among clusters, obtaining new information about the nature of the queries and the URLs. To evaluate our method we use a part of the query log of the Yahoo! search engine, with 2.8 million queries with at least one clicked URL and 4.9 million different URLs. For each query, the data includes information on which URLs were clicked and with which frequency. The quality of the inferred semantic relations is evaluated both with a tf-idf (term frequency-inverse document frequency) derived score and against data from Open Directory Project (ODP).

The rest of the paper is organized as follow. In Section 2 we discuss previous work on query similarity and knowledge extraction from queries. In Section 3 we describe the cover graph and its properties. In Section 4 we describe the clustering method and we present the modularity score. In Section 5 we analyze a large graph and we evaluate our approach. We end with some final remarks and future work.

2   Previous Work

Most of the work on query similarity is related to query expansion or query clustering. Here we mention only the main related papers.

Wen *et al* [6] proposed to cluster similar queries to recommend URLs to frequently asked queries of a search engine. They used four notions of query distance: (1) based on keywords or phrases of the query; (2) based on string matching of keywords; (3) based on common clicked URL's; and (4) based on the distance of the clicked documents in some pre-defined hierarchy. As the average number of words in queries is small (about two) and the number of clicks in the answer pages is also small [7], notions (1) and (2) generate distance matrices that are very sparse. Notion (4) needs a concept taxonomy and requires the clicked documents to be classified into the taxonomy as well, something that cannot be done in a large scale. Although (3) is also sparse, this sparsity can be diminished by using large query logs. Befferman and Berger [8] also proposed a query clustering technique based on (3) and Zaiane and Strilets [9] used variants of (1) and (3) as well as other simpler features.

Baeza-Yates *et al.* [10,11] used the content of clicked Web pages to define a term-weight vector model for a query. They consider terms in the URLs clicked after a query. Each term is weighted according to the number of occurrences of the query and the number of clicks of the documents in which the term appears. Then, the similarity of two queries is equivalent to the similarity of their vector representations, using the cosine distance function. This notion of query similarity has several advantages. First, it is simple and easy to compute. On the other hand, it makes it possible to relate queries that happen to be worded differently but stem from the same topic. Therefore, semantic relationships among queries are captured.

In the work by Chuang *et al.* [12,13,14,15] they use query logs to build a query taxonomy to also cluster answers. However they do not use any user feedback, like user clicks. This idea of building a taxonomy based on queries is extended in [16], but this is not the same as building a taxonomy of the queries, which is what we would call a query taxonomy. Later, Dupret and Mendoza [17] used the rank of clicked URLs to define relations among queries. They recommend better queries by generating query relations that can be associated to parts of a query taxonomy. Recently, Baeza-Yates and Tiberi [2] used (3) and a very large query log to define semantic relations such as equivalence or specificity based on different set conditions among the set of clicked URLs. Using the ODP they found a precision up to 83% on the relations discovered and also that the ones not found were too specific to appear in ODP. Our work can be viewed as a followup to this paper.

3   The Cover Graph

In this section we define the cover graph G that arises naturally from the bipartite graph of queries and URLs, based on the notion of common clicked URLs [18,19]. Let $\mathcal{Q}$ be the set of queries and $\mathcal{U}$ be the set of URLs. Given a query $q \in \mathcal{Q}$, the cover of q is the set of URLs clicked by q. Let $\mu : \mathcal{Q} \to 2^{\mathcal{U}}$ be a function that maps each query q to its cover set $\mu(q) \subseteq \mathcal{U}$.

The *cover graph* $G = (V, E)$ is an undirected and unweighted graph with queries as vertices and where exists an edge between two queries whenever they share at least one common clicked URL. Formally, $V = \mathcal{Q}$ and $E \subseteq V \times V$ is such that $(q_1, q_2) \in E$ if and only if $\mu(q_1) \cap \mu(q_2) \neq \emptyset$.

For the part of the query log of the Yahoo! search engine analyzed in this paper, the full cover graph is very large with more than 359 million edges (first row of Table 1). Since many URLs are clicked with a very low frequency, this graph is also very noisy. Hence, we will filter the edges in order to remove the noise and reduce the graph size.

Let $\mathcal{W} : \mathcal{Q} \times \mathcal{U} \to [0, 1]$ be a function such that $\mathcal{W}(q, u)$ is the ratio with which the URL u was clicked for the query q. Thus, given a ratio threshold w, the filtered cover graph $G = (V, E)$ is such that $V = \mathcal{Q}$ and $(q_1, q_2) \in E$ if and only if $\{u \in \mu(q_1) \mid \mathcal{W}(q_1, u) \geq w\} \cap \{u \in \mu(q_2) \mid \mathcal{W}(q_2, u) \geq w\} \neq \emptyset$. Note that this type of filtering is different from previous methods used to filter edges [2], where each query has a frequency weight vector associated and edges are weighted according to the cosine similarity. Both approaches are related since high frequency URLs increase the cosine similarity and also increase the confidence we have in the fact that queries joined by a given edge are truly related. However, with our approach, we can easily find cliques in G as we describe in the next section.

In Table 1 we give details about the cover graph for different values of w. For $w = 0.5$ we successfully reduce the size of the cover graph, since the filtered graph has only 36% of the edges in the full graph. Higher values of w reduce even more the size of the graph and the size of the giant connected component. The degree distributions of studied cover graphs follow a power law behavior, as exemplified

Table 1. Details of the studied cover graphs, where CC is the set of connected components, S is the set of singleton vertices, gC is the giant component and $|V| = 2,822,337$

w	$|E|$	$2|E|/|V|$	$|E|/|V|\log|V|$	$|CC|/|V|$	$|S|/|V|$	$|gC|/|V|$
0.0	359,881,327	255.023	8.584	0.556	0.499	0.348
0.5	129,915,749	92.062	3.099	0.697	0.620	0.145
0.7	70,487,699	49.949	1.681	0.785	0.718	0.063
0.8	35,324,706	25.032	0.842	0.859	0.806	0.002
1.0	30,695,828	21.752	0.732	0.890	0.847	0.002

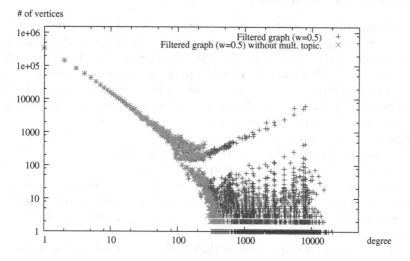

Fig. 1. Cover graph degree distribution before and after multitopical URL removal

later in Figure 1 for the case $w = 0.5$ the power law has an exponent of 1.51. We note also that an abrupt change occurs in the size of the giant connected component, as is clear in Table 1 when w changes from 0.7 and 0.8. Another interesting fact is that for $w \geq 0.8$, even for $w = 1.0$, the number of edges does not decrease as much as we may expect. That happens because there are many navigational queries, i.e., queries with just one clicked URL, and many of them refer to the same URL. In the next section we will study the cover graph for $w = 0.5$ in detail and analyze the existing semantic relations among queries.

Given $0 \leq w \leq 1$, the cover graph can be computed efficiently. Let $\bar{N}$ and $\hat{N}$ be the average and the maximum number of URLs covered per query, respectively. Let also $\bar{M}$ and $\hat{M}$ be the average and the maximum number of queries that cover an URL, respectively. First, we sort the URLs for each query in $O(|Q|\bar{N}\log\hat{N})$ time. Filtering the URLs for each query takes linear time with respect to the number of URLs per query, i.e., takes $O(|Q|\bar{N})$ time. We also need to compute the list of queries for each URL. This can be done in $O(|Q|\bar{N})$ time by transposing the list of URLs for each query, a procedure that can be performed while filtering.

Finally, we compute the adjacency list for each query by merging the list of queries for each URL in the URL list of the current query. Given the list of URLs for some query and given that the lists of queries and URLs are sorted, we can do the merge using a priority queue where we store the head of each URL query list. Thus, the merging takes $O(\bar{N}\hat{M} \log \hat{N})$ time per query. Given these complexities, the cover graph can be computed in $O(|\mathcal{Q}|\bar{N}\hat{M} \log \hat{N})$ where $\bar{N}$, $\hat{N}$ and $\hat{M}$ are small compared to $|\mathcal{Q}|$ and $|\mathcal{U}|$. For the data we study in this paper and for $w = 0.5$, $\bar{N}$ is 1.370, $\hat{N}$ is 4 and $\hat{M}$ is 7,974.

4   Clique Analysis and Clustering

We are interested in studying overlapping clusters of G in order to identify query relationships and extract semantic information from them. In this paper we specifically study the cliques of G and how they overlap. Previous work has been done on the identification of overlapping clusters from overlapping cliques [20,3]. In our approach, we use a clustering method similar to clique percolation as introduced by Palla et al. [3], where clusters are formed by joining cliques that overlap above a given threshold k.

Since computing maximal cliques is computationally hard [21], we will use *a priori* induced cliques. Let $\mu^{-1} : \mathcal{U} \to 2^{\mathcal{Q}}$ be the "inverse" function of μ that maps each u to its coverable set of queries $\mu^{-1}(u) \subseteq \mathcal{Q}$. Clearly, since every query q in the set $\mu^{-1}(u)$ share at least the URL u, the set $\mu^{-1}(u)$ induces a clique in the graph G. If we are dealing with a filtered version of G, the cliques are induced by the sets $\{q \in \mu^{-1}(u) \mid \mathcal{W}(q, u) \geq w\}$, where w is the frequency threshold.

Given the induced cliques and a threshold k, the clustering method works by merging every clique which overlaps in more than k vertices. Note that each URL has a (filtered) list of queries that are the vertices of the induced clique. Therefore we must intersect each URL list with all other URL lists. The running time is $O(|\mathcal{U}|^2\bar{M})$, where $\bar{M}$ is the average number of queries per URL and, for the studied graph with $w = 0.5$, $\bar{M}$ is 0.784. Given that we are filtering by k, the number of URLs is usually much smaller than $|\mathcal{U}|$.

We compute an extension of the modularity measure for overlapping clusters to evaluate the quality of the clustering. Given a non-overlapping clustering $\mathcal{C} = \{C_1, ..., C_n\}$ of G, the *modularity* Q is defined [4,5] as

$$Q = \frac{1}{2|E|} \sum_{p,q \in V} \left[A_{pq} - \frac{d_p d_q}{2|E|} \right] \delta(c_p, c_q), \tag{1}$$

where $1 \leq c_i \leq n$ denotes the cluster where vertex i belongs, A is the adjacency matrix of G, d_i is the degree of vertex i and the δ-function is such that $\delta(i, j) = 1$ if $i = j$ and $\delta(i, j) = 0$ otherwise. Note that the above sum runs over all possible pairs of vertices. Therefore each edge is summed twice. If we split the sum in two terms,

$$\frac{1}{2|E|} \sum_{p,q \in V} A_{pq}\delta(c_p, c_q) \quad \text{and} \quad \frac{1}{2|E|} \sum_{p,q \in V} \frac{d_p d_q}{2|E|}\delta(c_p, c_q), \tag{2}$$

the first term is the fraction of edges that fall within the clusters and the second term is the expected fraction of edges within the clusters, if the edges were randomly distributed while respecting the vertices degrees. In particular, if the edges were randomly placed as mentioned, $d_p d_q / |E|$ is the probability of the existence of an edge between vertices p and q.

Thus, modularity measures the fraction of edges that connect vertices in the same component minus the expected value of the same quantity in a graph with the same components but random connections between the vertices [4]. Values near 1, the maximum value of Q, indicate strong community structure. Typically, values for graphs underlying common networks with known community structure are in the range from 0.3 to 0.7.

However, if $\mathcal{C}$ has overlapping clusters, the measure needs refinement because it was designed for non-overlapping clusters. In this work, we extend the definition of modularity by weighting edge contributions with the cluster overlap centrality of the vertices. The *cluster overlap centrality* ν of a vertex $q \in V$ is the number of clusters that contain q. It is a generalization of the clique overlap centrality proposed by Everett and Borgatti [20]. Therefore, we extend the modularity definition as

$$Q = \frac{1}{2|E|} \sum_{p,q \in V} \left[A_{pq} - \frac{d_p d_q}{2|E|} \right] \frac{\delta(c_p, c_q)}{\nu_p \nu_q}. \tag{3}$$

This definition of modularity is a particular case of a more general extension proposed recently by Nicosia *et al.* [22]. Notice that this definition is equivalent to equation 1 when the clustering does not contain overlapping clusters.

5   Experimental Evaluation

We studied several cover graphs for different values of w. In this section we present the experiments with $w = 0.5$, as they exemplify our approach and provide interesting insights with respect to semantic relations among queries. We built the filtered cover graph as described in Section 3. Then, we applied the clustering method described in Section 4 for different values of k and we computed the modularity score to evaluate the quality of clustering. The clustering obtained with $k = 266$ (Figure 2) has the highest modularity score, $Q = 0.667$, a value that indicates the existence of well defined clusters. We found that these clusters are induced by 67 URLs with a high number of queries, the biggest one having 7,974 queries. It is interesting to note that these 67 URLs are all multitopical web pages.

Since multitopical web pages usually introduce noise as they relate very different queries, we generate the graph without the above 67 URLs. We note that a previous method was proposed by Baeza-Yates and Tiberi [2] to remove multitopical URLs. It is interesting that, although different, both approaches reduce the number of edges to similar values. The filtered graph has less than 14 million edges and Figure 1 depicts the degree distribution before and after the

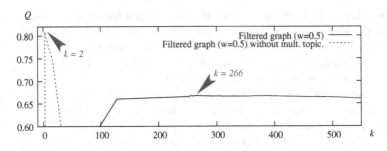

Fig. 2. Modularity score for different values of k

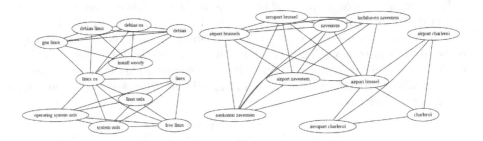

Fig. 3. Cluster intersections for queries "linux os" and "airport brussel"

URL removal. Both distributions follow power laws with coefficients of 1.51 and 1.70, respectively. We have applied the clustering method to this new graph and the results are rather interesting. The clustering with highest score, $Q = 0.809$, was obtained with $k = 2$ (Figure 2). This implies that the original graph structure was almost defined by a few large cliques and, after the multitopical URL removal, the result is a graph with a structure defined by cliques of size 3 (percolation of size 2 results from the intersection of cliques with more than 3 vertices). This clustering has 116, 044 clusters and covers 25% of the original queries. Note that 62% of the original queries are isolated vertices and, therefore, the clustering covers 66% of non singleton queries. More interesting, this clustering covers 79% of the queries in the giant component for the filtered graph with $w = 0.5$.

After obtaining this clustering, we examined the relations within and among clusters. Figure 3 contains two examples of the observed clusters and cluster intersections. In the first example the query "linux os" appears in two contexts: one related to the Debian distribution and the other to Unix operating systems. In the second example we see that the technique can find related queries even in different languages (English and Flemish). By analyzing the clusters we were able to find different contexts for given queries. We were also able to find equivalences between queries, e.g. "ge" and "general electrics". However, it would be interesting to have some method to classify queries and terms in the clusters and extract relevant semantic information as in [2].

For this, we have processed the clusters as follows. First we enumerated the terms for each cluster and then we computed the tf-idf score for each term. We

Table 2. Cluster tf-idf average, maximum and terms with tf-idf score ≥ 0.5

C.id	tf-idf	Max	|C|	Terms
1	5.658	5.831	3	synaptics, touchpad
2	2.374	2.569	4	erricson, ericcson, ericson, ericsson
3	2.336	4.998	4	charleroi, airport, brussel, aeroport
4	1.612	5.282	6	velux, skylite, www.velux.com, skylight, windows
5	1.423	2.881	6	debian, linux, os, gnu, woody, install
6	1.369	3.588	7	zaventem, brussel, airport, luchthaven, aankomst
7	1.329	2.359	3	linux, os, xp, servers
8	1.327	2.991	3	hfs, whfs, 99.1, whfstival, 105.7, dc
9	1.099	2.691	5	slackware, linux, kernel, slackware.com, 2.6, 9.0
10	1.024	2.143	6	unix, linux, linex, system, operating, os
11	1.019	2.778	5	longhorn, windows, screenshots
12	0.899	2.359	3	linux, wine, windows
13	0.848	2.105	4	cooking, wine, recipes, good, food
14	0.757	1.616	9	longhorn, steakhouse, horn, long, steak
15	0.662	1.262	5	spirits, liquor, pa, wine, pawineandspirits.com
16	0.591	1.727	7	baseball, longhorn, texas
17	0.585	1.054	7	union, credit, federal, teachers, teacher, tfcu
18	0.464	1.994	8	scorpios, scorpio, meaning, sign
19	0.091	0.970	51	windows, xp
20	0.072	0.838	76	delta, flights
21	0.069	0.594	104	yahooligans, games, kids

ranked the clusters by tf-idf average and, for each cluster, we also ranked the terms by tf-idf score. In Table 2 we provide tf-idf values for some clusters.

The analysis of the cluster ranking provides interesting information with respect to the nature of the queries. Clusters with navigational queries appear at the top of the ranking. Usually these queries have few terms, although very informative. The top ten clusters in Table 2 are examples of clusters with navigational queries, e.g., "synaptics touchpad" that is one of the 3 queries in the first cluster or "www.velux.com" that appears as term and as query in the second cluster. But we can generalize this analysis and verify that when a user knows what to search the queries appear in a higher ranked cluster. An example is the query "install woody" in Figure 3 and also in the fifth cluster of Table 2. Although this query does not refer Debian or Linux, it clearly refers to the installation of the Debian Linux distribution named Woody. Note that navigational queries are also an example of queries where the user knows what he wants.

The importance of a given term within a cluster can be inferred from the ranking of terms for each cluster. In Table 2 the terms are sorted from left to right, with the maximum tf-idf being the score of the first term. Thus, those terms are the most relevant in each cluster and can be seen as a description of the cluster.

By studying the overlap of terms among clusters, we can improve the semantic information obtained from cluster overlapping and we are able to identify context

Table 3. Distribution of queries with respect to ODP score

σ	≥ 0.25	≥ 0.33	≥ 0.50	≥ 0.66	≥ 0.75	$= 1.00$
%	90.5	85.6	69.7	56.5	49.6	34.7

and polysemy. In Table 2 we can identify clusters where a given term appears in the same context and others where the context is different. For example, the term "windows" appears in cluster three with the usual meaning and it appears in the other clusters as the Microsoft operating system. The term "wine" is an example of polysemy. It appears in cluster eleven as the Windows api for Unix-like operating systems and in the other clusters as a beverage.

In spite of the above interesting results, evaluating the quality of semantic relations is difficult. We would like to know if, from the user perspective, the queries in each cluster are truly related and contain valuable semantic information. Thus, we systematically evaluated a sample of clusters against data from ODP[1]. The Open Directory is a comprehensive human-edited directory of the Web and is constructed and maintained by a community of volunteer editors.

We randomly selected $1,000$ clusters among the $116,044$ clusters obtained. Then, we submitted each query in these clusters to the ODP and we obtained a set of categories matches in form of paths between directories. Note that these categories are ordered by relevance. For instance, the query "airport brussel" would provide the category "Regional: Europe: Belgium: Transportation" as the most relevant. To measure the similarity between queries, we measure the similarity between categories [2]. Thus, given two queries, we select the two most similar categories d_1 and d_2 as provided by ODP. The similarity score σ is defined as follows

$$\sigma(d_1, d_2) = \frac{|\pi(d_1, d_2)|}{\max\{|d_1|, |d_2|\}}, \qquad (4)$$

where $\pi(d_1, d_2)$ is the longest common prefix and $|\cdot|$ is the directory path length. The ODP similarity for two queries is the value of equation 4 for the most similar categories between them, i.e., the maximum among all possible pairs of categories for those queries. For each cluster we computed the ODP score as average of ODP similarity over all pairs of queries for which ODP provides at least one category. Note that, for pairs with queries with 0 categories, equation 4 is undefined.

Although the similarity score has values from near 0 to 1, we get an average of 0.7 among the $1,000$ clusters and 35% of the clusters have a score of 1.0. We provide the ODP score distribution in Table 3. Thus, we can infer that the clusters found reflect relevant semantic relations. We also verify that small clusters usually have higher values, which is to be expected, given the focus of their queries as discussed above. But, as depicted in Figure 4, the cluster ranking is much different from the tf-idf ranking and the score decreases much slower. It

---

[1] http://www.dmoz.org/

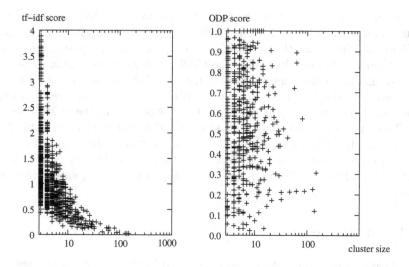

Fig. 4. Tf-idf and ODP scores for different cluster sizes

Table 4. Cluster ODP scores and relevant categories

C.id	Score	$|C|$	ODP category (most significant)
1	1.000	3	Computers: Software: Operating Systems: Linux: Hardware ...
3	1.000	4	Regional: Europe: Belgium: Transportation
11	1.000	5	Computers: Software: Operating Systems: Microsoft Windows: ...
9	1.000	5	Computers: Software: Operating Systems: Linux: Distributions: ...
4	1.000	6	Business: Construction and Maintenance: Materials and Supplies: ...
17	0.976	7	Business: Financial Services: Banking Services: Credit Unions
5	0.889	6	Computers: Software: Operating Systems: Linux: Projects: ...
13	0.722	4	Home: Cooking
16	0.722	7	Sports: Baseball: College and University: NJCAA
10	0.629	6	Computers: Software: Operating Systems: Linux
19	0.621	51	Computers: Software: Operating Systems: Microsoft Windows: ...
12	0.600	3	Computers: Emulators: Intel x86 Architecture: DOS and Windows
15	0.535	5	Recreation: Food: Drink: Liquor
21	0.477	104	Kids and Teens: Games: Online: Collections
7	0.467	3	Computers: Software: Shareware: Networking
14	0.443	9	Business: Hospitality: Restaurant Chains: Steakhouses
6	0.400	7	Regional: Europe: Belgium: Regions: Brussels: Travel ...
20	0.312	76	Recreation: Aviation: Experience Flights
8	0.056	3	Arts: Radio: Regional: North America: United States: Maryland
2	0.048	4	Regional: ...
18	0.024	8	Society: Religion and Spirituality: Divination: Astrology: ...

is also important to note that there are small clusters with very low tf-idf and ODP score. This happens because these clusters usually have few queries with many terms and usually these are more specific queries.

It is also interesting that, with the ODP score we were able to evaluate large clusters and distinguish which ones have more relevant information, as shown in Figure 4. Note that the score is not correlated to the cluster size and is better than tf-idf similarity. With the tf-idf score we were unable to perform this discrimination since larger clusters have smaller scores. In fact, if we also had considered the best ranked terms by tf-idf in each cluster, we would have improved the ranking of some larger clusters (although not the ODP score).

In Table 4 we provide the obtained scores for the clusters given in Table 2. We also provide the ODP category obtained with the two or three most significant terms for each cluster given in Table 2. Note that this is done by combining the tf-idf score and the ODP score, thus providing an interesting categorization of clusters.

6   Final Remarks

The efficient graph mining techniques proposed in this paper can be applied to very large graphs obtained from query logs, and the results show that these techniques are effective at obtaining semantic relations between queries. In the concrete case studied, the semantic relations discovered are useful and have provided interesting insights about implicit knowledge contained in queries submitted to a search engine.

The quality of the results can be improved by incorporating more data, i.e., by using larger logs, since more data will consolidate the relations obtained. Further analysis of the structure of the graph will also be important to unveil more relations. One possibility is to extend the similarity analysis to all the clusters and use edges with $w < 0.5$ to find weaker relations among the clusters to infer a possible taxonomy. The efficiency of the proposed methods makes them applicable to much larger graphs, thus making them useful for the extraction of semantic relations from less frequent queries.

References

1. Surowiecki, J.: The Wisdom of Crowds: Why the Many are Smarter than the Few and How Collective Wisdom Shapes Business, Economies, Societies and Nations. Little and Brown (2004)
2. Baeza-Yates, R.A., Tiberi, A.: Extracting semantic relations from query logs. In: Proc. of the 13th ACM SIGKDD International Conference on Knowledge Discovery and Data Mining, pp. 76–85. ACM, New York (2007)
3. Palla, G., Dernyi, I., Farkas, I., Vicsek, T.: Uncovering the overlapping community structure of complex networks in nature and society. Nature 435 (2005)
4. Newman, M.E.J., Girvan, M.: Finding and evaluating community structure in networks. Physical Review Letters 69 (2004)
5. Newman, M.E.J.: Modularity and community structure in networks. Proc. of the National Academy of Sciences 103 (2006)
6. Wen, J., Mie, J., Zhang, H.: Clustering user queries of a search engine. In: Proc. of the 10th International World Wide Web Conference, W3C (2001)

7. Baeza-Yates, R.: Applications of web query mining. In: Losada, D.E., Fernández-Luna, J.M. (eds.) ECIR 2005, vol. 3408, pp. 7–22. Springer, Heidelberg (2005)
8. Beeferman, D., Berger, A.: Agglomerative clustering of a search engine query log. In: Proc. of the 5th ACM SIGKDD International Conference on Knowledge Discovery and Data Mining. ACM, New York (1999)
9. Zaiane, O.R., Strilets, A.: Finding similar queries to satisfy searches based on query traces. In: Proc. of the International Workshop on Efficient Web-Based Information Systems (EWIS) (2002)
10. Baeza-Yates, R., Hurtado, C., Mendoza, M.: Query clustering for boosting web page ranking. In: Favela, J., Menasalvas, E., Chávez, E. (eds.) AWIC 2004. LNCS (LNAI), vol. 3034, pp. 164–175. Springer, Heidelberg (2004)
11. Baeza-Yates, R., Hurtado, C., Mendoza, M.: Query recommendation using query logs in a search engine. In: Lindner, W., Mesiti, M., Türker, C., Tzitzikas, Y., Vakali, A.I. (eds.) EDBT 2004, vol. 3268, pp. 588–596. Springer, Heidelberg (2004)
12. Chuang, S.-L., Chien, L.-F.: Automatic query taxonomy generation for information retrieval applications. Online Information Review 27(5) (2003)
13. Chuang, S.L., Chien, L.F.: Enriching web taxonomies through subject categorization of query terms from search engine logs. Decision Support System 30(1) (2003)
14. Chuang, S.L., Chien, L.F.: Towards automatic generation of query taxonomy: A hierarchical query clustering approach. In: Proc. of the 2002 IEEE International Conference on Data Mining. IEEE, Los Alamitos (2002)
15. Pu, H.T., Chuang, S.L., Yang, C.: Subject categorization of query terms for exploring web users' search interests. Journal of the American Society for Information Science and Technology (JASIST) 53(8) (2002)
16. Cheng, P.J., Tsai, C.H., Hung, C.M., Chien, L.F.: Query taxonomy generation for web search (poster). In: CIKM (2006)
17. Dupret, G., Mendoza, M.: Automatic query recommendation using click-through data. In: Symposium on Profesional Practice in Artificial Intelligence, 19th IFIP World Computer Congress, WCC 2006. Springer, Heidelberg (2006)
18. Wen, J., Mie, J., Zhang, H.: Clustering user queries of a search engine. In: Proc. of the 10th international conference on World Wide Web, pp. 162–168. ACM, New York (2001)
19. Baeza-Yates, R.A.: Graphs from search engine queries. In: van Leeuwen, J., Italiano, G.F., van der Hoek, W., Meinel, C., Sack, H., Plášil, F. (eds.) SOFSEM 2007. LNCS, vol. 4362, pp. 1–8. Springer, Heidelberg (2007)
20. Everett, M.G., Borgatti, S.P.: Analyzing clique overlap. Connections 21 (1998)
21. Karp, R.: Reducibility among combinatorial problems. In: Proc. of a Symposium on the Complexity of Computer Computations, Plenum Press (1972)
22. Nicosia, V., Mangioni, G., Carchiolo, V., Malgeri, M.: Extending modularity definition for directed graphs with overlapping communities (2008) arXiv.org:0801.1647

Out of the Box Phrase Indexing

Frederik Transier[1,2] and Peter Sanders[2]

[1] SAP NetWeaver EIM TREX, SAP AG, Walldorf, Germany
[2] University of Karlsruhe, Karlsruhe, Germany
{transier,sanders}@ira.uka.de

Abstract. We present a method for optimizing phrase search based on inverted indexes. Our approach adds selected (two-term) phrases to an existing index. Whereas competing approaches are often based on the analysis of query logs, our approach works out of the box and uses only the information contained in the index. Also, our method is competitive in terms of query performance and can even improve on other approaches for difficult queries. Moreover, our approach gives performance guarantees for arbitrary queries. Further, we propose using a phrase index as a substitute for the positional index of an in-memory search engine working with short documents. We support our conclusions with experiments using a high-performance main-memory search engine. We also give evidence that classical disk based systems can profit from our approach.

1 Introduction

Searching in huge collections of text documents is a pervasive feature of modern life. Many search engines are deployed to execute queries on collections of documents or records taken from the web, from personal computers, or from large corporate intranets and databases. Typically, users of search engines have several options for defining their queries. As well as Boolean operators like *AND*, *OR* and *NOT*, they can enter their queries as short phrases. The advantage of entering a phrase is that it specifies certain relations between the words in the phrase, and this constrains the result set more narrowly than individual search terms can do.

Examining query logs, one observes that often *hype words* appear for a while and quickly disappear again [1]. This dynamical aspect of user behavior regarding search terms may be unpredictable and makes it difficult to optimize search engines for (phrase) queries.

The most popular data structure for text search engines is the inverted index (see, e.g., [2]). Use of an inverted index to answer Boolean queries containing single terms is efficient, but phrase queries are computationally more expensive. Attempts have been made to speed up phrase queries on inverted indexes by adding further information to their data structures (see Section 2). These attempts have the disadvantage that they consume more space. This disadvantage is especially critical for high-performance main-memory search engines. Although

A. Amir, A. Turpin, and A. Moffat (Eds.): SPIRE 2008, LNCS 5280, pp. 200–211, 2008.

modern hardware configurations feature ever more random access memory, memory capacity is still significantly more restricted and more expensive than disk capacity in classical search engine systems.

In this paper, we present a method for selecting phrases to add to an existing inverted index. Whereas competing approaches have often been based on the analysis of query logs, our approach works out of the box and uses only the information contained in the index of single terms. Our approach has the advantages over the other approaches that query logs (a) are not available for many real-world applications and (b) are just a snapshot of user behavior, which often changes unpredictably. Despite this, our method is competitive in terms of query performance and can even improve on the other approaches for *difficult* queries. Also, our approach enables us to derive theoretical performance guarantees for arbitrary queries. Our main focus is on search engines that hold all their data structures in main memory, but we give evidence that our results can be applied to classical disk-based systems as well.

The remainder of the paper is organized as follows. We give an overview of related work in Section 2. Section 3 describes our phrase indexing scheme. Section 4 gives an experimental evaluation using real-world benchmark data. Section 5 summarizes our results and outlines possible future work.

2 Related Work

Decades ago, researchers began to create inverted indexes by indexing phrases instead of single terms (see, e.g., [3]). But building indexes that cover all phrases within a (large) document collection takes enormous amounts of space and time. So authors designed a variety of methods for selecting (short) phrases. All of these approaches focus on choosing phrases that are as *meaningful* as possible. They use syntactical and statistical analyses of a given text to extract suitable candidate phrases for the index [3,4,5]. The main aim of such analyses is to improve the retrieval quality in the absence of a full-text index.

More recently, ways have been found to speed up phrase query times for (full-text) inverted indexes. Williams et al. [6] proposed *nextword indexes*, in which for each term or *firstword*, a list of all successors is stored together with the positions at which they occur as a consecutive pair. With this approach, phrase queries can be answered four times faster than by a classical inverted index. However, this technique requires an additional 50% to 60% of the space occupied by a standard inverted index. This can be reduced to about 40% to 50% using the compression techniques of [7] at little cost in query performance. To achieve further reductions in memory consumption, Bahle et al. introduced *partial nextword indexes* [8]. A partial nextword index contains only the most common words as firstwords. It can be used in combination with a conventional inverted index acting as fall-back for query terms that are not listed in the nextword index. A partial nextword index enables phrase queries to be evaluated in half the time using 10% more space compared to an inverted index.

A combination of partial nextword index, partial phrase index, and inverted index was proposed in [9]. An existing query log is analyzed to optimize the indexes. This reduces the average query time to a quarter by indexing the three most common words as firstwords and the 10 000 most common phrase queries. The space overhead for the *three-way index* combination is about 26%.

Another method, based on the analysis of query logs, was introduced by Chang and Poon [10]. Their approach divides the vocabulary into two sets: *rare* terms and *common* terms. Common terms are the most frequent terms found in the query log; all others are rare terms. For each common term, there is a tree whose root-to-leaf paths contain all phrases from the query log starting with that term and ending with a *terminal* term. Terminal terms are defined by a lexical analysis of the query log: any term that is not a preposition, adverb, conjunction, article, or pronoun is a terminal term. Each leaf points to the inverted list for its phrase. The resulting *common phrase index* shows 5% improvement in average query time over a partial nextword index with only a 1% extra storage cost. For long query phrases the query time improvement can reach 20%.

All these approaches are designed for systems that hold major parts of their indexes on disks. So the cited values apply for disk-based systems.

Another index data structure held mostly in main memory and capable of answering phrase queries efficiently is the suffix array. The array stores all suffixes of a given text and can therefore answer queries consisting of concatenated terms very quickly. Also, suffix arrays can be efficiently compressed. However, they cannot compete with inverted indexes in classical document-based search scenarios at present [11,12].

3   Two-Term Phrase Indexes

We propose a phrase selection scheme confined to a subset of the two-term phrases contained in a document collection to be indexed. Two-term phrases are the most widely used form of text queries [13]. They are also the slowest to resolve [11]. Once they are in the index, they can be used to speed up queries of arbitrary length greater than two [14]. We can easily generalize our approach to phrases consisting of more than two terms.

The idea behind our indexing scheme is simple. We build up an inverted index I and independently make a list of the two-term phrases that occur. We define a function $C((s,t))$ estimating the real costs caused by evaluating a term pair query $s \cdot t$ through I. We require C to be determined by properties that are also available at query time. We choose a suitable threshold T_I and add just the phrases $s \cdot t$ such that $C((s,t)) \geq T_I$ to the index I.

We can exploit this to speed up the handling of some phrase queries with empty result sets. We estimate the cost of each two-term phrase occurring in the query and check whether it is above the threshold for the index. If it is, and if there are results for the phrase, the phrase is in the index and we return the results. Otherwise, we know the result set is empty and we are done.

3.1   Selection Criteria

There are various options to define suitable cost functions for a phrase $s \cdot t$. One approach would be to take the sum of the lengths of their inverted lists $C((s,t)) = \|s\| + \|t\|$ as this reflects the amount of memory to be retrieved from an external storage for intersecting them. However, on main-memory systems the minimum $C((s,t)) = \min\{\|s\|, \|t\|\}$ may be more closely related to the cost of the intersection when a form of *per-list* indexes is used (see, e.g., [15]). Other criteria can be used. For example, the phrase query is often implemented as a (fast) AND query followed by a check for consecutive term positions. In this case, a cost function can be defined as the number of entries in the result set of the AND query. Queries that do not profit from the phrase index can be still evaluated in nearly the same time.

Our experiments show that these selection methods differ only marginally in average run-time. However, the sum-based cost function is of theoretical importance.

Proposition 1 (Performance guarantee). *Let I be a two-term phrase index and $C((s,t)) = \|s\| + \|t\| \geq T_I$ its selection criterion. Then, the cost of a two-term query $t_1 \cdot t_2$ on I is $O(T_I)$.*

3.2   Querying Phrases

The two-term phrase index can be used to speed up queries consisting of an arbitrary number of terms. Phrase queries can be evaluated efficiently, as follows [14,8]. Let $t_1 \cdot t_2 \cdot ... \cdot t_n$ be a phrase query consisting of n terms and $l_1, l_2, ..., l_n$ their inverted lists. Starting at l_1, we replace each list l_i with the inverted list of the phrase $t_i \cdot t_{i+1}$ where possible, because $\|t_i \cdot t_{i+1}\| \leq \|t_i\|$. Then, we sort the resulting set of inverted lists by increasing length. Taking the smallest list as the initial result set, the others are intersected (taking phrase offsets into account) with it successively. We skip inverted lists of two-term phrases that are already covered by previous lists during the intersections.

3.3   A Substitute for Positional Indexes

Positional indexes for text collections with small documents consisting of just a few words are not very space efficient [11]. However, business application scenarios often involve huge amounts of short business texts like memos or product descriptions that require fast access. With such scenarios in mind, we propose an alternative index design adapted to the characteristics of small text documents. Instead of storing the position of each term in a positional index, we store the sequence of term IDs for each document. Our experiments show that for small documents this requires less space than a positional index. Note that in scenarios where the search system has to answer its queries with actual document contents, this information is required anyway. A phrase query can then be implemented in two steps. In step 1, preselect possible documents by executing an AND query. In step 2, select the results from them by using the string matching algorithm of [16] on their term ID lists. Because this implementation is generally slower than evaluating a phrase query using a positional index, we add a phrase index that makes the preselection process more precise. This gives a better trade-off between compression and query performance (see Section 4). We call this modified index design a *short-text index*.

4   Experiments

We evaluated the performance of a two-term phrase index against the query log based three-way index combination of [9] mentioned in Section 2. To fill our indexes, we used the GOV2 [17] test collection from TREC. This collection is a crawl of a large proportion of the websites from the .gov domain in early 2004. This corpus contains 25 million documents filling 426 GB.

We used four different query logs. The first two consist of the queries from the TREC Efficiency task topics from the years 2005 [18] and 2006 [19]. We call them 05eff and 06eff respectively. They were extracted from a real-world query log to compare the performance of different search engine systems on the GOV2 test collection. In fact, the 2005 query log does not match GOV2 very well, so it can be processed much more quickly than more realistic queries because it contains many terms that occur either rarely or not at all in GOV2 (or in the documents of the .gov domain). For this reason, the 2006 queries were selected more carefully to fit the data [20,21].

For our third query log, we used the queries from the Excite log [1]. We extracted queries that were explicitly indicated as phrases by means of quotation marks. Again, these queries were originally addressed to the whole web and not just to the sites of the .gov domain, so this query log is also a rather unrealistic sample. Indeed, there are many very selective queries on the GOV2 corpus with empty or very small result sets that can be processed very quickly.

Finally, the Random query log was built using the pseudo real-world query generation method of [11]. We selected random hits from the GOV2 test corpus varying in their length of between two and ten terms. This can be seen as a counterweight to the two easy query logs 05eff and Excite. The algorithm selects many queries containing very frequent terms so that the result sets are likely to be larger than they would be in real life.

The three-way index combination needs a training set of queries. Hence, we split all the query logs into two parts and used the first one for training purposes. All our measurements were made on the second parts.

4.1   Results for Main-Memory Systems

We implemented the two-term phrase index and the three-way index combination as an extension to the main-memory search engine of [11] using C++. All our code was compiled by the gnu C++ compiler 4.1.2 with optimization level -O3. Timing was done using PAPI 3.5.0 [22]. The experiments were done on one core of an Intel Core 2 Duo E6600 processor clocked at 2.4 GHz with 4 GB main memory and 2 × 2 MB L2 cache, running OpenSuSE 10.2 (kernel 2.6.18).

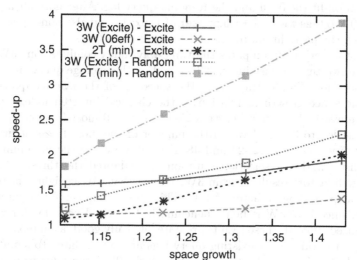

(a) Excite and Random logs on different index configurations

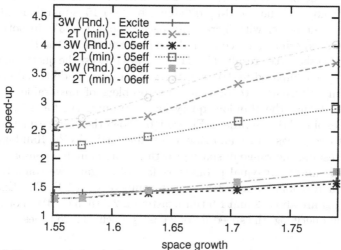

(b) Two-term index (2T) vs. Random-trained three-way index combination (3W)

Fig. 1. Average query time

We split the GOV2 corpus into 64 roughly equal-sized and contiguous parts to obtain the amount of data normally assigned to a single processing core of an in-memory text search engine based on a cluster of low-cost machines. Assigning more than 1–2 GB of data to one core would skew hardware investment too heavily toward RAM rather than processing power. For our measurements, we randomly selected one of the parts. It would have been much more expensive to average over all the parts, and the sizes of the result sets suggested that the chosen portion fit the 06eff and the Random query log. After normalization (removing HTML tags with the HTMLparser of libxml2 and also non-alphanumeric characters), the part shrank to about 1.8 GB of plain text.

Our first experiment compares the three-way index combination (3W) from [9] and the two-term phrase index (2T) based on the average query time for the Excite query log. Figure 1(a) shows the speed-up of the average query times against the space growth in relation to the classical inverted index, indexing only single words. Following [9], we added the 10 000 most frequent queries of the training log to 3W and varied the number of the most frequent firstwords between the values 3, 6, 12, 24 and 48. For 2T, we chose the minimum of the inverted list lengths as selection criterion and adapted the threshold for the space growth factors resulting from 3W. We see that 3W is slightly faster than 2T up to a space growth factor of 1.38. This observation should be interpreted with care, since here 3W is being benchmarked on the Excite log for which it was optimized. Indeed, these results are unrealistically good, as in most cases no query logs are available or existing query logs are out of date. To simulate the more usual situation, we also trained 3W with the 06eff query log instead of the Excite log and compared the outcomes. As Figure 1(a) shows, this transforms the outcome. The intersection with the 2T Excite line slips down to about 1.15 on the space axis and the speed-up difference below 1.15 is marginal. Moreover, 2T seems to scale better with increasing space factors than 3W in both cases.

In a second experiment we compared the performance on the Random query log. Figure 1(a) shows the result. The 2T index can process the query log in about 70% of the time compared to 3W over the entire range of additional memory.

A tempting alternative to overcome the problem of unsuitable or missing query logs is to use the Random query log as training set for 3W. Our next experiment explored this approach. Figure 1(b) shows the average querying times for different query logs on 2T and Random-trained 3W. As the Random queries yield more results, the range of space growth factors is higher compared to our previous experiments at equal parameters for 3W. Again, we come to a clear conclusion: a Random-trained 3W cannot compete with 2T. The Excite and 05eff query log are about 2 and 1.6 times faster on 2T than on 3W. For the more difficult 06eff query log, the speed-up of 2T is as much as 2.25 times higher than that of 3W.

Indexes for Small Documents. We also implemented the *short-text index* of Section 3.3 and used the two-term phrase index as a substitute for the positional information within the index. This somewhat unconventional application of two-term phrase indexes can reduce the memory consumption of an index.

For these experiments, we used the WT2g.s test collection introduced in [15]). It contains short documents and is derived from the WT2g corpus [23] by interpreting each sentence of the plain text as a single document. The total size of the text of WT2g.s is about 1.5 GB. The average document size of this test collection is 25.6 terms. However, about 2000 of its more than 10 million documents were much larger than that. They occupied more than 30 MB in total, so we truncated them to a size of 1024 terms each.

Table 1. Plain positional index vs. plain phrase index on WT2g.s

	pos. index	phrase index				no phrases
size [KB]	1 460 683	×1.0	×0.9	×0.8	×0.7	×0.6
avg. [μs] 05eff	614	186	260	386	592	987
06eff	687	120	178	296	487	804
Excite	1 193	298	475	765	1 179	2 200
Random	23 985	943	1 249	2 097	5 062	63 118
w.-c. [μs] 05eff	133 715	125 715	125 707	124 203	125 721	424 127
06eff	88 123	13 445	19 181	67 683	81 101	366 231
Excite	211 793	17 699	22 039	45 536	108 050	697 359
Random	1 917 584	109 618	109 575	109 605	178 414	4 276 635

We compared the short-text index with the classical positional index. Table 1 shows the average query times as well as the worst-case query times for different query logs on both indexes. The index size given for the positional index contains *bags of words* for each document. The bags are sorted lists of the term IDs for each document (without duplicates). They are used in the implementation of [11] for reconstructing document contents from the indexes. But if we add term ID sequences to the index as described in Section 3.3, the bags become superfluous. So all other size values contain just the space needed for these sequences. The short-text index without any additional phrases needs about 60% of the size of the positional index. We added phrases to the short-text index and measured the query log running times at different thresholds. As Table 1 shows, if we use about 17% more memory for the phrases, the short-text index performs better than the positional index.

4.2   Evaluation for External Memory Systems

For our tests, we did not have a disk-based external memory implementation of our search engine. However, to get an idea of performance on such systems, we measured the memory size of the inverted lists that were accessed for each query. We believe that this is a good estimation for the performance of disk-based systems as it reflects the costs for retrieving the lists from external storage.

The phrase index construction can run as follows. Write all term pairs during the standard indexing process to disk. Iterate through them and check if a phrase has to be added. If so, look up the phrase in the index. If the phrase is not yet present, do the search and add the outcomes.

We report results that correspond to those of our main-memory experiments. As cost function for 2T, we used the sum of the sizes of the inverted lists for the two terms. Figure 2(a) shows the average amount of memory touched during query execution. Again, we normalized all values against those of the classical inverted index. First of all, we can see that a 06eff-trained 3W loses by a constant factor compared to an Excite-trained 3W. The characteristics are similar to those we observed in the main-memory case. However, even the worse-trained 3W can process the (easy) Excite query log with access to less memory than 2T can. But

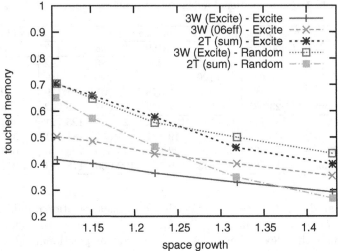

(a) Excite and Random logs on different index configurations

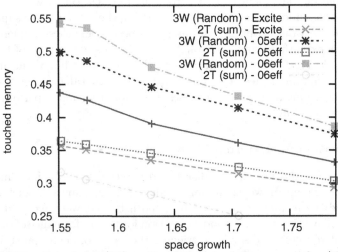

(b) Two-term index (2T) vs. Random-trained three-way index (3W)

Fig. 2. Average memory per query required to retrieve from external storage

the advantage diminishes as the space growth factor increases. For the Random log, we see again that 2T beats 3W. Also, Figure 2(b) shows that 2T touches less memory than a Random-trained 3W for all query logs.

Because we were aware that the amount of data used in our previous experiments was unrealistically small for external memory search engines, we performed additional experiments with larger data sets. Figure 3 shows the amount of touched memory for data sets of different sizes using about 22% of additional

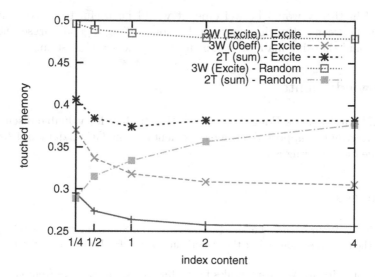

Fig. 3. Touched memory for different data set sizes

memory for both 2T and 3W. We indexed 25%, 50%, and 200% of the amount of data considered in Figure 2(a). For smaller inputs, we averaged over all pieces of the largest input. For the Excite query log, each curve converges to a flat line that is independent of the input size. So the conclusions for small inputs seem to remain valid for larger inputs.

5 Conclusion

Our experiments have shown that our two-term phrase indexing scheme can be used to achieve significant query time speed-up without the assistance of query logs. The run times of more expensive and difficult queries are halved when 13% more space is made available for the indexes on our main-memory search engine. Given that amount of space, an approach based on query logs can only achieve a run-time reduction of about 10%. Approaches based on query logs admittedly perform better on simple and very selective queries. However, we believe that it is more important to speed up difficult queries that fit the data than to reduce the costs of queries that are fast anyway, and this is suggested by our results.

In systems based on external memory, the reduction of query costs is even higher. On this point, our estimations for those systems coincide with previous work. For such systems, we arrive at similar conclusions as for main-memory systems. Approaches based on query logs can process easy queries slightly more quickly, but our two-term phrase index is clearly faster for difficult queries.

It would be interesting to try our approach on a disk-based implementation of a search engine to avoid relying on approximate estimations. Other interesting future work would be to investigate the query evaluation heuristics of Section 3.2. The previous work [14] is a useful experimental study but work grounded in theory would be welcome. It would be useful to explore the dependency between optimizing query evaluation, finding a good phrase selection criterion, and choosing main or disk memory for the search system.

Acknowledgments

We would like to thank Franz Färber and Holger Bast for valuable discussions. Markus Hartmann supported our measurements on the GOV2 data and Andrew Ross tidied up our English.

References

1. Spink, A., Wolfram, D., Jansen, B., Saracevic, T.: Searching the web: The public and their queries. Journal of the American Society for Information Science 52(3) (2001)
2. Zobel, J., Moffat, A.: Inverted files for text search engines. ACM Computing Surveys 38(2) (2006)
3. Salton, G., Yang, C.S., Yu, C.T.: A theory of term importance in automatic text analysis. Journal of the American Society for Information Science 26 (1975)
4. Fagan, J.: Automatic phrase indexing for document retrieval. In: Proceedings of the 10th Annual International Conference on Research and Development in Information Retrieval SIGIR 1987. ACM, New York (1987)
5. Gutwin, C., Paynter, G., Witten, I., Nevill-Manning, C., Frank, E.: Improving browsing in digital libraries with keyphrase indexes. Decision Support Systems 27(1-2) (1999)
6. Williams, H.E., Zobel, J., Anderson, P.: What's next? - index structures for efficient phrase querying. In: Proceedings of the 10th Australasian Database Conference ADC 1999. Springer, Heidelberg (1999)
7. Bahle, D., Williams, H.E., Zobel, J.: Compaction techniques for nextword indexes. In: Proceedings of the Symposium on String Processing and Information Retrieval SPIRE 2001. IEEE Computer Society, Los Alamitos (2001)
8. Bahle, D., Williams, H.E., Zobel, J.: Efficient phrase querying with an auxiliary index. In: Proceedings of the 25th Annual International Conference on Research and Development in Information Retrieval SIGIR 2002. ACM, New York (2002)
9. Williams, H.E., Zobel, J., Bahle, D.: Fast phrase querying with combined indexes. ACM Transactions on Information Systems 22(4) (2004)
10. Chang, M., Poon, C.K.: Efficient phrase querying with common phrase index. In: Lalmas, M., MacFarlane, A., Rüger, S.M., Tombros, A., Tsikrika, T., Yavlinsky, A. (eds.) ECIR 2006. LNCS, vol. 3936, pp. 61–71. Springer, Heidelberg (2006)
11. Transier, F., Sanders, P.: Compressed inverted indexes for in-memory search engines. In: Proceedings of the 10th Workshop on Algorithm Engineering and Experiments ALENEX 2008. SIAM, Philadelphia (2008)
12. Puglisi, S.J., Smyth, W.F., Turpin, A.: Inverted files versus suffix arrays for locating patterns in primary memory. In: Crestani, F., Ferragina, P., Sanderson, M. (eds.) SPIRE 2006. LNCS, vol. 4209. Springer, Heidelberg (2006)
13. Jansen, B.J., Spink, A., Bateman, J., Saracevic, T.: Real life information retrieval: a study of user queries on the web. SIGIR Forum 32(1) (1998)

14. Bahle, D., Williams, H.E., Zobel, J.: Optimised phrase querying and browsing in text databases. In: Proceedings of the Australasian Computer Science Conference (2001)
15. Sanders, P., Transier, F.: Intersection in integer inverted indices. In: Proceedings of the 9th Workshop on Algorithm Engineering and Experiments ALENEX 2007. SIAM, Philadelphia (2007)
16. Knuth, D.E., Morris Jr., J.H., Pratt, V.R.: Fast pattern matching in strings. SIAM Journal on Computing 6(2) (1977)
17. Clarke, C., Soboroff, I., Craswell, N.: GOV2 test collection (2004), http://ir.dcs.gla.ac.uk/test_collections/gov2-summary.htm
18. Clarke, C., Scholer, F., Soboroff, I.: TREC 2005 efficiency topics (2005), http://trec.nist.gov/data/terabyte/05/05.efficiency_topics.gz
19. Büttcher, S., Clarke, C., Soboroff, I.: TREC 2006 efficiency topics (2006), http://trec.nist.gov/data/terabyte/06/06.efficiency_topics.tar.gz
20. Clarke, C., Scholer, F., Soboroff, I.: The TREC 2005 terabyte track (2005)
21. Büttcher, S., Clarke, C., Soboroff, I.: The TREC 2006 terabyte track (2006)
22. Mucci, P.: Performance API (2005), http://icl.cs.utk.edu/papi/
23. Hawking, D., Voorhees, E., Craswell, N., Bailey, P.: TREC-8 web track (1999), http://ir.dcs.gla.ac.uk/test_collections/access_to_data.html

Approximated Pattern Matching
with the L_1, L_2 and L_∞ Metrics

Ohad Lipsky and Ely Porat

Bar-Ilan University*

Abstract. Given an alphabet $\Sigma = \{1, 2, \ldots, |\Sigma|\}$ text string $T \in \Sigma^n$ and a pattern string $P \in \Sigma^m$, for each $i = 1, 2, \ldots, n - m + 1$ define $L_d(i)$ as the d-norm distance when the pattern is aligned below the text and starts at position i of the text. The problem of pattern matching with L_p distance is to compute $L_p(i)$ for every $i = 1, 2, \ldots, n - m + 1$. We discuss the problem for $d = 1, \infty$. First, in the case of L_1 matching (pattern matching with an L_1 distance) we present an algorithm that approximates the L_1 matching up to a factor of $1 + \varepsilon$, which has an $O(\frac{1}{\varepsilon^2}n\log m log|\Sigma|)$ run time. Second, we provide an algorithm that approximates the L_∞ matching up to a factor of $1 + \varepsilon$ with a run time of $O(\frac{1}{\varepsilon}n\log m log|\Sigma|)$. We also generalize the problem of String Matching with mismatches to have weighted mismatches and present an $O(n\log^4 m)$ algorithm that approximates the results of this problem up to a factor of $O(\log m)$ in the case that the weight function is a metric.

1 Introduction

The last few decades have prompted the evolution of pattern matching from a combinatorial solution of the exact string matching problem [FP74, KMP77] to an area concerned with approximate matching of various relationships motivated by a wide range of scientific and business applications. To this end, two new paradigms were needed - *"Generalized matching"* and *"Approximate matching"*.

In generalized matching the input is still a text and pattern but the *"matching"* relation is defined differently. The output is all locations in the text where the pattern *"matches"* under the new definition of *"match"*. The different applications define the matching relation. An early generalized matching was the string matching with don't cares defined by Fischer and Paterson [FP74]. Another example of a generalized matching problem is the less-than matching [AF95] problem defined by Amir and Farach. In this problem both texts and patterns are numbers. All text locations are sought where every pattern number is less than its corresponding text number. Amir and Farach showed that the less-than-matching problem can be solved in $O(n\sqrt{m\log m})$ time. More examples of generalized matching are parameterized matching [AFM94] and shift matching [CH02]. Lower bound results on generalized matching can be found in [MR95].

---

* Research supported in part by US-Israel Binational Science Foundation.

A. Amir, A. Turpin, and A. Moffat (Eds.): SPIRE 2008, LNCS 5280, pp. 212–223, 2008.

The second important pattern matching paradigm and the one with which we are concerned this paper is approximate matching. In approximate matching the distance metric between the objects (e.g. strings, matrices) is defined and all locations in the text are sought where the distances between them and the patterns are within the tolerated bounds, under the given distance function. Some examples that motivate approximate matching include finding a "close" mutation in computational biology, adjusting noise transmission in communications, allowing common typing errors in texts and adjusting for lossy compressions, occlusions, scaling, affine transformations or dimension loss in multimedia.

A well known distance function concerning strings of characters is the Hamming distance. The Hamming distance is defined as the number of character changes needed to convert one string to the other. The string matching with mismatches is defined as computing the Hamming distance between the pattern and every substring of the same length of the text. By applying methods similar to those of Fischer and Paterson [FP74] we can show that the string matching with mismatches problem can be solved in $O(\min(|\Sigma|, m)n \log m)$ time. For given finite alphabets the time needed is $O(n \log m)$. Abrahamson [Abr87] developed an algorithm that solves this problem for general alphabets in $O(n\sqrt{m \log m})$ time. It is an open question whether this bound is tight or it can be reduced. Karloff [Kar93] introduced an algorithm that approximate the Hamming distance up to a factor of $1 + \epsilon$ and works in time $O(\frac{1}{\epsilon^2} n \log^3 m)$.

Hamming distance gives a mismatch error a fixed weight, independent of the characters that were exchanged. We discuss a generalization for the problem of String Matching with Mismatches. The problem of *String Matching with Weighted Mismatches* defined in [Mut95] is informally, summing up, for each text location i, the weighted mismatches that occur when the pattern is compared to $t_i, t_{i+1}, \ldots, t_{i+m-1}$. We show that the String Matching with Weighted Mismatches can be solved deterministically in $O(n|\Sigma| \log m)$ time, and can be approximated using techniques of embedding up to a factor of $O(\log m)$, in a time of $\widetilde{O}(n)$, if the weight function is a metric.

In many aspects of science we deal with strings of numbers, which are also referred to as *time-series data*. In meteorology we measure temperature , air pressure, wind speed, etc. over time. In the Stock Market or in the Foreign Exchange market we keep a record of the stock price or the exchange rate over time. Another area which deals with such strings is Music Information Retrieval (MIR) [Per00]. If we wish to seek patterns in such strings of numbers, it would be non realistic to seek the exact same values, rather than "close" instances of a pattern. When dealing with strings of numbers the most common distance metrics are the L_1, L_2 and L_∞ Minkowsky norm metrics. The 1-norm distance is more colorfully called the taxicab norm or Manhattan distance, because it is the distance a car would drive in a city laid out in square blocks. The 2-norm distance is the famous Euclidean distance, which is a generalization of the Pythagorean theorem to more than one coordinates. It is what would be obtained if the distance between two points were measured with a ruler: the "intuitive" idea of distance. L_∞ is known as the Maximum metric.

Therefore, we propose new pattern matching problems: The string L_1 distance matching problem (L_1 matching, for short) to compute the L_1 distance between the pattern and every substring of the same length of the text. Similarly we define the String L_∞ Distance Matching (L_∞ Matching) problems.

The problem of searching for patterns among time-series data has been extensively studied for building time-series databases. In this case, the problem is pre-processing the data in such a way that will enable us to run fast queries for patterns [FRM94, CW99].Another variant is looking for some periodicity along the data [IKM00]. Most of the algorithms used for these variants apply L_2 as the distance function. Moreover, most of them are also based on heuristics, and do not have any "worse time" analysis.

In this paper we provide a deterministic algorithm for the L_1 matching problem, based on methods similar to those of Fischer and Paterson [FP74] which works in $O(\min(|\Sigma|, m)n \log m)$ time can be constructed according to a method similar to the one used in [ILLP04]. For given finite alphabets, the runtime is $O(n \log m)$, but for large alphabets (i.e. $|\Sigma| = n$) it is worse than the naive algorithm which works in $O(nm)$ time. This time complexity leads us to assume that the main difficulty in constructing an efficient algorithm (i.e. $\widetilde{O}(n)$) is with the alphabet size. Therefore, we have searched for a method to reduce the alphabet size. In [LP01] techniques for dividing the alphabet into frequent and non-frequent symbols, similar to those in [Abr87] provided an algorithm that works in $O(n\sqrt{m \log m})$ time. We show in [LP05] that the string matching with mismatches problem can be linearly reduced to the L_1 matching problem. This fact has led us to look for an approximation algorithm rather than an exact solution that seems to be difficult to construct in $\widetilde{O}(n)$ time. A randomized algorithm that approximates the L_1 matching results up to a factor of $1 + \varepsilon$ and an err with a probability bounded by δ can be constructed based on [Ind00], with a run time of $O(\frac{1}{\varepsilon^2}n \log \frac{1}{\delta} \log m)$. We present a deterministic algorithm for approximating L_1 which has a runtime of $O(\frac{1}{\varepsilon^2}n \log m \log |\Sigma|)$.

We also present an algorithm that approximates the L_∞ matching results up to a factor of $1 + \varepsilon$ and works, as well, in $O(\frac{1}{\varepsilon}n \log m \log |\Sigma|)$ time. All algorithms, except the approximation method for String Matching with Weighted Mismatches, are based on convolutions, and therefore can easily be extended to include *don't care* symbols within the alphabet.

This paper is organized as follows: section 2 provides definitions and preliminaries, section 3 presents the algorithms for string matching with weighted mismatches. Section 4 deals with the problem of L_1 Matching - presents an approximation algorithm for the L_1 matching problem. Finally section 5 discusses the problem of L_∞ Matching and presents an approximation algorithm for the problem.

2   Preliminaries and Problem Definition

All algorithms in this paper, except those in section 3, deal with alphabets of the form $\Sigma = \{1, 2, \ldots |\Sigma|\}$. We assume the RAM model of computation, which allows arithmetic on $\log N$ bit numbers in $O(1)$ time, where N is the order of the maximum problem size. In addition we use n to denote $|T|$ and m to denote $|P|$. We can assume w.l.o.g $n \leq 2m$, otherwise we can use the method of cutting the text into n/m overlapping segments, each with a length of $2m$.

Let $x = (x_1, x_2, \ldots, x_n)$ and $y = (y_1, y_2, \ldots, y_n)$ be two vectors over Σ. Then the Minkowsky L_d distance metric between x and y is defined as: $L_d(x, y) = \sqrt[d]{\sum_{i=1}^{n} |x_i - y_i|^d}$ The L_1, L_2 and L_∞ metrics are the well-known Manhattan, Euclidean and Max metrics, respectively. The approximation algorithms we construct for L_1 and L_∞ Matching have an approximation factor of $1 \pm \varepsilon$ and are easily changed to have a factor of $1 + \varepsilon$ by running the algorithm with $\frac{\varepsilon}{2}$ and multiplying the results by $1 + \frac{\varepsilon}{2}$.

Convolutions

The convolution vector of two vectors t, p, denoted by $t \otimes p$ is defined as vector w such that

$$w[i] = \sum_{j=1}^{m} t[i + j - 1]p[j].$$

The convolution can be computed in $O(n \log m)$ time, in a computational model with a word size of $O(\log m)$, by using the Fast Fourier Transform (FFT) [CLR92].

2.1 Problem Definitions

We provide more convenient definitions of the problems that are equivalent to the form in which they were introduced above.

The *String Matching with L_1 Distance* (*L_1 Matching* for short) is defined as follows:

Input: Text vector $t = t_1, t_2, \ldots, t_n$, pattern vector $p = p_1, p_2, \ldots, p_m$ where $t_i, p_j \in \{1, 2, \ldots, |\Sigma|\}$.

Output: results vector $O[1, \ldots, n - m + 1]$, where for every i $O[i] = \sum_{j=1}^{m} |t_{i+j-1} - p_j|$.

The *Approximated String Matching with L_1 Distance* (*Approximated L_1 Matching*) is defined as follows:

Input: Text vector $t = t_1, t_2, \ldots, t_n$, pattern vector $p = p_1, p_2, \ldots, p_m$, exactness parameter $0 < \epsilon < 1$.

Output: results vector $\widehat{O}[1, \ldots, n - m + 1]$, s.t. $O[i] \leq \widehat{O}[i] \leq (1 + \epsilon)O[i]$, where $O[i] = \sum_{j=1}^{m} |t_{i+j-1} - p_j|$.

The *String Matching with L_∞ Distance* (*L_∞ Matching*) is defined for the same form of input, but the Output vector is defined differently.

Output: results vector $O[1, \ldots, n - m + 1]$ where for every i $O[i] = Max_{j=1}^{m} |t_{i+j-1} - p_j|$.

The *Approximated String Matching with L_∞ Distance* (*Approximated L_∞ Matching*) is defined as follows:

Input: Text vector $t = t_1, t_2, \ldots, t_n$, pattern vector $p = p_1, p_2, \ldots, p_m$, exactness parameter $0 < \epsilon < 1$.

Output: results vector $\hat{O}[1,\ldots,n-m+1]$, s.t. $O[i] \leq \hat{O}[i] \leq (1+\epsilon)O[i]$, where $O[i] = Max_{j=1}^m |t_{i+j-1} - p_j|$.

The *String Matching with Weighted Mismatches* is defined for an arbitrary alphabet as follows:

Input: Given some distance function $f : \Sigma \times \Sigma \to \Re$, Text vector $t = t_1, t_2, \ldots, t_n$ and pattern vector $p = p_1, p_2, \ldots, p_m$, where $t_i, p_j \in \Sigma$.

Output: result vector $O[1, \ldots, n-m+1]$, s.t. $O[i] = \sum_{j=1}^m f(t_{i+j-1}, p_j)$.

3   String Matching with Weighted Mismatches

Note that the alphabet distance function does not have to be a metric. The time complexity of the algorithm is $O(|\Sigma| n \log m)$.

This algorithm is a direct extension of the algorithm given by Fischer and Paterson in [FP74]. They examined a case where the distance function defined on the alphabet is:

$$f(\sigma_1, \sigma_2) = \begin{cases} 1 & \sigma_1 \neq \sigma_2 \\ 0 & \sigma_1 = \sigma_2 \end{cases}$$

We generalize the problem to any given distance function $f : \Sigma \times \Sigma \to \Re$.

Algorithm steps:

1. For each $\sigma \in \Sigma$
   (a) Create t^σ by replacing every σ by 1 and every other symbol by 0.

   (b) Create p^σ by replacing every $p_j \neq \phi$ by $f(p_j, \sigma)$, and ϕ by 0.

   (c) Compute $O_\sigma = t^\sigma \otimes p^\sigma$
2. Compute $O \leftarrow \sum_{\sigma \in \Sigma} O_\sigma$

It is easily seen that the algorithm computes the desired results. The time complexity of this algorithm is derived from the time needed to compute a convolution, and the size of the alphabet. Therefore the overall time needed for this algorithm is $O(|\Sigma| n \log m)$. For given finite alphabets, the time needed is $O(n \log m)$, but for large alphabets (i.e. $|\Sigma| = m$) it is worse than the naive algorithm which works in $O(nm)$ time.

If the alphabet distance function f is a metric we can approximate the results up to a factor of $\log m$ in the $\tilde{O}(n)$ time algorithm, using embedding techniques [Ind01] into L_1 and then using the L_1 Matching approximation algorithm.

4   String Matching with L_1 Distance

The problem of Exact L_1 Matching can be solved in $O(|\Sigma| n \log m)$ time using the algorithm described in section 3. An $O(n\sqrt{m \log m})$ time algorithm can be designed using methods of dividing the alphabet into frequent and non-frequent symbols, as in [ILLP04, ALPU05, LP01]. Therefore, the current best known time for Exact L_1 Matching is $O(n\sqrt{\log m}\min(|\Sigma|\sqrt{\log m}, \sqrt{m}))$.

The problem of String Matching with Mismatches can be linearly reduced to the problem of Exact L_1 Matching [LP05]. This fact led us to search for an approximation algorithm rather than an exact one, which seems hard to design in a better time complexity.

We have designed an algorithm that approximates the results of L_1 Matching up to a factor of $1 + \varepsilon$ and works in a run time of $O(\frac{1}{\varepsilon^2} n \log m \log |\Sigma|)$.

4.1 L_1 Approximation Algorithm

We begin by describing an exact algorithm that solves the L_1 Matching problem in $O(n|\Sigma|\log m \log |\Sigma|)$ and showing its correctness. Then we modify it and insert the exactness parameter ε in order to reduce the time complexity. The total time complexity of our algorithm for Approximated L_1 matching is $O(\frac{1}{\varepsilon^2} n \log m \log |\Sigma|)$. We emphasize the fact that the exact algorithm is not efficient, and is used only for methodological reasons, to construct the approximation algorithm.

We use a specific alphabet distance function to construct our algorithm. Given some k, define $k - 2k$ *String Matching* to be String Matching with Weighted Mismatches with the specific weight function: $f : \Sigma \times \Sigma \to \Re$ defined as:

$$f(\sigma_1, \sigma_2) = \begin{cases} 0 & |\sigma_1 - \sigma_2| < k \\ \lceil |\sigma_1 - \sigma_2| - k \rceil & k \le |\sigma_1 - \sigma_2| \le 2k \\ \lceil k \rceil & 2k < |\sigma_1 - \sigma_2| \end{cases}$$

For $\sigma_1 \neq \phi$ and $\sigma_2 \neq \phi$, and $f(\sigma_1, \sigma_2) = 0$ otherwise. Note that this problem can be solved in time $O(|\Sigma|n \log m)$, because it is a specific case of the String Matching with Weighted Mismatches.

Exact L_1 Matching: We provide an algorithm to solve L_1 Matching that uses $k - 2k$ Matching as a given procedure.

Algorithm Steps:

1. Initialize array $O[1, \dots, n - m + 1]$ to zeros.
2. $k = \frac{1}{2}$
3. While $k < |\Sigma|$ do:
   (a) Run $k - 2k$ Matching with t,p and k and get a result vector $V[1, \dots, n - m + 1]$
   (b) For $i = 1, \dots, n - m + 1$ do $O[i] \leftarrow O[i] + V[i]$
   (c) $k = k * 2$

Algorithm Correctness: By problem definition of L_1 Matching, we need to show that for every i, $O[i] = \sum_{j=1}^{m} |t_{i+j-1} - p_j|$. It is suffice to prove that for every i, j we add $|t_{i+j-1} - p_j|$ to $O[i]$. For some i, j call $d = |t_{i+j-1} - p_j|$. Concerning d, in every iteration of the while loop the $k - 2k$ Matching checks whether

$d < k, k \le d \le 2k$ or $2k < d$, and adds $0, d - k$ or k respectively. In the first while iteration 1 is added if $d \ne 0$, in the following iterations it adds $1, 2, 4$, and so on up to 2^x where $2^{x+1} \le d < 2^{x+2}$. In the next iteration, where $k = 2^{x+1}$, $d - k = d - 2^{x+1}$ is added. In the remaining iterations $d < k$ and therefore nothing is added to $O[i]$. In the cases where $d > 0$, the summary of the additions we have is:$1 + 1 + 2 + \cdots + 2^x + d - 2^{x+1} = d$. If $d = 0$ all the iterations do not add anything to $O[i]$ since $d < k$ for all k.

Time: We have $\log |\Sigma|$ iterations of the while loop, each using one time $k - 2k$ Matching and $O(n)$. If it takes $f(n, m)$ time to compute the $k - 2k$ Matching then the time complexity of the L_1 algorithm is $O((f(n, m) + n) \log |\Sigma|)$. The current best known time for the $k - 2k$ Matching is $O(|\Sigma|n \log m)$, therefore the time is $O(|\Sigma|n \log m \log |\Sigma|)$.

First Modification of the Exact Algorithm. Consider the exact algorithm described above. Note that in fact we add every $d = |t_{i+j-1} - p_j|$ to $O[i]$ by iterations that check whether $d > \frac{1}{2}$, $d > 1$, $d > 2$, $d > 4$, $\cdots$, $d > 2^x$, $\cdots$. Therefore d is made from the sum $1 + 1 + 2 + 4 + \ldots + 2^x + (d - 2^{x+1}) = d$. The idea behind the following modification is that in case we drop the first elements of this sum, where d is large enough, we do not greatly err. Now consider that we are given some $0 < \varepsilon < 1$. Our first modification is to reduce the alphabet size as follows. For each time we run $k - 2k$ Matching we run it with t', p' instead of t, p where t' and p' are defined as:

$$t'_i \leftarrow t_i \ mod \ \frac{4k}{\varepsilon} \quad 1 \le i \le n$$
$$p'_j \leftarrow p_j \ mod \ \frac{4k}{\varepsilon} \quad 1 \le j \le m$$

The error gained by this modification is if d is "very large" compared to k. In such cases, in order to err in this iteration of $k - 2k$ Matchingm it is required that $d mod \frac{4k}{\varepsilon} < 2k$. In details, consider $d = |t_{i+j-1} - p_j|$, if $d > \frac{4k}{\varepsilon}$ then perhaps where $d \ mod \ \frac{4k}{\varepsilon} < 2k$ then less than k is added to $V[i]$ at this iteration of $k - 2k$ (It can only err by adding less than needed). This is acceptable, because if we err with at most k in this iteration and $d > \frac{4k}{\varepsilon}$ then our error is less than $\varepsilon/4$. Summing all the iterations we can err, and $1, 1, 2, \ldots, \frac{\varepsilon}{4}d$ is bounded by $\frac{\varepsilon}{2}d$. The time complexity of this modified version of the algorithm is derived from the alphabet size in each of the iterations of the $k - 2k$ procedure, which leads to: $\Sigma_{k=\frac{1}{2}, 1, 2, 4, \ldots, \frac{|\Sigma|}{2}} \frac{4k}{\varepsilon} n \log m = O(\frac{1}{\varepsilon}|\Sigma|n \log m)$. This time complexity is not satisfactory leading us to the next modification.

Second Modification of the Exact Algorithm. If we keep in mind that every $d = |t_{i+j-1} - p_j|$ is computed from the sum of $1 + 1 + 2 + 4 + \ldots + 2^x + (d - 2^{x+1})$ and we already allowed error within the small elements, we can reduce the alphabet even more, in such a way that cause a small error in the biggest element in the sum, $d - 2^{x+1}$, where $2^{x+1} < d < 2^{x+2}$. We now run the stage of $k - 2k$ Matching with t', p' which has an alphabet size of $\frac{4k}{\varepsilon}$. The Second Modification will be as follows:

For each time we run $k - 2k$ Matching we run it with t'', p'', k'' instead of t', p', k where t'', p'' and k'' are defined as:

$$t''_i \leftarrow t'_i \ div \ \frac{k\varepsilon}{2} \quad 1 \leq i \leq n$$
$$p''_j \leftarrow p'_j \ div \ \frac{k\varepsilon}{2} \ 1 \leq j \leq m$$
$$k'' \leftarrow \frac{2}{\varepsilon}$$

And we change step 3(b) in the exact algorithm to:

For $i = 1, \ldots, n - m + 1$ do $O[i] \leftarrow O[i] + V[i]\frac{k\varepsilon}{2}$.

If we consider some $d = |t_{i+j-1} - p_j|$, let $d_k = |t_{i+j-1} div \frac{k\varepsilon}{2} - p_j div \frac{k\varepsilon}{2}|$. This method would not make any differrence for any of the elements in the sequence with a sum of d except for the largest one. We can concentrate on three groups of elements in the sequence: where $d > 2k, k \leq d \leq 2k$ and $d < k$.

In the iterations where $d > 2k$, it is easily seen that $d_k > \frac{4}{\varepsilon}$ and therefore $\frac{4}{\varepsilon} \times \frac{k\varepsilon}{2} = 2k$ is added to $O[i]$. In the iteration where $d < k$, then $d_k < \frac{2}{\varepsilon}$ and thus nothing is added to $O[i]$. The only element that is affected from the modification is where $k \leq d \leq 2k$, i.e. $d - 2^{x+1}$. If $k \leq d \leq 2k$ then $\frac{2}{\varepsilon} \leq d_k \leq \frac{4}{\varepsilon}$. Therefore, $(d_k - \frac{2}{\varepsilon})\frac{k\varepsilon}{2} = d_k\frac{k\varepsilon}{2} - k$ is added to O[i]. We use the following inequaliy to assure that our error is bounded by $\frac{k\varepsilon}{2}$:

Lemma 1. *For any integers x, y s.t. $k \leq |x - y| \leq 2k$:*

$$|x - y|(1 - \varepsilon) \leq |x \ div \ k\varepsilon - y \ div \ k\varepsilon|k\varepsilon \leq |x - y|(1 + \varepsilon)$$

The total error from both modifications is bounded by $k\varepsilon$ for $k \leq d \leq 2k$, and therefore is within the tolerated distance of up to a factor of ε. Note that $L_1[i]$ is defined as $\Sigma_{j=1}^m |t_{i+j-1} - p_j|$, so if we err in each pair (t_{i+j-1}, p_j) up to a factor of $1 + \varepsilon$ the total error is also bounded by this factor.

Proof (of lemma). If we divide the natural numbers into groups of c elements i.e. $g_1 = (1, 2, \ldots, c), g_2 = (c + 1, c + 2, \ldots, 2c), \cdots$ (this is equivalent to the div c operation) and the distance between each two elements is assumed to be the distance in groups multiplied by c, then our error is bounded by the sum of the distances of each of these two elements from the center of the group, which is $\frac{c}{2} + \frac{c}{2} = c$. If $c = k\epsilon$ we find that:

$$|x - y| - k\varepsilon \leq |x \ div \ k\varepsilon - y \ div \ k\varepsilon|k\varepsilon \leq |x - y| - k\varepsilon$$

Given that $k \leq |x - y| \leq 2k$ the inequality holds.

The fact that in each iteration we reduce our alphabet first by $mod\frac{4k}{\varepsilon}$ and then by $div\frac{k\varepsilon}{2}$ leads to an alphabet size of $O(\frac{1}{\varepsilon^2})$. The time needed to run the $k - 2k$ procedure is $O(\frac{1}{\varepsilon^2} n \log m)$. The total time for this algorithm is $O(\frac{1}{\varepsilon^2} n \log m \log |\Sigma|)$.

5   String Matching with L_∞ Distance

This problem differs from the problems of L_1 and L_2 Matching since the distance function defined between strings is not the sum of distances between symbols, but rather is defined for all pairs of symbols. We first construct an $O(n|\Sigma|\log m)$ time algorithm, and then construct an algorithm that approximates L_∞ matching up to a factor of $1 + \varepsilon$ and works within a run time of $O(\frac{1}{\varepsilon}n\log m\log|\Sigma|)$.

5.1 $O(n|\Sigma|\log(m+|\Sigma|))$ Algorithm [ALPU05]

The method in this algorithm is to encode the text and the pattern in such a way that we find the results in one convolution, and a linear time pass on the convolution result.

Key idea: If we look at one text number, t, and one pattern number p. We encode both of them into $|\Sigma|$ long binary strings. The encoding of t is all 0's except the t-th bit which is 1, and similarly with p, which is encoded to all 0's except the p-th bit. Let $c(i)$ denote the encoded i. Now, we begin by aligning $c(p)$ below $c(t)$ and starting at position $-|\Sigma|$ (where $c(t)$ is fixed to start at position 1). We move $c(p)$ to the right till both 1-bits are one below the other. At this position, the distance between the starting position of $c(t)$ and the starting position of $c(p)$ equals the difference $|t - p|$. An example depicting this process is given in Figure 1. If we look at $r = c(t) \otimes c(p)$ we see either $r[-|t - p|] = 1$ or $r[|t - p|] = 1$. Extending this idea to encode strings of numbers requires us to add leading (or tracing) zeros between the encoded numbers.

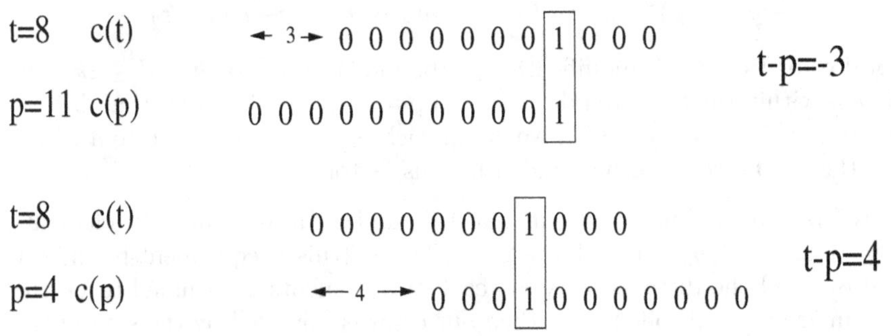

t=8   c(t)   ← 3 → 0 0 0 0 0 0 0 | 1 | 0 0 0   t-p=-3

p=11 c(p)   0 0 0 0 0 0 0 0 0 0 0 | 1

t=8   c(t)   0 0 0 0 0 0 0 | 1 | 0 0 0   t-p=4

p=4   c(p)   ← 4 → 0 0 0 | 1 | 0 0 0 0 0 0 0

Fig. 1. $c(p)$ moved below $c(t)$ till the 1-bits are aligned

In detail: first, define $\chi_{\neq 0}(x) = 1$ if $x \neq 0$ and otherwise 0. Next, define for every $x \in \Sigma = \{1, \ldots, n\}$, $c^t(x) = c^t(x)_1, \ldots, c^t(x)_{2|\Sigma|}$ where $c^t(x)_i = 1$ if $i = |\Sigma| + x$ and otherwise 0. Similarly define $c^p(x) = c^p(x)_1, \ldots, c^p(x)_{2|\Sigma|}$ where $c^p(x)_i = 1$ if $i = x$ and otherwise 0.

Algorithm Steps

1. Construct $c^t(T) = c^t(t_1)\cdots c^t(t_n)$
2. Construct $c^p(P) = c^p(p_1)\cdots c^p(p_m)$
3. Compute $R = c^t(T) \otimes c^p(P)$
4. For $i = 1, \ldots, n - m + 1$
   $O[i] \leftarrow \max_{s=-|\Sigma|}^{|\Sigma|} \chi_{\neq 0}(R[(2i - 1)|\Sigma| + 1 + s])|s|$

Claim: At the end of the algorithm $O[i] = \max_{j=1}^{m} |t_{i+j-1} - p_j|$.

Time: The time needed to convolve 2 strings of size $n|\Sigma|$ and $m|\Sigma|$ is $O(n|\Sigma| \log n)$. The computation of $O[i]$ takes $2|\Sigma|$ steps , and $i = 1, 2, \ldots, n - m + 1$, thus this step takes $O(n|\Sigma|)$. Both steps together take $O(n|\Sigma| \log n)$. We can slightly improve the time to a total time of $O(n|\Sigma| \log(m + |\Sigma|))$ by using the technique of cutting the text into n/m overlapping segments, each with a length of $2m$.

5.2 L_∞ Approximation Algorithm

Below we present an algorithm that approximates the L_∞ Matching up to a factor of $1 + \varepsilon$.

Algorithm Steps:

1. Initialize array $O[1, \ldots, n - m + 1]$ to zeros.
2. For $q = 1, 2, 4, \ldots |\Sigma|$ do:
   (a) Create t' from t, where $t'_i = (t_i \ div \ q\varepsilon) mod \frac{4}{\varepsilon}$.
   (b) Create p' from p, where $p'_i = (p_i \ div \ q\varepsilon) mod \frac{4}{\varepsilon}$.
   (c) Run L_∞ exact algorithm with t', p' and attain a result vector $V[1, \ldots, n - m + 1]$.
   (d) For $i = 1, \ldots, n - m + 1$ do $O[i] \leftarrow V[i] q\varepsilon$ if $V[i] \geq \frac{1}{\varepsilon}$ or leave unchanged otherwise.

Algorithm Correctness. For each $O[i]$, the maximal q where $V[i] > \frac{1}{\varepsilon}$ mean that $L_\infty(p, t[i, \ldots, i + m - 1]) > q$, but also that $L_\infty(p, t[i, \ldots, i + m - 1]) < 2q$. Now since we reduce our alphabet by $divq\varepsilon$ we receive the value computed with an error of $\pm q\varepsilon$, which is acceptable since the exact value is between q and $2q$.

Time. For each q it takes $O(\frac{1}{\varepsilon} n \log n)$. since we reduce the alphabet to be of size $\frac{4}{\varepsilon}$. We run it with $q = 1, 2, 4, \ldots, |\Sigma|$, therefore $O(\frac{1}{\varepsilon} n \log n \log |\Sigma|)$. If we apply the technique of cutting the text into n/m overlapping segments, we get the total time of $O(\frac{1}{\varepsilon} n \log m \log |\Sigma|)$.

6   Open Problems and Conclusions

We have presented an $O(\frac{1}{\varepsilon^2} n \log m \log |\Sigma|)$ time $1 + \epsilon$ approximation algorithm for the L_1 matching problem. L_2 matching is related to convolution computation, therefore any improvement in the time for one of the problems, directly means improvement for the other as well. The current best time for both L_2 and convolution is $O(n \log m)$. The most interesting case is L_∞. In this case we have an $\tilde{O}(n|\Sigma|)$ algorithm, which means its run time is worse than the naive for large alphabets. The problem of solving the *exact* L_∞ matching within a better-than-naive time for large alphabets still remains open. We have presented a $1 + \epsilon$ approximation algorithm for this problem in $O(\frac{1}{\varepsilon} n \log m \log |\Sigma|)$.

For the *String Matching with Weighted Mismatches* problem we have provided an $O(n|\Sigma|\log m)$-time exact algorithm. Nonetheless, whether the *exact* solution can be improved or a better approximation ratio can be shown with the $\tilde{O}(n)$-time algorithm remains an open question.

References

[Abr87]   Abrahamson, K.: Generalized string matching. SIAM J. Computing 16(6), 1039–1051 (1987)

[AF95]   Amir, A., Farach, M.: Efficient 2-dimensional approximate matching of half-rectangular figures. Information and Computation 118(1), 1–11 (1995)

[AFM94]   Amir, A., Farach, M., Muthukrishnan, S.: Alphabet dependence in parameterized matching. Information Processing Letters 49, 111–115 (1994)

[ALPU05]   Amir, A., Lipsky, O., Porat, E., Umanski, J.: Approximate matching in the l_1 metric. In: Apostolico, A., Crochemore, M., Park, K. (eds.) CPM 2005, vol. 3537, pp. 91–103. Springer, Heidelberg (2005)

[CH02]   Cole, R., Hariharan, R.: Verifying candidate matches in sparse and wildcard matching. In: Proceedings of the thiry-fourth annual ACM symposium on Theory of computing, pp. 592–601. ACM Press, New York (2002)

[CLR92]   Cormen, T.H., Leiserson, C.E., Rivest, R.L.: Introduction to Algorithms. MIT Press and McGraw-Hill (1992)

[CW99]   Kam Wing Chu, K., Hon Wong, M.: Fast time-series searching with scaling and shifting. In: Symposium on Principles of Database Systems, pp. 237–248 (1999)

[FP74]   Fischer, M.J., Paterson, M.S.: String matching and other products. In: Karp, R.M. (ed.) Complexity of Computation, SIAM-AMS Proceedings, vol. 7, pp. 113–125 (1974)

[FRM94]   Faloutsos, C., Ranganathan, M., Manolopoulos, Y.: Fast subsequence matching in time-series databases. In: Proceedings 1994 ACM SIGMOD Conference, Mineapolis, MN, pp. 419–429 (1994)

[IKM00]   Indyk, P., Koudas, N., Muthukrishnan, S.: Identifying representative trends in massive time series data sets using sketches. In: International Conference on Very Large Data Bases (VLDB), pp. 363–372 (2000)

[ILLP04]   Indyk, P., Lewenstein, M., Lipsky, O., Porat, E.: Closest pair problems in very high dimensions. In: Díaz, J., Karhumäki, J., Lepistö, A., Sannella, D. (eds.) ICALP 2004, vol. 3142, pp. 782–792. Springer, Heidelberg (2004)

[Ind00]   Indyk, P.: Stable distributions, pseudorandom generators, embeddings and data stream computation. In: Foundations of Computer Science (FOCS), pp. 189–197 (2000)

[Ind01]   Indyk, P.: Algorithmic applications of low-distortion geometric embeddings. In: Foundations of Computer Science, FOCS (2001)

[Kar93]   Karloff, H.: Fast algorithms for approximately counting mismatches. Information Processing Letters 48(2), 53–60 (1993)

[KMP77]   Knuth, D.E., Morris, J.H., Pratt, V.R.: Fast pattern matching in strings. SIAM J. Computing 6, 323–350 (1977)

[LP01]   Lipsky, O., Porat, E.: Efficient l1-matching algorithm (manuscript, 2001)

[LP05]   Lipsky, O., Porat, E.: L_1 pattern matching lower bound. In: Consens, M.P.,
           Navarro, G. (eds.) SPIRE 2005, vol. 3772, pp. 327–330. Springer, Heidel-
           berg (2005)
[MR95]   Muthukrishnan, S., Ramesh, H.: String matching under a general matching
           relation. Information and Computation 122(1), 140–148 (1995)
[Mut95]   Muthukrishnan, S.: New results and open problems related to non-standard
           stringology. In: Combinatorial Pattern Matching Conference, pp. 298–317
           (1995)
[Per00]   Perttu, S.: Combinatorial pattern matching in musical sequences. Master's
           thesis, Department of Computer Science, University of Helsinki (2000)

Interchange Rearrangement: The Element-Cost Model

Oren Kapah[2], Gad M. Landau[1], Avivit Levy[3,4], and Nitsan Oz[1]

[1] Department of Computer Science
University of Haifa, Haifa 31905, Israel
`landau@cs.haifa.ac.il, nitskeus@gmail.com`
[2] Department of Computer Science
Bar Ilan University, Ramat Gan 52900, Israel
`kapaho@cs.biu.ac.il`
[3] CRI, University of Haifa, Haifa 31905, Israel
`avivitlevy@gmail.com`
[4] Shenkar College, Anna Frank 12, Ramat Gan, 52526, Israel

Abstract. Given an input string S and a target string T when S is a permutation of T, the *interchange rearrangement problem* is to apply on S a sequence of interchanges, such that S is transformed into T. The *interchange* operation exchanges the position of the two elements on which it is applied. The goal is to transform S into T at the minimum cost possible, referred to as the distance between S and T. The distance can be defined by several cost models that determine the cost of every operation. There are two known models: The *Unit-cost model* and the *Length-cost model*. In this paper, we suggest a natural cost model: The *Element-cost model*. In this model, the cost of an operation is determined by the elements that participate in it. Though this model has been studied in other fields, it has never been considered in the context of rearrangement problems. We consider both the special case where all elements in S and T are distinct, referred to as a *permutation string*, and the general case, referred to as a *general string*. An efficient optimal algorithm for the permutation string case and efficient approximation algorithms for the general string case, which is $\mathcal{NP}$-hard, are presented.

Keywords: Interchange rearrangement, Cost models.

1 Introduction

The problem of defining the distance or similarity between two strings S and T has been studied extensively over the years. There are many known and established methods, such as the *Edit distance* and the *Hamming distance* [13]. The Edit distance allows three operations (substitution, insertion or deletion) upon the input string. There are several generalizations of the basic Edit distance (also referred to as the *Levenshtein distance*), which defines a unit-cost for every operation. One is the *the operation-weight edit distance*, which gives a unit-cost for every type of operation. Another is the *alphabet-weight edit-distance*, which

A. Amir, A. Turpin, and A. Moffat (Eds.): SPIRE 2008, LNCS 5280, pp. 224–235, 2008.

defines a cost for every operation depending on the elements participating in the specific operation.

These string metrics deal with errors of data appearing in the text and give a measure of either similarity or distance between an input string S to a target string T. The order of the elements is assumed to be correct. However, address errors may also be considered ([1,2,3,4]). In these types of errors, elements in S may only be mispositioned. It is commonly assumed that the input string is a permutation of the target string in order to have a finite distance. In a rearrangement problem, it is assumed that only address errors have occurred. The goal is to apply a sequence of legal operations on S, such that S is transformed into T at the minimum cost possible, referred to as the distance between S and T.

The interchange rearrangement problem was studied by Cayley [10]. Cayley solved this problem for permutation strings under the Unit-cost model and left open the problem of general strings. Recently, Amir et al. solved Cayley's open problem by showing it is $\mathcal{NP}$-hard and giving a 1.5-approximation algorithm. In addition, they extended this problem by examining it under the Length-cost model [4]. In this paper, we further extend this problem on both permutation strings and general strings by examining it under the Element-cost model.

We begin with formal definitions of the interchange operator and the Element-cost model.

Definition 1. *Let* $S = s_1, \ldots, s_n$ *be a string. An interchange of elements* s_i *and* s_j, $i < j$, *transforms* S *into* $S' = s_1, \ldots, s_{i-1}, s_j, s_{i+1}, \ldots, s_{j-1}, s_i, s_{j+1}, \ldots, s_n$.

There are two known cost models in the context of rearrangement problems. In the Unit-cost model (UCM) each operation is given a unit cost, so the problem is to transform S into T with a minimum number of operations. In the Length-cost model (LCM) [4,7], the cost of an operation depends on its length characteristic. Other characteristics may be considered in the rearrangement problem. For example, some elements may be heavier than other elements. In such cases, moving light elements is preferable to moving heavy elements. This observation motivated researchers to explore the Element-cost model (ECM). In [12], Gupta and Kumar considered the problem of sorting and selection in the comparison model for structured costs. In their work, it is assumed that every element has a weight and that the cost of a comparison is defined by a function applied to the weight of the elements that participate in the comparison. They gave approximations for the optimal solution for families of structured functions such as summation, multiplication, etc. Recently, [5] addressed the same problem of sorting and selection for random costs. However, this paper is the first to consider the ECM for dealing with rearrangement problems.

Definition 2. *Let* $w : \Sigma \to \mathbb{R}^+$ *be a weight function, which assigns a non-negative weight to every element in* Σ. *Let* $g : \Sigma \times \Sigma \to \mathbb{R}^+$ *be a function defining the interchange cost. The function* g *is called a* general function *if it satisfies the following conditions:*

1. $\forall x, y \in \Sigma : g(x, y) = g(y, x)$.
2. $\forall x, y, z \in \Sigma : w(y) \leq w(z) \Leftrightarrow g(x, y) \leq g(x, z)$.

The summation function $g(x, y) = w(x) + w(y)$ and the multiplication function $g(x, y) = w(x) \cdot w(y)$ are two examples of intuitive general functions.

If all elements in S are distinct, a unique bijection $f : S \to \{1, \ldots, n\}$ can be defined such that $f(s_i)$ equals the position of the element s_i in T. Thus S can be represented by $\pi = f(s_1), f(s_2), \ldots, f(s_n)$ and T by $1, \ldots, n$. For this case the term permutation string is used. The input string is then assumed to be π, i.e, a permutation of $1, \ldots, n$. Under this assumption the rearrangement problem is simply a sorting problem, i.e. the distance is the minimum cost for sorting π. Problems of sorting a permutation string have been studied extensively [6,8,9,11,14,15]. For the general case in which S may have repetitions of elements, the term general string is used.

Our main results are:

1. $O(n)$ time algorithm for the interchange rearrangement problem for permutation strings for any general function.
2. Two approximation algorithms for the general strings case, which is $\mathcal{NP}$-hard:
   (a) $O(n)$ time 3-approximation algorithm for any general function.
   (b) $O(n \cdot \lg |\Sigma|)$ time 1.72-approximation algorithm for the summation function.

In addition, we give some minor results, which are presented in the full version of the paper, considering the *transposition rearrangement problem* under the ECM, UCM and the LCM for general strings and permutation strings. A *transposition* of an element s_i, ℓ positions forward transforms the string S into the string $S' = s_1, \ldots, s_{i-1}, s_{i+1}, \ldots, s_{i+\ell}, s_i, s_{i+\ell+1}, \ldots, s_n$ and a transposition of an element s_i, ℓ positions backward transforms the string S into the string $S' = s_1, \ldots, s_{i-\ell-1}, s_i, s_{i-\ell}, \ldots, s_{i-1}, s_{i+1}, \ldots, s_n$. Table 1 summarizes the known and new results.

The paper is organized as follows. Section 2 gives additional preliminaries and notations. Section 3 presents an algorithm for the interchange rearrangement problem for permutation strings for any general function. Section 4 present an approximation algorithm for the interchange rearrangement problem for general

Table 1. A Summary of Results

	UCM	ECM	LCM
Interchanges			
Permutation strings	$O(n)$ [10]	$O(n)$ (general functions)	$O(n)$ [4]
General Strings	$\mathcal{NP}$-hard [4] $O(n \cdot \lg \|\Sigma\|)$ 1.5-approximation [4]	General function: $O(n)$ 3-approximation Summation function: $O(n \cdot \lg \|\Sigma\|)$ 1.72-approximation	$O(n)$ [4]
Transpositions			
Permutation strings	$O(n \lg n)$ [14]	$O(n \lg n)$	$O(n \lg n)$
General Strings	$O(n^2)$	$O(n^2)$	$O(n \lg n)$

strings for any general function and an improved approximation algorithm for
the summation function.

2 Preliminaries and Notations

Given an input string S and a target string T, we define a multi-graph $G_{S,T} = (V, E)$ in the following way: $V = \{v \in \Sigma : v$ appears in S and $T\}$ and $E = \{(t_i, s_i), 1 \leq i \leq n\}$. In other words, every distinct character has a vertex and for every index $1 \leq i \leq n$ there is an edge connecting the vertex representing t_i with the vertex of s_i, meaning that by the end of the rearrangement process, s_i will be moved and replaced by a t_i character. Since S and T have the same quantities of each element of Σ, the number of incoming edges of every vertex equals the number of its outgoing edges, which is the number of occurrences of the vertex's character in S (and hence in T). Therefore, $G_{S,T}$ is an Eulerian directed graph and by definition can be decomposed into edge-disjoint directed cycles. If S is a permutation string, every vertex has exactly one incoming edge and one outgoing edge and therefore, $G_{S,T}$ can be uniquely decomposed into edge-disjoint directed cycles. This fact is not true for general strings. Furthermore, there might be an exponential number of ways to decompose $G_{S,T}$ into edge-disjoint directed cycles. However, once such a decomposition of $G_{S,T}$ is given, it uniquely defines a labeling of the elements of S and T such that every element appears exactly once. An edge-disjoint directed cycle in a given decomposition is also called a permutation cycle. We use the following notations:

- $d(\pi)$: The distance in the permutation string case (the minimum cost for sorting π) and $d(S, T)$ in the general string case (the minimum cost for transforming S into T).
- $e \leftrightarrow f$: Denotes the operation of interchanging elements e and f. Note that if e and f appear in the same cycle, interchanging them splits their cycle into two cycles. If e and f appear in different cycles, interchanging them unites their cycles into one cycle.
- S_{min}: Denotes the minimum cost element in S. If the input string is a permutation string we substitute this notation with π_{min}.
- $\tilde{S}$: Denotes the multi-set of elements that are not in place. For example, if $T = abcab$ and $S = bbaca$ then $\tilde{S} = \{a, a, b, c\}$.

The following notations apply directly to a permutation string. However, given a decomposition of $G_{S,T}$ into edge-disjoint directed cycles in the case of a general string, these notations may be also applied. We use the notation G_π instead of $G_{S,T}$ for the case of a permutation string:

- For a cycle C:
  - $|C|$: Denotes the number of elements in C (the size of C). We use the term ℓ-cycle for a cycle of size ℓ.
  - C_{min}: Denotes the minimum cost element in C.
- $c(\pi)$: Denotes the number of cycles in G_π.

3   Sorting a Permutation String

In this section we present an algorithm for the interchange rearrangement problem when the input string is a permutation string for any general function under the ECM. This problem is defined as follows:

Definition 3. *Let π be a permutation string and let $g : \Sigma \times \Sigma \to \mathbb{R}^+$ be a general function. Compute the minimum cost for sorting π by interchanges when $g(x, y)$ is the cost of interchanging elements x and y.*

Cayley [10] studied this problem under the UCM. He showed that given a permutation π of $1, \ldots, n$, the minimum number of interchanges needed for sorting π, is $n - c(\pi)$. This is achieved by interchanging only elements that share a cycle until there are no such elements (the permutation is sorted). When the ECM is used, one might also be inclined to apply a minimum number of interchanges. This inclination implies that one would be making interchanges only within cycles. Any interchange between elements of different cycles would result in an increase in the number of interchanges needed for sorting π and probably in the total cost for sorting π. However, this inclination is incorrect. Moreover, there might be cases in which the optimal solution would be to increase the number of interchanges needed for sorting π in order to decrease the total cost. We will describe an algorithm for sorting a permutation string by interchanges under ECM, and then prove that it yields the optimal cost, i.e., the distance $d(\pi)$.

3.1   The $O(n)$ Time Algorithm

The basic idea of the CEA_{ps} algorithm (Fig. 2) is quite simple. In order to sort the permutation π at the minimum cost, either the cheapest element in some cycle is used to sort all the other elements including itself, or (if the cheapest element in the cycle is not cheap enough) the cost for introducing into the cycle the cheapest element in π is "paid" by interchanging it with the cheapest element of the cycle. Doing so unites the cycle with the cycle of the minimum cost element of π. Then the cheapest element of π can be used to sort all the other elements in the cycle. We call this algorithm *"The Cheapest Employee Algorithm" (CEA)*.

Definition 4. *Let C be a cycle in G_π, define:*

- $\alpha_{in}(C) = \sum_{x \in C \setminus \{C_{min}\}} g(C_{min}, x) = \sum_{x \in C} g(C_{min}, x) - g(C_{min}, C_{min})$

This represents the case in which a cycle C is sorted within itself, i.e. by using only interchanges of elements within C. This is done by repeatedly interchanging C_{min} with the other elements in C as shown in Fig. 1(a) until all C's elements including C_{min} are sorted.

- $\alpha_{out}(C) = \sum_{x \in C} g(\pi_{min}, x) + g(\pi_{min}, C_{min})$

This represents the case in which in order to sort the elements of C, π_{min} is introduced into C by interchanging C_{min} with π_{min}. The result of this interchange

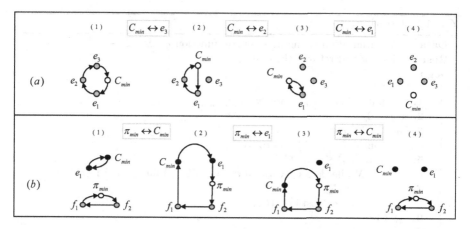

Fig. 1. In (a) the sorting is done within the cycle using its minimum cost element, C_{min}. In (b) the sorting is done by introducing the cycle to the minimum cost element, π_{min}. Note that after the interchange $C_{min} \leftrightarrow \pi_{min}$ the elements of C form a connected path in the new cycle (the black vertices path) and π_{min} is positioned at the tail of this path (white vertex).

is that the elements of C in the new united cycle form a connected path and π_{min} is positioned at the tail of this path. Then π_{min} is interchanged with all the elements of C in order to sort them in the same manner described for $\alpha_{in}(C)$ (see Fig. 1(b)).

- $\alpha(C) = \min\{\alpha_{in}(C), \alpha_{out}(C)\}$

The minimum cost method for sorting C.

Step 1 of the CEA_{ps} algorithm (Fig. 2) computes the permutation cycles of π. This is done by a left to right traversal of π. In addition, the minimum cost element for every cycle and for the whole permutation string is computed. Then, in steps $3-13$, each cycle is tested separately for the cheapest sorting method and this method is applied.

3.2   Correctness of the Algorithm

In this subsection we show that the CEA_{ps} algorithm is optimal, i.e., returns the distance $d(\pi)$. The cost returned by the CEA_{ps} algorithm defines an upper bound for the distance, which is:

$$d(\pi) \leq \sum_{1 \leq i \leq c(\pi)} \alpha(C_i)$$

We now show that it matches the lower bound.

Lemma 1. *Let π be a permutation string and let $C_1, \ldots, C_{c(\pi)}$ be the cycles of G_π, then:*

$$d(\pi) \geq \sum_{1 \leq i \leq c(\pi)} \alpha(C_i)$$

CEA_{ps} algorithm

Data : A permutation string π, a general function $g : \Sigma \times \Sigma \to \mathbb{R}^+$
Result : Sorts π and returns the cost
begin

    1. Compute $C_1, \ldots, C_{c(\pi)}$ and $\pi_{min},, C_{1_{min}}, \ldots, C_{c(\pi)_{min}}$
    2. cost$\leftarrow 0$
    3. For $1 \le i \le c(\pi)$ do
    4.   Compute $\alpha_{in}(C_i)$ and $\alpha_{out}(C_i)$
    5.   If $\alpha_{in}(C_i) \le \alpha_{out}(C_i)$
    6.   While $\exists e \in C_i$ with an edge $(e, C_{i_{min}})$ and $|C_i| \ne 1$ do
    7. $C_{i_{min}} \leftrightarrow e$
    8.   cost $\leftarrow$ cost$+\alpha_{in}(C_i)$
    9.   Else
    10. $C_{i_{min}} \leftrightarrow \pi_{min}$
    11.   While $\exists e \in C_i$ with an edge (e, π_{min}) do
    12. $\pi_{min} \leftrightarrow e$
    13.   cost$\leftarrow$ cost$+\alpha_{out}(C_i)$
    14. return $cost$

end

Fig. 2. Algorithm for sorting a permutation string by interchanges under ECM

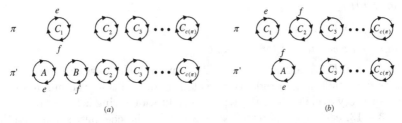

(a)   (b)

Fig. 3. In case 1 - (a), $e, f \in C_1$ and after the interchange $e \leftrightarrow f$: $e \in A$ and $f \in B$. In case 2 - (b), $e \in C_1$ and $f \in C_2$ and after the interchange $e \leftrightarrow f$: $e, f \in A$.

Proof. By induction on the number of interchanges performed by the optimal solution. The case in which the optimal solution performs 0 operations is trivial (a sorted permutation). Assume that the lemma applies for a permutation that can be optimally sorted in $k - 1$ interchanges. We prove that the lemma also applies for a permutation that can be optimally sorted in k interchanges. Let π be a permutation of $1, \ldots, n$ with cycles $C_1, \ldots, C_{c(\pi)}$, which can be optimally sorted in k interchanges. Suppose that the first interchange of this solution is $e \leftrightarrow f$. Then the resulting permutation after performing this interchange is a permutation π', which can be optimally sorted in $k - 1$ operations. Thus π' satisfies the induction hypothesis. The cost for sorting π is: $d(\pi) = d(\pi')+g(e, f)$. There are two cases to consider:

Case 1: e and f in π belong to the same cycle. Assume w.l.o.g. that $e, f \in C_1$ and after performing the interchange, $e \in A$ and $f \in B$ (see Fig. 3 (a)). The distance is:

$$d(\pi) = d(\pi') + g(e, f) \geq \alpha(A) + \alpha(B) + \sum_{2 \leq i \leq c(\pi)} \alpha(C_i) + g(e, f)$$

This case implies four subcases depending on the values of $\alpha(A)$ and $\alpha(B)$.

Case 2: e and f in π belong to different cycles. Assume w.l.o.g. that $e \in C_1$ and $f \in C_2$ and after performing the interchange $e, f \in A$ (see Fig. 3 (b)). The distance is:

$$d(\pi) = d(\pi') + g(e, f) \geq \alpha(A) + \sum_{3 \leq i \leq c(\pi)} \alpha(C_i) + g(e, f)$$

This case implies two additional subcases depending on the value of $\alpha(A)$.

The proof of these subcases is omitted and is fully detailed in the full version of the paper. $\qquad\square$

Theorem 1 immediately follows from the upper bound of the algorithm and Lemma 1.

Theorem 1. *Let π be a permutation string and let $C_1, \ldots, C_{c(\pi)}$ be the cycles of G_π. Then the minimum cost for sorting π by interchanges under ECM for any general function is $d(\pi) = \sum_{1 \leq i \leq c(\pi)} \alpha(C_i)$.*

Complexity: By Theorem 1, the CEA_{ps} algorithm computes the distance $d(\pi)$. Since computing the permutation cycles can be done in linear time by a left to right traversal and since testing all the cycles is done in linear time, the CEA_{ps} algorithm runs in linear time in the size of G_π and thus linear in n.

4   Rearranging General Strings

In the previous section we showed a linear time algorithm that computes the distance in the interchange rearrangement problem when the input string is a permutation string and for every general function. In this section we consider the following problem:

Definition 5. *Let S be the input string and T be the target string, when S is a permutation of T and let $g : \Sigma \times \Sigma \to \mathbb{R}^+$ be a general function. Compute the minimum cost for transforming S into T by interchanges when the cost of interchanging elements x and y is given by $g(x, y)$.*

The interchange rearrangement problem under the UCM for general strings is $\mathcal{NP}$-hard [4]. Hence, as the UCM is a special case of ECM where all elements have equal weights, Corollary 1 follows:

Corollary 1. *The interchange rearrangement problem under ECM for general strings is $\mathcal{NP}$-hard.*

In the following subsections we present an $O(n)$ time, 3-approximation algorithm for any general function. In addition, we present an $O(n \cdot \lg |\Sigma|)$ time 1.72-approximation algorithm for the summation function.

4.1 $O(n)$ Time 3-Approximation Algorithm for General Functions

The hardness of this problem is due to the difficulty of pairing each element in S with an identical element in T (converting the problem into a permutation string problem) in a way that gives the minimum distance. As explained in Section 2, pairing elements from S with elements in T is equivalent to performing an edge-disjoint decomposition of $G_{S,T}$ into directed cycles. Since S is a permutation of T, $G_{S,T}$ is an Eulerian directed graph and such a decomposition exists. The CEA_{gs} algorithm (Fig. 4) arbitrarily decomposes $G_{S,T}$ into cycles and then applies the CEA_{ps} algorithm (Fig. 2). We prove the following theorem:

Theorem 2. *The CEA_{gs} algorithm gives a 3-approximation ratio for any general function.*

Proof. We first observe that any solution for the problem implies a decomposition of $G_{S,T}$ into edge-disjoint directed cycles. This is true because any solution implies a pairing of identical elements of S and T, which is equivalent to performing such a decomposition. Assume that the optimal solution implies a decomposition of $G_{S,T}$ into cycles $C_1, \ldots, C_k$. Then by Theorem 1:

$$d(S,T) = \sum_{i=1}^{k} \alpha(C_i)$$
$$= \sum_{i=1}^{k} min\{ \sum_{x \in C_i} g(C_{i_{min}}, x) - g(C_{i_{min}}, C_{i_{min}}) ,$$
$$\sum_{x \in C_i} g(S_{min}, x) + g(C_{i_{min}}, S_{min}) \}$$

Since $w(S_{min}) \leq w(C_{i_{min}})$ then by decreasing the weight of $C_{i_{min}}, \forall 1 \leq i \leq k$ to $w(S_{min})$ the total cost may only decrease:

CEA_{gs} algorithm

Data   : Input string S, target string T, a general function $g : \Sigma \times \Sigma \to \mathbb{R}^+$
Result : Transform S into T
begin

1. Compute $G_{S,T}$.
2. Compute a decomposition D of $G_{S,T}$ as follows:
3. $D \leftarrow \emptyset$.
4.   Add to D all the 1-cycles of $G_{S,T}$ and remove their edges.
5.   Add to D an arbitrary decomposition of the remaining edges.
6. Apply the CEA_{ps} algorithm on D.

end

Fig. 4. 3-approximation algorithm for the interchange rearrangement problem under ECM for general strings for a general function g

$$d(S,T) \geq \sum_{i=1}^{k}(\sum_{x \in C_i} g(S_{min}, x) - g(S_{min}, C_{i_{min}}))$$

Define $Z = \sum_{x \in \tilde{S}} g(S_{min}, x) = \sum_{i=1}^{k}\sum_{x \in C_i} g(S_{min}, x)$. The expression $\sum_{i=1}^{k} g(S_{min}, C_{i_{min}})$ is bounded by the case when all cycles are 2-cycles. Since for every 2-cycle, C, with elements x and C_{min}: $g(S_{min}, C_{min}) \leq \frac{1}{2}(g(S_{min}, x) + g(S_{min}, C_{min}))$, it follows that $\sum_{i=1}^{k} g(S_{min}, C_{i_{min}}) \leq \frac{1}{2}Z$. Therefore, a lower bound for the distance of the optimal solution is:

$$d(S,T) \geq Z - \frac{1}{2}Z = \frac{1}{2}Z$$

We now show an upper bound on the distance computed by the CEA_{gs} algorithm, denoted by d_{alg}. Consider a modified version of the CEA_{gs} algorithm that sorts each cycle in the decomposition D with the α_{out} sorting method. Since the CEA_{ps} applied in step 6 of the CEA_{gs} is optimal, the distance computed by the CEA_{gs} algorithm may only be lower than the distance computed by the modified version. Let $C_1, \ldots, C_l$ be the cycles arbitrarily decomposed by the CEA_{gs} algorithm. We therefore have:

$$d_{alg} \leq \sum_{i=1}^{l}(\sum_{x \in C_i} g(S_{min}, x) + g(S_{min}, C_{i_{min}}))$$
$$\leq Z + \frac{1}{2}Z = 1\frac{1}{2}Z$$

The ratio between d_{alg} and $d(S,T)$ is: $\frac{d_{alg}}{d(S,T)} \leq \frac{1\frac{1}{2}Z}{\frac{1}{2}Z} = 3$. $\qquad\square$

Complexity: Since a $G_{S,T}$ computation and an arbitrary decomposition of $G_{S,T}$ can be computed in linear time and since the CEA_{ps} algorithm is a linear time algorithm, the CEA_{gs} algorithm runs in linear time.

4.2 $O(n \cdot \lg|\Sigma|)$ Time 1.72-Approximation Algorithm for the Summation Function

In this subsection we consider the special case of the summation function, i.e, $g(x,y) = w(x) + w(y)$. The $\alpha_{in}(C), \alpha_{out}(C)$ for a given cycle are therefore defined as follows:

- $\alpha_{in}(C) = \sum_{x \in C \setminus \{C_{min}\}} g(C_{min}, x) = \sum_{x \in C} w(x) + (|C| - 2) \cdot w(C_{min})$
- $\alpha_{out}(C) = \sum_{x \in C} g(S_{min}, x) + g(S_{min}, C_{min}) = \sum_{x \in C} w(x) + (|C| + 1) \cdot w(S_{min}) + w(C_{min})$

We show that applying the CEA_{ps} algorithm on the decomposition presented by [4] gives a 1.72-approximation ratio. The decomposition presented by [4] is basically the same as the decomposition of the CAE_{gs} except that it contains a maximum number of 2-cycles. This difference is represented by step 5 of the CAE_{gs}^+ (Fig. 5). We use the following lemma:

Lemma 2. [4] *Given an Eulerian directed graph $G = (V, E)$, then for every 2-cycle, C, in G there exists a decomposition of E into a maximum number of edge-disjoint directed cycles, in which C appears as a cycle in the decomposition.*

CEA_{gs}^+ algorithm

Data   : Input string S, target string T
Result : Transform S into T
begin

> **1.** Compute $G_{S,T}$.
> **2.** Compute a decomposition D of $G_{S,T}$ as follows:
> **3.** $D \leftarrow \emptyset$.
> **4.**   Add to D all the 1-cycles of $G_{S,T}$ and remove their edges.
> **5.**   Add to D a maximum number of 2-cycles from $G_{S,T}$
>   and remove their edges.
> **6.**   Add to D an arbitrary decomposition of the remaining edges.
> **7.** Apply the CEA_{ps} algorithm on D.

end

Fig. 5. 1.72-Approximation algorithm for the interchange rearrangement problem under ECM for general strings for the summation function

By Lemma 2 there exists a decomposition of $G_{S,T}$ into a maximum number of edge-disjoint directed cycles that contains a maximum number of 2-cycles. This can be shown inductively by repeatedly finding a 2-cycle and removing it from $G_{S,T}$ until there are no more 2-cycles. By lemma 2 in every stage, there exists a decomposition into a maximum number of edge-disjoint directed cycles that contains the chosen 2-cycle. Removing it results in a new graph G', which is also an Eulerian directed graph. Therefore, the same can be applied for G'. We prove the following theorem (the proof is omitted and given in the full version of the paper):

Theorem 3. *The CEA_{gs}^+ algorithm gives a 1.72-approximation ratio.*

Complexity: The CEA_{gs}^+ algorithm differs from the CEA_{gs} algorithm only in step 5 of CEA_{gs}^+. Since finding a maximum number of 2-cycles in $G_{S,T}$ can be done in $O(n \cdot \lg(|\Sigma|))$ time, the CEA_{gs}^+ algorithm runs in $O(n \cdot \lg(|\Sigma|))$ time.

References

1. Amir, A., Aumann, Y., Benson, G., Levy, A., Lipsky, O., Porat, E., Skiena, S., Vishne, U.: Pattern matching with address errors: Rearrangement distances. In: Proc. of the 17th annual ACM-SIAM Symposium on Discrete Algorithm (SODA), pp. 1221–1229 (2006)
2. Amir, A., Aumann, Y., Indyk, P., Levy, A., Porat, E.: Efficient computations of ℓ_1 and ℓ_∞. In: Ziviani, N., Baeza-Yates, R. (eds.) SPIRE 2007. LNCS, vol. 4726, pp. 39–49. Springer, Heidelberg (2007)
3. Amir, A., Aumann, Y., Kapah, O., Levy, A., Porat, E.: Approximate string matching with address bit errors. In: Proc. of the 19th Annual Symposium on Combinatorial Pattern Matching (CPM), pp. 118–130 (2008)

4. Amir, A., Hartman, T., Kapah, O., Levy, A., Porat, E.: On the cost of interchange rearrangement in strings. In: Arge, L., Hoffmann, M., Welzl, E. (eds.) ESA 2007. LNCS, vol. 4698, pp. 99–110. Springer, Heidelberg (2007)
5. Angelov, S., Kunal, K., McGregor, A.: Sorting and selection with random costs. In: Laber, E.S., Bornstein, C., Nogueira, L.T., Faria, L. (eds.) LATIN 2008. LNCS, vol. 4957, pp. 48–59. Springer, Heidelberg (2008)
6. Bafna, V., Pevzner, P.A.: Sorting by transpositions. SIAM Journal on Discrete Mathematics 11, 224–240 (1998)
7. Bender, M.A., Ge, D., He, S., Hu, H., Pinter, R.Y., Skiena, S., Swidan, F.: Improved bounds on sorting with length-weighted reversals. In: Proc. of the 15th annual ACM-SIAM Symposium on Discrete Algorithm (SODA), pp. 919–928 (2004)
8. Berman, P., Hannenhalli, S.: Fast sorting by reversal. In: Proc. 8th Annual Symposium on Combinatorial Pattern Matching (CPM), vol. 1075, pp. 168–185 (1996)
9. Carpara, A.: Sorting by reversals is difficult. In: Proc. 1st Annual Intl. Conf. on Research in Computational Biology (RECOMB), pp. 75–83 (1997)
10. Cayley, A.: Note on the theory of permutations. Philosophical Magazine 34, 527–529 (1849)
11. Christie, D.A.: Sorting by block-interchanges. Information Processing Letters 60, 165–169 (1996)
12. Gupta, A., Kumar, A.: Sorting and selection with structured costs. In: Proc. of the 42nd Symposium on Foundations of Computer Science (FOCS), pp. 416–425 (2001)
13. Gusfield, D.: Algorithms on strings, trees, and sequences: Computer science and computational biology. Cambridge University Press, Cambridge (1997)
14. Heath, L.S., Vergara, J.P.C.: Sorting by bounded block-moves. Discrete Applied Mathematics 88(1-3), 181–206 (1998)
15. Heath, L.S., Vergara, J.P.C.: Sorting by short swaps. Journal of Computational Biology 10(5), 775–789 (2003)

$\delta\gamma$ − Parameterized Matching

Inbok Lee[1], Juan Mendivelso[2], and Yoan J. Pinzón[2]

[1] School of Electronic, Telecommunication, and Computer Engineering
Hankuk Aviation University, Republic of Korea
`inboklee@hau.ac.kr`
[2] Department of System Engineering and Industrial Engineering
Research Group on Algorithms and Combinatorics (ALGOS-UN)
National University of Colombia, Colombia
`{jcmendivelsom,ypinzon}@unal.edu.co`

Abstract. This paper defines a new pattern matching problem by combining two paradigms: $\delta\gamma$–matching and parameterized matching. The solution is essentially obtained by a combination of bitparallel techniques and a reduction to a graph matching problem. The time complexity of the algorithm is $O(nm)$, assuming text size n, pattern size m and a constant size alphabet.

Keywords: combinatorial algorithms, δ–matching, $\delta\gamma$–matching, parameterized matching, bipartite matching.

1 Introduction

String searching is inarguably one of the foremost and most basic and useful computational primitives [6]. More formally, the input to the string matching problem consist of two strings: the pattern $P_{1..m}$ and the text $T_{1..n}$. The output should list all occurrences of the pattern string in the text string. The symbols in the strings are chosen from some set which is called an *alphabet*. An alphabet could be any collection of symbols and it is normally drawn from a set of pre-existing characters which is habitually designated as the common ASCII[1] code set. Nonetheless, in many real computing situations instead, the alphabet is drawn from a set of integer values. These integer strings are normally found in cipher text, financial data [44], meteorology data, image data, and music data [16, 23], to name some. If we were to seek for patterns in those strings of numbers, it would prove unrealistic and ineffective to seek for exactly the same values, but rather ought to search for a close instance of this pattern.

Delta-Gamma Matching. δ–matching algorithms are very effective in searching for all similar but not necessarily identical occurrences of a given pattern. In the δ–matching problem two integer strings of the same length match if the corresponding integers differ by at most a fixed bound δ. We also consider $\delta\gamma$–matching, where γ is a bound on the total sum of differences. Many kinds

---

[1] American Standard Code for Information Interchange.

A. Amir, A. Turpin, and A. Moffat (Eds.): SPIRE 2008, LNCS 5280, pp. 236–248, 2008.

of algorithms have been put forward to resolve this problem (*see* for instance [15, 19, 24, 25, 26]). According to our research, Cambouropoulos *et.al.* [15] was perhaps the first to propose an algorithm in this context, possibly motivated by Crawford *et.al.* [22] where some open problems were posed for applications that arise in bioinformatics, computer vision, but mainly, music information retrieval. Recently, several variants to this problem have been developed in order to allow for *don't care* symbols [21, 46], transposition-invariant [39] and gaps [17, 18, 28], among others. On the other hand, δ– and $\delta\gamma$–matching are closely related but not identical to the most common distance metrics L_1 and L_∞ also referred to as the Manhattan and Chevyshev Distances, respectively. For recent work in this direction, see *e.g.* [2, 4, 40, 41, 42, 43, 47].

Parameterized Matching. In this variant, two equal length strings (but not necessarily integer strings) parameterized-match if there exist a bijective function π from the alphabet for which every text symbol in one string is equal to the image under π of the corresponding symbol in the other string. In 1993, Brenda Baker [10] was the first researcher to have addressed this problem, and many others [3, 8, 11, 12, 13, 20, 29, 31, 36, 45, 48] since have followed Baker's work. She did, indeed, open up a wide-field of extensive research. Over the years, other lines of research that have been pursued are: parameterized matching under edit distance [14], parameterized matching under Hamming distance [7, 30], parameterized matching under LCS distance [34], multiple parameterized matching [33], 2-dimensional parameterized matching [1] and function matching [1, 5]. This accelerated research could only be justified by the usefulness of its practical applications such as in software maintenance [10], plagiarism detection [13], image processing [9, 49] and computational biology [1], to name some.

Our contribution. In this paper, we show that one can extend Baker's theory of parameterized string matching [10] to algorithms that support the δ– and $\delta\gamma$–distance. For a given pattern $P_{1..m}$ and a text $T_{1..n}$, we provide an algorithm for the δ–*Approximate Parameterized Matching* problem that takes time $O(nm)$, based on a bitparallel technique known as the SHIFT-AND algorithm [15] and a reduction to the *Maximum–Size Bipartite Matching* (MSBM) problem. For the MSPM problem, the classic solution is an $O(\sqrt{|V|}|E|)$ algorithm by Hopcroft and Karp [32]. However we use an improved algorithm by Feder and Motwani [27] that runs in $O((\sqrt{|V|}|E|)/k(|V|, |E|))$, where $k(x, y) = \log(x)/(\log(x^2/y))$. We furthermore give an $O(nm)$–time algorithm for the $\delta\gamma$–*Approximate Parameterized Matching* problem based on the SHIFT-PLUS bitparallel algorithm [15] and a reduction to the *Maximum–Weight Bipartite Matching* (MWBM) problem. The classic solution for the MWBM problem is the $O(|V|^3)$-time algorithm by Kuhn [37] (*a.k.a.* the *Hungarian Algorithm*). Nevertheless, we use a recent result by Kao *et.al.* [35] that runs in $O((\sqrt{|V|}W)/k(|V|, \frac{W}{N}))$, where N is the largest weight of any edge and W is the total weight of the graph. Therefore, there are two main contributions in this paper: (1) A formalization of the δ– and $\delta\gamma$-approximate parameterized matching problems and (2) simple but cost-effective solutions to both of these problems. All time complexities above assume constant alphabet size (*cf.* Section 3.1, 4.1 for asymptotically tighter bounds).

The outline of the paper is as follows: Some preliminaries are described in §2. An algorithm for the *δ–approximate parameterized matching* problem is presented in §3. We follow this in §4 by an algorithm for the *δγ–approximate parameterized matching* problem. Conclusions and further remarks are drawn in the last section.

2   Preliminaries

A string is a sequence of zero or more symbols from an alphabet Σ; the string with zero symbols is denoted by ϵ. (Recall that in the original definition of [10] for parameterized matching, the alphabet is supposed to be divided in *constants* and *parameters*. However, for our purposes, it is more convenient to consider the alphabet as being composed of parameters only. The extension to the general case is straightforward). The set of all strings over the alphabet Σ is denoted by Σ^*. Throughout the paper, the *alphabet* Σ is assumed to be an interval of integers and considered to be $\Sigma = \{1, 2, ..., \sigma\}$, $\sigma = |\Sigma|$. A *text* $T = T_{1..n}$ is a string of length n defined on Σ. T_i is used to denote the *i-th element* of T, $T_{i..j}$ is used as a notation for the *substring* $T_iT_{i+1}\cdots T_j$ of T, where $i, j \in \{1..n\}$. Similarly, a *pattern* $P = P_{1..m}$ is a string of length m defined on Σ. For easy notation, we use T^i to denote the substring of T of length m starting at position i, thus $T^i = T_{i..i+m-1}$.

For the *string comparison problem*, let m denote the (equal) length of two strings X and Y in Σ. Let δ, γ be two given numbers ($\delta, \gamma \in \mathbb{N}$). Then, X and Y are said to be *δ–matched* (denoted as $X \overset{\delta}{=} Y$), iff $\max_{j=1}^m |X_j - Y_j| \leq \delta$. Additionally, X and Y are said to be *δγ–matched* (denoted as $X \overset{\delta\gamma}{=} Y$), iff $\sum_{j=1}^m |X_j - Y_j| \leq \gamma$ and $X \overset{\delta}{=} Y$. For the same strings X and Y, we say that X *δ–parameterized-matches* Y (denoted as $X \overset{\delta}{\approx} Y$) if there exist a bijective function $\pi : \Sigma \to \Sigma$ such that $\max_{j=1}^m |X_j - \pi(Y_j)| \leq \delta$, for some permutation π of Σ. We also say that there is a *δγ–parameterized-match* between X and Y (denoted as $X \overset{\delta\gamma}{\approx} Y$) if there exist a bijective function $\pi : \Sigma \to \Sigma$ such that $\sum_{j=1}^m |X_j - \pi(Y_j)| \leq \gamma$ and $X \overset{\delta}{\approx} Y$. To give an example, let us assume that $\delta = 1$, $\Sigma=\{1,2,3,4,5\}$, $X=\{2,2,1,3,4,3,4,5,2,2\}$, and $Y=\{3,5,3,4,1,2,1,2,5,4\}$ (*cf.* Fig. 1). Then, $X \overset{\delta}{\approx} Y$ iff there exist a bijective function $\pi : \Sigma \to \Sigma$ such that $\max_{j=1}^{10} |X_j - \pi(Y_j)| \leq 1$. Note that each possible ordering/permutation of the set Σ corresponds to a specific bijection π, and there are $\sigma!$ different permutations for a σ-set. Hence, naively, one could compute all possible permutations of Σ to see whether there exist a permutation satisfying $\max_{j=1}^{10} |X_j - \pi(Y_j)| \leq 1$. For our running example, Fig. 1(d) lists all 120 possible permutations of given Σ. Only three of them (shown bold-underlined) make $X \overset{\delta}{\approx} Y$. Fig. 1(a,b,c) present the corresponding matching for each of them. Thus, one can safely conclude that $X \overset{\delta}{\approx} Y$. If your aim is also to check if $X \overset{\delta\gamma}{\approx} Y$ for, let us say, same δ and $\gamma = 6$, then only permutation $(5, 4, 1, 3, 2)$ (*cf.* Fig. 1(d)) will make $X \overset{\delta\gamma}{\approx} Y$.

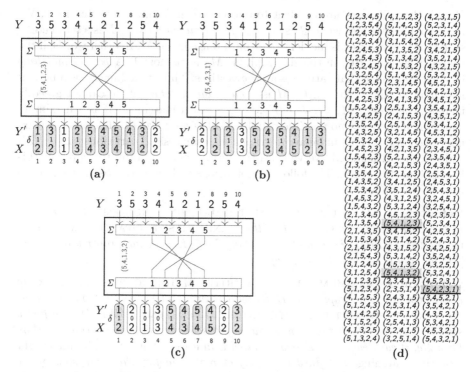

Fig. 1. $\delta-$ and $\delta\gamma-$parameterized matching example for $\Sigma=\{1,2,3,4,5\}$, $\delta=1$, $\gamma=6$, $X=\{2,2,1,3,4,3,4,5,2,2\}$, and $Y=\{3,5,3,4,1,2,1,2,5,4\}$, **(a)** permutation (5,4,1,2,3): $\delta-$equal but not $\delta\gamma-$equal, **(b)** permutation (5,4,2,3,1): $\delta-$equal but not $\delta\gamma-$equal, **(c)** permutation (5,4,1,3,2): both $\delta-$equal and $\delta\gamma-$equal, **(d)** All 120 possible permutations of Σ. Best permutations yielding a $\delta-$parameterized match are shown bold-underlined.

The main problems studied in this paper are for the *string pattern matching problem*, we formally defined the problems of $\delta-$ and $\delta\gamma-$approximate parameterized matching as follows:

Definition 1 ($\delta-$APPROXIMATE PARAMETERIZED MATCHING PROBLEM). *For a given text T, pattern P and integer δ, the $\delta-$APPROXIMATE PARAMETERIZED MATCHING PROBLEM (DAPM) is to calculate the set of all indices $i \in \{1..n-m+1\}$ satisfying the condition $P \overset{\delta}{\sim} T^i$. Note that the best permutation π yielding $P \overset{\delta}{=} \pi(T^i)$ is not necessarily the same at every position.*

Definition 2 ($\delta\gamma-$APPROXIMATE PARAMETERIZED MATCHING PROBLEM). *For a given text T, pattern P and integers δ and γ, the $\delta\gamma-$APPROXIMATE PARAMETERIZED MATCHING PROBLEM (DGAPM) is to calculate the set of all indices $i \in \{1..n-m+1\}$ satisfying the condition $P \overset{\delta\gamma}{\sim} T^i$. Note that the best permutation π yielding $P \overset{\delta\gamma}{=} \pi(T^i)$ is not necessarily the same at every position.*

In the sequel we also make use of the following graph-theoretic notions: An undirected graph $G(V,E)$ is *bipartite* if we can partition V into two sets L and

R such that all edges are incident to one vertex in L and one vertex in R. We henceforth sometimes write G as $G(L, E, R)$. A *matching* $M \subseteq E$ is a set of edges such that every vertex is incident to at most one edge in M. In other words, if the degree of every vertex in the subgraph (V, M) is at most 1. A *perfect matching* is a matching such that saturates all the vertices. Thus, matching M is a perfect matching iff $|L| = |R| = |M|$. M is a *maximal matching* if we cannot greedily increase the size of M, i.e. $\forall e \in E, M \cup e$ is not a matching. M is a maximum matching if there are no possible matchings of greater size. In this paper, we are interested in matching in bipartite graphs with same-sized partitions ($|L| = |R|$). Unless otherwise stated, we will use the term bipartite graph to represent this kind of bipartite graph. The following definitions will be central to the techniques used in this paper.

Definition 3 (MAXIMUM–SIZE BIPARTITE MATCHING PROBLEM). *For an undirected bipartite graph* $G(L, E, R)$, *the* MAXIMUM–SIZE BIPARTITE MATCHING PROBLEM[2] *(MSBM) is to find a matching* $M \subseteq E$ *with the property of* $|M'| \leq |M|$ *for any other matching* M' *of* G.

Definition 4 (MAXIMUM–WEIGHT BIPARTITE MATCHING PROBLEM). *For an undirected bipartite graph* $G(L, E, R)$ *with positive integer weights on the edges, the* MAXIMUM–WEIGHT BIPARTITE MATCHING PROBLEM[3] *(MWBM) is to find a set* $M \subseteq E$ *for which the sum of the weights of the edges is maximum. More formally, if each edge* $e_i \in E$ *is associated with a weight* w_i. *A* maximum–weight bipartite matching *is defined as a perfect matching for which the sum of the weights* w_i *associated with the edges in the matching has a maximal value, i.e. the perfect matching* M *that maximizes* $\sum_i w_i | e_i \in M$.

The following theorems are standard and crucial to our algorithms. They hold for both definitions of bipartite matching above.

Theorem 1 (Complexity of the MSBM problem). *For an undirected bipartite graph* $G(L, E, R)$, *the MSBM problem can be solved accurately and efficiently in* $O(\frac{\sqrt{|V|}|E|}{k(|V|,|E|)})$, *where* $k(x, y) = \frac{\log(x)}{\log(\frac{x^2}{y})}$, *by applying the Feder-Motwani algorithm [27].*

Theorem 2 (Complexity of the MWBM problem). *For an undirected bipartite graph* $G(L, E, R)$ *with positive integer weights on the edges. Let* N *be the largest weight of any edge. Let* W *be the total weight of* G. *Then, the MWBM problem can be solved accurately and efficiently in* $O(\frac{\sqrt{|V|}W}{k(|V|,\frac{W}{N})})$, *where* $k(x, y) = \frac{\log(x)}{\log(\frac{x^2}{y})}$, *via the Kao et.al. algorithm [35].*

For the bit-parallel operations we adopt the following notation. A machine word has w bits, numbered from the least significant bit to the most significant bit. We use C-like notation for the bitwise operations; & is bitwise AND, and | is bitwise OR.

---

[2] *a.k.a.* Maximum Cardinality Bipartite Matching Problem.

[3] *a.k.a.* Assignment Problem.

3 δ – Approximate Parameterized Matching

We begin by limiting our attention to the special case of *string comparison*, thus, when the two strings to be compared have the same length. Then we will extend this solution to the general case where the two strings are unequal in length, *i.e.* the *string pattern matching* case. In this case, the two strings are called: the pattern (shorter) string and the text string.

The core idea of our algorithm consists of three basic steps; they are as follows:

Step 1. [**Preprocessing Step**] Compute bitvector $\mathcal{D}$: $\forall \alpha \in \Sigma$, we set $\mathcal{D}[\alpha] = r$, where $r = r_\sigma..r_2 r_1$ is a σ-length binary number with $r_\ell = (\alpha \overset{\delta}{=} \ell ? 1 : 0)$, $\forall \ell \in \{1..\sigma\}$. The preprocessing step is carried out only once for the entire alphabet.

Step 2. [**Filter Step**] Once the $\mathcal{D}$ bitvector is computed, we compute bitvector $\mathcal{E}$ and $\mathcal{F}$ as follows: $\forall j \in \{1..m\}$, we set $\mathcal{E}[j] = \mathcal{D}[X_j]$, and $\forall \alpha \in \Sigma$, $\mathcal{F}[\alpha]$ is set as the bitwise AND operation among all $\mathcal{E}[j]$ with $Y_j = \alpha$.

Step 3. [**Matching Step**] Construct a balanced bipartite graph $G_\delta = (L, E, R)$ with bipartition (L, R) as follows: $L = R = \Sigma$. There is an edge $(u, v) \in E$ with weight 1 iff $\mathcal{F}[u]_v = 1$, $u, v \in \Sigma$. Then, we run a MSBM algorithm over G_δ to find a *maximum-size matching* $M \subseteq E$. We conclude that $X \overset{\sim}{\delta} Y$ iff $|M| = \sigma$ (*i.e.*, M is a perfect matching).

Step 1–2 identify candidate symbols for a successful renaming function from Y to X. Let us say that x and y are two corresponding symbols taken, *resp.*, from X and Y, and that y appears once more in Y but with corresponding symbol $x' \neq x$. Then, the permitted renamings (to make $x \overset{\delta}{=} y$ and $x' \overset{\delta}{=} y$) for y are clearly $\{x-\delta..x+\delta\} \cap \{x'-\delta..x'+\delta\}$. For example, if $\sigma = 10, \delta = 2, x = 4, x' = 6$ and $y = \star^4$, then $\pi(\star) \in \{2,3,4,5,6\} \cap \{4,5,6,7,8\} = \{4,5,6\}$. Hence, in order to solve the problem correctly, we need to find the permitted renamings for each symbol in Y. Note that $\mathcal{E}[j] = \mathcal{D}[X_j]$ allows us to know (independently) which are the permitted renamings for each Y_j using a bit number[5]. Since all sets are stored as binary words, the set-intersection operation can be perform as an $O(1)$-time bitwise AND operation. For our example, 0000111110 & 0011111000 = 0000111000. That is why, table $\mathcal{F}[\alpha]$ stores the intersection of all possible renamings of symbol α in Y. Therefore, $\mathcal{F}[\alpha]_\ell = 1$, iff ℓ is a valid renaming of α in Y such that if we rename all occurrences of α for ℓ, all these renamings will δ-math with their corresponding symbol in X. Since each $\alpha \in \Sigma$ in Y can have more than one renaming we can easily reduce this problem to a bipartite matching problem [7]. Having built the required bipartite graph G_δ as outlined in Step 3, it immediately follows that $X \overset{\sim}{\delta} Y$ iff there exist a perfect matching in G_δ.

Example 1. Let us assume $\Sigma = \{1,2,3,4,5\}$ and that we want to know whether $X=\{2,2,1,3,4,3,4,5,2,2\}$ is δ–parameterized equal to $Y=\{3,5,3,4,1,2,1,2,5,4\}$, for

---

[4] The symbol '$\star$' denotes any symbol in the alphabet.

[5] This underlying idea is similar, but not identical, to that in [15].

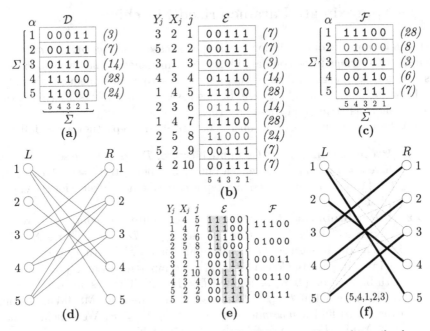

Fig. 2. (a) $\mathcal{D}$ bitvector, (b) $\mathcal{E}$ bitvector, and (c) $\mathcal{F}$ bitvector, if we arbitrarily choose to set $\Sigma = \{1, 2, 3, 4, 5\}$, $\delta = 1$, $X=\{2,2,1,3,4,3,4,5,2,2\}$ and $Y=\{3,5,3,4,1,2,1,2,5,4\}$. (d) Corresponding G_δ bipartite graph. (f) A perfect matching in G_δ. (e) An illustrative example of how one can use bitvector $\mathcal{E}$ to get bitvector $\mathcal{F}$. The numbers in parenthesis are the bit numbers represented as decimal numbers.

Algorithm 1. DAPM algorithm

Input: P, T, δ, Σ ▷ $\sigma=|\Sigma|, m=|P|, n=|T|, G_\delta(L,E,R), L=R=\Sigma$

Output: $\{i \in \{1..n - m + 1\} : T^i \overset{\sim}{\delta} P\}$

1. $\mathcal{D}[\alpha]_\ell \leftarrow \begin{cases} 1, & \text{if } \alpha \overset{\delta}{=} \ell \\ 0, & \text{otherwise} \end{cases}, \forall \alpha, \ell \in \Sigma$

2.   **for** $i \leftarrow 1$ **to** $n - m + 1$ **do**
3. $\mathcal{F}[\alpha] \leftarrow (2^\sigma - 1), \forall \alpha \in \Sigma$
4.   **for** $j \leftarrow 1$ **to** m **do** $\mathcal{F}[T_j^i] \leftarrow \mathcal{F}[T_j^i]$ & $\mathcal{D}[P_j]$
5. $E \leftarrow \emptyset$
6. $G_\delta.add\_edge(u, v)$ iff $\mathcal{F}[u]_v = 1, \forall u, v \in \Sigma$
7. $M \leftarrow \text{MSBM}(G_\delta)$
8.   **if** $|M| \leq \delta$ **then** OUTPUT(i)

Fig. 3. DAPM Algorithm

$\delta = 1$. We first compute bitvector $\mathcal{D}$ (*see* Fig. 2(a)), $\mathcal{E}$ (*see* Fig. 2(b)) and $\mathcal{F}$ (*see* Fig. 2(c)). Note, for instance, that $\mathcal{E}[3] = \mathcal{D}[X_3] = \mathcal{D}[1] = 00011$. Also notice how $\mathcal{F}[2] = \mathcal{E}[6]$ & $\mathcal{E}[8] = 01000$, because $Y_6 = Y_8 = 2$ (*see* Fig. 2(e) for an illustrative example of how to get bitvector $\mathcal{E}$ out from $\mathcal{F}$). Fig. 2(d) graphically depicts the bipartite graph generated by the algorithm in Step 3.

Fig. 2(f) shows a perfect matching solution of G_δ. Therefore, we conclude that $X \overset{\sim}{\delta} Y$ because Y can be renamed to $Y'=\{1,3,1,2,5,4,5,4,3,2\}$ using renaming function $\pi : (1,2,3,4,5) \rightarrow (5,4,1,2,3)$ (*cf.* Fig. 2(e)) that makes $X \overset{\delta}{=} Y'$.

In order to generalize this algorithm for δ–approximate parameterized matching we need to use these simple algorithmical ideas, $O(n)$ times, to check whether $P \overset{\sim}{\delta} T^i$, $i \in \{1..n - m + 1\}$. Fig. 3 shows the main steps of the algorithm and Section 3.1 analyzes its time complexity.

3.1   Time Complexity Analysis

In Line 1, bitvector $\mathcal{D}$ can be computed in $O(\sigma\lceil\frac{\sigma}{w}\rceil)$ time/space, where w is the computer size word. This time complexity can be achieve, for instance, by creating a binary word $z = 2^{2\delta+1} - 1$ with $2\delta + 1$ one-bits, and then shifting z accordingly to the left using a $O(\lceil\frac{\sigma}{w}\rceil)$-time bitwise Shift-Left operation. The body of the loop of line 2 is executed $O(n)$ times and corresponds to lines 3 to 8. Line 3 initializes the bitvector $\mathcal{F}$ in $O(\sigma\lceil\frac{\sigma}{w}\rceil)$ time/space. Line 4 takes $O(m\sigma\lceil\frac{\sigma}{w}\rceil)$ time. Note that Algorithm 1 dispenses with the use of bitvector $\mathcal{E}$. Bitvector $\mathcal{E}$ was used above to partially explain (for a better understanding) the computation of bitvector $\mathcal{F}$. Line 5 takes constant time. Note that each binary number in $\mathcal{F}$ is comprised of bitwise-AND operations on binary numbers with at most $2\delta + 1$ one-bits, so the total number of bits set to 1 in $\mathcal{F}$ is $O(\sigma\delta)$, hence line 6 can be implemented in $O(\sigma\delta)$ time (proportional to the number of bits set to 1 in $\mathcal{F}$). This is possible since $\log_2(b)$ determines the location of the most significant bit in a binary word b, then, in constant time, we can remove that bit and proceed to find the next significant bit on b and so on so forth until b becomes zero, thus, it is possible to extract all 1's from b in time proportional to the number of bits set to 1 in b. In line 7, the time complexity to find the maximum–size bipartite matching on G_δ using Theorem 1, $|V| = O(\sigma)$ and $|E| = O(\delta\sigma)$ is $O(\frac{\sqrt{\sigma}\sigma\delta}{k(\sigma,\sigma\delta)}) = O(\frac{\sigma^{1.5}\delta}{k(\sigma,\sigma\delta)})$, where $k(x,y) = \log(x)/\log(\frac{x^2}{y})$. Therefore, the total time complexity of Algorithm DAPM is

$$O\left(\sigma\lceil\tfrac{\sigma}{w}\rceil + n\left(\sigma\lceil\tfrac{\sigma}{w}\rceil + m\sigma\lceil\tfrac{\sigma}{w}\rceil + \sigma\delta + \tfrac{\sigma^{1.5}\delta}{k(\sigma,\sigma\delta)}\right)\right),$$

which is $O(nm)$ if we assume a constant alphabet. The overall space complexity is bounded by $O(\sigma\lceil\frac{\sigma}{w}\rceil) + \sigma + \sigma\delta)$.

4 $\delta\gamma$–Approximate Parameterized Matching

This algorithm follows the same main steps as the previous algorithm in Section 3. The basic steps are as follows:

Step 1. [**Preprocessing Step**] We need to compute the $\mathcal{D}$ bitvector as we did before, and the $\mathcal{G}$ bitvector: $\forall\alpha \in \Sigma$, we set $\mathcal{G}[\alpha] = s$, where $s = s_\sigma..s_2s_1$, $s_\ell = |\alpha-\ell|$ if $\alpha \overset{\delta}{=} \ell$, otherwise, $s_\ell = 0, \forall\ell \in \Sigma$. Each s_ℓ is in turn a binary number of d bits, $d = \lceil\log(\delta m)\rceil$. The preprocessing step is carried out only once for the entire alphabet.

Step 2. [**Filter Step**] Once the $\mathcal{G}$ bitvector is computed, we can use it to compute a second bitvector $\mathcal{H}$. $\forall \alpha \in \{1..\sigma\}$, we set $\mathcal{H}[\alpha] = \sum_{j=1}^{m}((Y_j = \alpha)\,?\,\mathcal{G}(X_j), 0)$.

Step 3. [**Matching Step**] We construct a balanced bipartite graph $G_{\delta\gamma} = G_\delta$ and assign weights to edges as follows: $\forall (u, v) \in E$, $w(u, v) = \delta m - \mathcal{H}[u]_v$. Then, we run a maximum-weight bipartite matching over $G_{\delta\gamma}$ to find a perfect matching M. Let $w(M)$ be the total weight of this match, then $X \mathbin{\overset{\sim}{\approx}} Y$ iff $(\sigma \delta m - w(M)) \leq \gamma$.

Apart from identifying candidate symbols for a successful renaming function from Y to X, we also need to know the total δ–differences we incur if we choose a candidate symbol. The main idea is by using bitparallelism: Each binary number in $\mathcal{H}$ will work like a collection of σ counters. We have to make sure that each counter will never overflow. That is why the size of each counter is set to $\log_2(m\delta)$. For example, if $X = Y\{3,3,3,3,3,3,3,3,3,3\}$ and $\delta = 2$, the maximum counter value will be 20, so we should use 5 bits capacity for each counter. This allow us to perform several additions in parallel, which is also the main idea used in [15]. Having bitvector $\mathcal{H}$, we can easily reduce this problem to a maximum-weight bipartite matching problem. Since we want to minimize the δ–differences, we built the required bipartite graph $G_{\delta\gamma}$ in a way such that the problem becomes of minimization. It immediately follows that $X \mathbin{\overset{\sim}{\approx}} Y$ iff there exist a matching M in $G_{\delta\gamma}$ of size σ and $w(M) \leq \gamma$.

Example 2. For our running example, suppose now that we are told that $\gamma = 6$ in order to prune the δ–match. If we use the permutation $(5, 4, 1, 2, 3)$ (*cf.* Fig. 2(e)) obtained by the previous algorithm, the sum of all errors is 8 (*cf.* Fig. 1(a)) and therefore we should conclude that X and Y do not $\delta\gamma$–parameterized match. If we use the above algorithm, we would get a different result. So let us see. We compute bitvector $\mathcal{G}$ (*see* Fig. 4(a)) and bitvector $\mathcal{H}$ (*see* Fig. 4(b)). Note, for instance, that $\mathcal{H}[2] = \sum_{j=1}^{10}((Y_j = 2)\,?\,\mathcal{G}[X_j], 0) = 0+0+0+0+0+\mathcal{G}[X_6]+0+ \mathcal{G}[X_8]+0+0 = \mathcal{G}[3]+\mathcal{G}[5] = 4112+4096 = 8209$ (0000 0010 0000 0001 0000). Fig. 4(c) graphically depicts the bipartite graph $G_{\delta\gamma}$ generated by the algorithm in Step 3. Note, for instance, that $w(2, 4) = \delta m - \mathcal{H}[2]_4 = 10 - 2 = 8$. Fig. 4(d) shows a perfect matching solution of $G_{\delta\gamma}$. Therefore, we conclude that $X \mathbin{\overset{\sim}{\approx}} Y$ because Y can be renamed to $Y'=\{1,2,1,3,5,4,5,4,2,3\}$ using renaming function $\pi : (1, 2, 3, 4, 5) \rightarrow (5, 4, 1, 3, 2)$ (found in Fig. 4(d)), and $X \overset{\delta\gamma}{=} Y'$ under this permutation.

In order to generalize this algorithm for $\delta\gamma$–approximate parameterized matching we need to use these simple algorithmical ideas, $O(n)$ times, to check whether $P \mathbin{\overset{\sim}{\approx}} T^i$, $i \in \{1..n - m + 1\}$. Fig. 5 shows the main steps of the algorithm and Section 4.1 analyzes its time complexity.

4.1   Time Complexity Analysis

The computation of $\mathcal{D}$ and $\mathcal{F}$ is just like in Algorithm 1. In Line 1, bitvector $\mathcal{G}$ can be computed in $O(\sigma\lceil\frac{\sigma d}{w}\rceil)$ space and $O(\delta + \sigma\lceil\frac{\sigma d}{w}\rceil)$ time, where w is the computer size word and $d = \log_2(m\delta)$. Line 3 initializes the bitvector $\mathcal{H}$ in $O(\sigma\lceil\frac{\sigma d}{w}\rceil)$

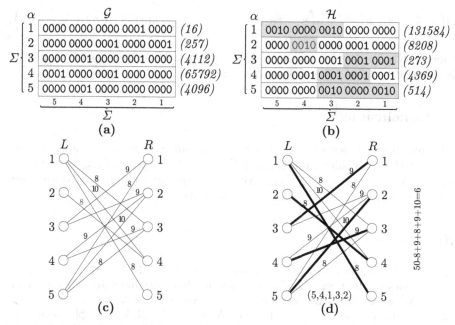

Fig. 4. An example of the **(a)** $\mathcal{G}$ bitvector and **(b)** $\mathcal{H}$ bitvector for $\Sigma = \{1, 2, 3, 4, 5\}$, $\delta = 1$, $\gamma = 6$, $X=\{2,2,1,3,4,3,4,5,2,2\}$ and $Y=\{3,5,3,4,1,2,1,2,5,4\}$. **(c)** Bipartite graph $G_{\delta\gamma}$. **(d)** A perfect matching solution.

Algorithm 2. DGAPM algorithm

Input: $P, T, \delta, \gamma, \Sigma$ $\triangleright$ $\sigma=|\Sigma|, m=|P|, n=|T|, G_{\delta\gamma}(L, E, R), L=R=\Sigma$

Output: $\{i \in \{1..n - m + 1\} : T^i \underset{\delta\gamma}{\approx} P\}$

1. $\mathcal{D}[\alpha]_\ell \leftarrow \begin{cases} 1, & \text{if } \alpha \overset{\delta}{=} \ell \\ 0, & \text{otherwise} \end{cases}, \mathcal{G}[\alpha]_\ell \leftarrow \begin{cases} |\alpha - \ell|, & \text{if } \alpha \overset{\delta}{=} \ell \\ 0, & \text{otherwise} \end{cases}, \forall \alpha, \ell \in \Sigma$

2. **for** $i \leftarrow 1$ **to** $n - m + 1$ **do**

3. $\mathcal{F}[\alpha] \leftarrow (2^\sigma - 1), \mathcal{H}[\alpha] \leftarrow 0, \forall \alpha \in \Sigma$

4.   **for** $j \leftarrow 1$ **to** m **do** $\mathcal{F}[T_j^i] \leftarrow \mathcal{F}[T_j^i]$ & $\mathcal{D}[P_j], \mathcal{H}[T_j^i] \leftarrow \mathcal{H}[T_j^i] + \mathcal{G}[P_j]$

5. $E \leftarrow \emptyset$

6. $G_{\delta\gamma}.add\_edge(u, v, \mathcal{H}[u]_v)$ iff $\mathcal{F}[u]_v = 1, \forall u, v \in \Sigma$

7. $M \leftarrow \text{MWBM}(G_{\delta\gamma})$

8.   **if** $((|M| \leq \delta)$ AND $(w(M) \leq \gamma))$ **then** OUTPUT(i)

Fig. 5. DGAPM Algorithm

---

time/space. Line 4 takes $O(m\sigma \lceil \frac{\sigma d}{w} \rceil)$ time. Line 6 can be implemented in $O(\sigma\delta)$ time. In line 7, the time complexity to find the maximum–weight bipartite matching using Theorem 2, $|V| = O(\sigma)$, $N = O(m\delta)$ and $W = O(m\delta|E|) = O(m\sigma\delta^2)$ is $O(\frac{\sqrt{\sigma}m\sigma\delta^2}{k(\sigma, \frac{m\sigma\delta^2}{m\delta})}) = O(\frac{m\sigma^{1.5}\delta^2}{k(\sigma, \delta^2)})$, where $k(x, y) = \log(x)/\log(\frac{x^2}{y})$. Therefore, the total time complexity of Algorithm DGAPM is

$$O\left(\sigma\lceil\tfrac{\sigma}{w}\rceil+\delta+\sigma\lceil\tfrac{\sigma d}{w}\rceil+n\left(\sigma\lceil\tfrac{\sigma}{w}\rceil+\sigma\lceil\tfrac{\sigma d}{w}\rceil+m\sigma\lceil\tfrac{\sigma}{w}\rceil+m\sigma\lceil\tfrac{\sigma d}{w}\rceil+\sigma\delta+\tfrac{m\sigma^{1.5}\delta^2}{k(\sigma,\delta^2)}\right)\right),$$

which is $O(nm)$ if we assume a constant alphabet. The overall space complexity is bounded by $O(\sigma\lceil\tfrac{\sigma}{w}\rceil)+\sigma\lceil\tfrac{\sigma d}{w}\rceil)+\sigma+\sigma\delta)$.

5 Conclusions

We have presented new $O(nm)$–algorithms that solve the δ– and $\delta\gamma$–approximate parameterized matching problem for two strings (the pattern $P_{1..m}$ and the text $T_{1..n}$). We believe that this complexity could be further improved since every alignment combines two previous alignments [38].

References

1. Amir, A., Aumann, Y., Cole, R., Lewenstein, M., Porat, E.: Function matching: algorithms, applications, and a lower bound. In: Proc. 30th International Colloquium on Automata, Languages and Programming, pp. 929–942 (2003)
2. Amir, A., Aumann, Y., Indyk, P., Levy, A., Porat, E.: Efficient computations of L_1 and L_∞ rearrangement distances. In: Proc. 14th String Processing and Information Retrieval, pp. 39–49 (2007)
3. Amir, A., Farach, M., Muthukrishnan, S.: Alphabet dependence in parameterized matching. Inform. Process. Lett. 49(3), 111–115 (1994)
4. Amir, A., Lipsky, O., Porat, E.: Approximate matching in the L_1 metric. In: Proc. 16th Annual Symposium on Combinatorial Pattern Matching, pp. 91–103 (2005)
5. Amir, A., Nor, I.: Generalized function matching. J. Discrete Algorithms 5(3), 514–523 (2007)
6. Apostolico, A., Galil, Z.: Pattern matching algorithms. Oxford University Press, Oxford (1997)
7. Apostolico, A., Erds, P.L., Lewenstein, M.: Parameterized matching with mismatches. J. Discrete Algorithms 5(1), 135–140 (2007)
8. Apostolico, A., Giancarlo, R.: Periodicity and repetitions in parameterized strings. Discrete Appl. Math. 156(9), 1389–1398 (2008)
9. Babu, G.P., Mehtre, B.M., Kankanhalli, M.S.: Color indexing for efficient image retrieval. Multimedia Tools and Applications 1(4), 327–348 (1995)
10. Baker, B.S.: A theory of parameterized pattern matching: algorithms and applications. In: Proc. 25th Annual Symposium on Theory of Computing, pp. 71–80 (1993)
11. Baker, B.S.: Parameterized pattern matching by Boyer-Moore-type algorithms. In: Proc. 6th Annual ACM-SIAM Symposium on Discrete Algorithms 1995, pp. 541–550 (1995)
12. Baker, B.S.: Parameterized pattern matching: algorithms and applications. J. Comput. Syst. Sci. 52(1), 28–42 (1996)
13. Baker, B.S.: Parameterized duplication in strings: algorithms and an application to software maintenance. SIAM J. Comput. 26(5), 1343–1362 (1997)
14. Baker, B.S.: Parameterized diff. In: Proc. 10th Symposium on Discrete Algorithms, pp. 854–855 (1999)

15. Cambouropoulos, E., Crochemore, M., Iliopoulos, C.S., Mouchard, L., Pinzon, Y.J.: Algorithms for computing approximate repetitions in musical sequences. Int. J. Comput. Math. 79(11), 1135–1148 (2002)

16. Cambouropoulos, E., Tsougras, C.: Influence of musical similarity on melodic segmentation: representations and algorithms. In: Proc. International Conference on Sound and Music Computing (2004)

17. Cantone, D., Cristofaro, S., Faro, S.: An efficient algorithm for δ-approximate matching with α-bounded gaps in musical sequences. In: Proc. 4th International Workshop on Efficient and Experimental Algorithms, pp. 428–439 (2005)

18. Cantone, D., Cristofaro, S., Faro, S.: On tuning the $(\delta;\alpha)$-sequential-sampling algorithm for δ-approximate matching with α-bounded gaps in musical sequences. In: Proc. 6th International Conference on Music Information Retrieval (2005)

19. Clifford, P., Clifford, R., Iliopoulos, C.S.: Faster algorithms for delta, gamma-matching and related problems. In: Proc. 16th Annual Symposium on Combinatorial Pattern Matching, pp. 68–78 (2005)

20. Cole, R., Hariharan, R.: Faster suffix tree construction with missing suffix links. In: Proc. 32nd ACM Symposium on Theory of Computing, pp. 407–415 (2000)

21. Cole, R., Iliopoulos, C.S., Lecroq, T., Plandowski, W., Rytter, W.: On special families of morphisms related to δ-matching and don't care symbols. In: Inform. Process. Lett., pp. 227–233 (2003)

22. Crawford, T., Iliopoulos, C.S., Raman, R.: String matching techniques for musical similarity and melodic recognition. Computers and Musicology 11, 72–100 (1998)

23. Cui, B., Jagadish, H.V., Ooi, B.C., Tan, K.: Compacting music signatures for efficient music retrieval. In: Proc. 11th international Conference on Extending Database Technology: Advances in Database Technology, pp. 229–240 (2008)

24. Crochemore, M., Iliopoulos, C.S., Lecroq, T., Plandowski, W., Rytter, W.: Three heuristics for δ-matching: δ-BM algorithms. In: Proc. 13th Annual Symposium on Combinatorial Pattern Matching, pp. 178–189 (2002)

25. Crochemore, M., Iliopoulos, C.S., Lecroq, T., Pinzon, Y.J., Plandowski, W., Rytter, W.: Occurence and substring heuristics for δ-matching. Fundam. Inf. 56(1-2), 1–21 (2003)

26. Crochemore, M., Iliopoulos, C.S., Navarro, G., Pinzon, Y., Salinger, A.: Bit-parallel (delta,gamma)-matching suffix automata. J. Discrete Algorithms 3(2-4), 198–214 (2004)

27. Feder, T., Montwani, R.: Clique partitions, graph compression and speeding-up algorithms. J. Comp. Sys. Sci. 51, 261–272 (1995)

28. Fredriksson, K., Grabowski, S.: Efficient Algorithms for (δ, γ, α) and $(\delta, k_\triangle, \alpha)$–matching. Int. J. Found. Comp. Sc. 19(1), 163–183 (2008)

29. Fredriksson, K., Mozgovoy, M.: Efficient parameterized string matching. Inform. Process. Lett. 100(3), 91–96 (2006)

30. Hazay, C., Lewenstein, M., Sokol, D.: Approximate parameterized matching. ACM Trans. Algorithms 3(3), 29 (2007)

31. Hazay, C.: Parameterized matching. Master's thesis, Bar-Ilan University (2004)

32. Hopcroft, J.E., Karp, R.M.: An $n^{5/2}$ algorithm for the maximum matching in bipartite graphs. SIAM J. Comp. 2, 225–231 (1973)

33. Idury, R.M., Schäffer, A.A.: Multiple matching of parameterized patterns. Theor. Comput. Sci. 154(2), 203–224 (1996)

34. Iliopoulos, C.S., Kubica, M., Rahman, M.S., Walen, T.: Algorithms for computing the longest parameterized common subsequence. In: Proc. 18th Annual Symposium on Combinatorial Pattern Matching, pp. 265–273 (2007)

35. Kao, M., Lam, T., Sung, W., Ting, H.: A decomposition theorem for maximum weight bipartite matchings. SIAM J. Comput. 31(1), 18–26 (2002)
36. Kosaraju, S.R.: Faster algorithms for the construction of parameterized suffix trees. In: Proc. 36th Annual Symposium on Foundations of Computer Science, pp. 631–637 (1995)
37. Kuhn, H.W.: The Hungarian methos for the assigment problem. Naval Res. Logist. Quart. 2, 83–97 (1955)
38. Landau, G.M., Vishkin, U.: Introducing efficient parallelism into approximate string matching. In: Proc. 18th ACM Symposium on Theory of Computing, pp. 220–230 (1986)
39. Lee, I., Clifford, R., Kim, S.K.: Algorithms on extended delta, gamma-matching. In: International Conference on Computational Science and its Applications, pp. 1137–1142 (2006)
40. Lipsky, O.: Efficient distance computations. Master's thesis, Bar-Ilan University (2003)
41. Lipsky, O., Porat, E.: Approximate pattern matching with the L_1, L_2 and L_∞ metrics (manuscript, 2002)
42. Lipsky, O., Porat, E.: Approximate matching in the L_∞ metric. Inf. Process. Lett. 105(4), 138–140 (2008)
43. Lipsky, O., Porat, E.: L1 pattern matching lower bound. Inf. Process. Lett. 105(4), 141–143 (2008)
44. Maasoumi, E., Racine, J.: Entropy and predictability of stock market returns. J. Econometrics 107(1), 291–312 (2002)
45. du Mouza, C., Rigauxb, P., Scholla, M.: Parameterized pattern queries. Data Knowl. Eng. 63(2), 433–456 (2007)
46. Pinzon-Ardila, Y.J., Christodoulakis, M., Iliopoulos, C.S., Mohamed, M.: Efficient (delta,gamma)-matching with don't cares. In: Proc. 16th Australasian Workshop on Combinatorial Algorithms, pp. 27–38 (2005)
47. Porat, E., Efremenko, K.: Approximating general metric distances between a pattern and a text. In: Proc. 9th Annual ACM-SIAM Symposium on Discrete Algorithms, pp. 419–427 (2008)
48. Salmela, L., Tarhio, J.: Sublinear Algorithms for Parameterized Matching. In: Proc. 17th Annual Symposium on Combinatorial Pattern Matching, pp. 354–364 (2006)
49. Swain, M., Ballard, D.: Color indexing. Int. J. Comput. Vis. 7(1), 11–32 (1991)

Pattern Matching with Pair Correlation Distance

Benny Porat, Ely Porat, and Asaf Zur

Department of Computer Science
Bar-Ilan University
52900 Ramat-Gan
Israel
{bennyporat,porately,zurasa}@cs.biu.ac.il
http://www.cs.biu.ac.il*

Abstract. In pattern matching with *pair correlation* distance problem, the goal is to find all occurrences of a pattern P of length m, in a text T of length n, where the distance between them is less than a threshold k. For each text location i, the distance is defined as the number of different kinds of mismatched pairs (α, β), between P and $T[i \ldots i+m]$. We present an algorithm with running time of $O\left(min\{|\Sigma_P|^2\, n \log m, n(m \log m)^{\frac{2}{3}}\}\right)$ for this problem. Another interesting problem is the *one-side pair correlation* distance where it is desired to find all occurrences of P where the number of mismatched characters in P is less than k. For this problem, we present an algorithm with running time of $O\left(min\{|\Sigma_P|\, n \log m, n \sqrt{m \log m}\}\right)$.

1   Introduction

Approximate pattern matching requires finding all occurrences of a pattern P in a text T where a match is defined by a distance metric and a threshold. The simplest distance metric is *Hamming distance*, where the distance in location i is the number of mismatches between the pattern and the sub-string $T[i \ldots i+m]$. Landau and Vishkin [LV86] used suffix trees and LCA queries to solve this problem in $O(nk)$. Amir et al. [ALP00] used Landau and Vishkin method and combined it with filtering and verification to get an algorithm that runs in $O\left(n\sqrt{k \log k}\right)$ and solves *Hamming distance* problem. A more generalized problem is the *edit distance problem*, which captures also insertion and deletion. It was presented by Levenshtein [Lev66] and a dynamic programming algorithm was presented by Lowrance and Wagner [LW75, Wag75]. Another distance metrics were defined in *parameterized matching* [Bak93, Bak96, Bak97, AFM94, HLS04], *function matching* [AAC+03], and *swap matching* [ALLL98]. The last three metrics are more sophisticated than the traditional distance metrics in the sense that they take into account relationships between mismatches. However, even those metrics don't take into account repetitive mismatches of the same kind, and the weight they give to each occurrence is equal for all mismatches.

---

* Research supported in part by US-Israel Binational Science Foundation.

A. Amir, A. Turpin, and A. Moffat (Eds.): SPIRE 2008, LNCS 5280, pp. 249–256, 2008.

In this paper we present a new metric, called *pair correlation distance*, which counts the number of different kinds of mismatched pairs. A mismatched pair $(\alpha, \beta), \alpha \in \Sigma_P, \beta \in \Sigma_T$, increases the distance only once, regardless of the number of times it occurs. Computational biology is a field where pair correlation is needed. The need for such a metric arises in computational biology. For example, if substance A is required for some reaction but it is missing, it will be replaced by some other substance B. This can be addressed by *pair correlation*. Such a case may occur in protein chain synthesis, when some amino-acid is replaced by another one due to a shortage or some malfunction (radiation that deteriorates the protein structure). The reason for the first substitute may cause another similar substitutes, hence it is very important to detect such connection between mismatches. Sometimes finding a connection between repeating mismatches can yield better explanations for experiments than traditional *edit distance*.

We also define another problem called *one-side pair correlation* where mismatches are counted by pattern characters only. This metric is used when we want to know that substance A is missing, but we are not interested which substance replaces it.

Pair correlation is well motivated by music retrieval [SYHC+99], stock market analysis [Nos04] and copy detection [SWW03], where mismatches influence each other.

2 Problem Definition

Following are some useful definitions that we use through this document.

Notation 1. *Let S_1 and S_2 be two equal length strings of size ℓ, over alphabets Σ_1 and Σ_2 respectively. The function $Occ(a, b)$ denotes the number of times the symbol $a \in \Sigma_1$ is aligned with the symbol $b \in \Sigma_2$. Formally,*
$Occ(a, b) = |\{1 \leq i \leq \ell : S_1[i] = a \land S_2[i] = b\}|.$

Definition 1. *Let S_1 and S_2 be two equal length strings, over alphabets Σ_1 and Σ_2 respectively. Pair Correlation Distance is $PC(S_1, S_2) = |\{(a, b) : Occ(a, b) \geq 1, a \neq b, a \in \Sigma_1, b \in \Sigma_2\}|$. In other words, $PC(S_1, S_2)$ counts the number of pairs $(a, b), a \in \Sigma_1, b \in \Sigma_2$ that are mismatched.*

Definition 2. *Let $T = t_1 \ldots t_n$ be a text, and $P = p_1 \ldots p_m$ be a pattern over alphabets Σ_T and Σ_P respectively, and let $k \in \mathbf{N}$. The Pair Correlation Distance problem of P and T with threshold k, is that of finding all locations $i = 1, ..., n$, where the Pair Correlation Distance of P and a prefix of $t_i \ldots t_n$ is no more than k, i.e. all locations where $PC(P, t_i \ldots t_{i+m-1}) \leq k$.*

Definition 3. *Let S_1 and S_2 be two equal length strings, over alphabets Σ_1 and Σ_2 respectively. One Side Pair Correlation Distance is $PC1(S_1, S_2) = |\{a : \exists b \in \Sigma_2, a \neq b, Occ(a, b) \geq 1\}|$. In other words, this metric counts the number of pattern symbols that caused at least one mismatch.*

Definition 4. *Let* $T = t_1 \ldots t_n$ *be a text, and* $P = p_1 \ldots p_m$ *be a pattern over alphabets* Σ_T *and* Σ_P *respectively, and let* $k \in \mathbf{N}$. *The* One-Side Pair Correlation Distance problem *of* P *and* T *with threshold* k, *is that of finding all locations* $i = 1, \ldots, n$, *where the* One-Side Pair Correlation Distance *of* P *and a prefix of* $t_i \ldots t_n$ *is no more than* k, *i.e. all locations where* $PC1(P, t_i \ldots t_{i+m-1}) \leq k$.

For example, consider the following two (equal length) strings:

abcaabbcd
fbeffbbee

Using traditional hamming distance, the number of mismatches is 6. Applying *two-side pair correlation distance*, the number of mismatches is only 3 because there are only three pairs that are mismatched (f, a), (e, c), (e, d). The *one-side pair correlation distance* gives only 2 mismatches because there are only two symbols that are mismatched: f and e. This example shows that *pair correlation distance* metric succeeds to detect the similarity between the two strings while under the traditional hamming metric they are not resemble each other. The result of this comparison helps a researcher to find out the real reason for the difference between those strings.

Remark. Algorithms that solve traditional pattern matching problems, usually find all text locations that match the pattern. The algorithms presented in this paper, reports the number of mismatches in each text location that is a match as well.

2.1   Naive Algorithms

The naive algorithm runs over all text alignments and compares each one of them to the whole pattern. It updates counters for each mismatch to calculate the moment distance in each alignment. The running time of the naive algorithm is $O(nm)$, because for each text alignment no more than m counters are updated.

2.2   Convolutions

In some cases we can improve the naive algorithm by using convolutions. The naive algorithm finds for each symbol in the pattern how many times it appears against each symbol in the text for each text location. For each symbol $a \in \Sigma_P$ in the pattern we make a convolution with every symbol $b \in \Sigma_T$ in the text (except the symbol a itself, since it is not a mismatch). These convolutions give all errors caused by each pattern symbol. Knowing how many mismatches each symbol caused it is possible to calculate *Pair Correlation Distance* for each location. Basically, this algorithm does what the naive algorithm does, but by using convolutions it achieves better running time than the naive algorithm when the alphabets are small.

The number of convolutions made is $O\left(|\Sigma_T| |\Sigma_P|\right)$, which gives a total running time of $O\left(|\Sigma_T| |\Sigma_P| n \log m\right)$.

For *one-side pair correlation* it is required to count how many errors each pattern symbol causes, regardless of the text symbols it is aligned with. Hence for each pattern symbol we make only one convolution to check how many errors it contributes. Therefore, the number of convolutions made is only $O\left(|\Sigma_P|\right)$, which result in running time of $O\left(|\Sigma_P|\, n \log m\right)$.

2.3   Filtering and Verification

In some cases it is possible to utilize some properties of the pattern and to divide the algorithm into two stages:

1. **Filtering -** In this stage a quick scan of the text is made in order to eliminate a considerable number of text locations.
2. **Verification -** In this stage each text location that passed the filtering stage is checked whether it is a match or not. Due to the filtering the number of locations to be verified is much lower than the total number of text locations.

In [ALP00] this method is used widely. However in our case, some changes are required as described in the next section.

3   Algorithm for One-Side

In this section we deal with *one-side pair correlation*. This is a simpler problem than *two-side pair correlation* since we are interested only in pattern symbols. For each pattern symbol we want to know whether it causes at least one mismatch or doesn't cause any. This can be achieved by using one convolution for each symbol, resulting in $|\Sigma_P|$ convolutions and running time of $O\left(|\Sigma_P|\, n \log m\right)$, as described in sub-section 2.2. This algorithm has reasonable running time when $|\Sigma_P|$ is smaller than m. However, when $|\Sigma_P| = O(m)$ the running time becomes $O(nm \log m)$ which is worse than the naive algorithm.

To improve this we define x to be a threshold such that if $|\Sigma_P| < x$ we make $|\Sigma_P|$ convolutions to solve the problem. The exact value of x will be determined later. From now on, we deal only with the case where $|\Sigma_P| \geq x$.

Definition 5. *A pattern symbol is called* frequent *if it appears more than* $\frac{m}{x+1}$ *times in the pattern. Otherwise it is called* rare.

The number of frequent symbols is no more than x, hence making a convolution for each frequent symbol results in running time of $O(nx \log m)$.

For rare symbols we use the filtering and verification method. In the filtering stage we look at the first occurrence of each of the pattern symbols. Following is the filtering algorithm:

The filtering stage actually counts how many first occurrences of pattern symbols are aligned with each text location. Each text location that got less than $|\Sigma_P| - k$ scores is discarded, since there are more than k mismatches. The number of locations that passed the filtering stage is at most $O\left(\frac{n}{|\Sigma_P| - k}\right)$. For each text

---

Input: a pattern P of length m and a text T of length n
1 Let `offset` be an array of all offsets of the first occurrence in the pattern of each symbol in Σ_P;
2 Let `score` be an array of length $|T|$, initialized to 0;
3 **for** $i = 1$ **to** n **do**
4   Let j=i-offset[T[i]];
5   **if** $j \geq 0$ **then** $score[i - j] \leftarrow score[i - j] + 1$;

Algorithm 1. Filtering Stage

location that passed the filtering stage we have to check only rare symbols, since the frequent symbols were counted by convolutions. Each rare symbol appears no more than $\frac{m}{x+1}$ times and there are no more than $|\Sigma_P|$ rare symbols, so for each text location we have to check at most $O\left(\frac{m|\Sigma_P|}{x+1}\right)$ locations in the pattern. We assume that $k \leq \frac{|\Sigma_P|}{2}$. Since there are no more than $O\left(\frac{n}{|\Sigma_P|-k}\right)$ text locations to check, total time for rare symbols is $O\left(\frac{nm}{x}\right)$. The total time for all symbols, rare and frequent is $O\left(xn \log m + \frac{nm}{x}\right)$. Optimizing over x values we get that the minimum is when $x = \sqrt{\frac{m}{\log m}}$, yielding a running time of $O\left(n\sqrt{m \log m}\right)$.

Conclusion: Total running time for the algorithm that solves *one-side pair correlation* is $O\left(\min\{|\Sigma_P| n \log m, n\sqrt{m \log m}\}\right)$.

In the above analysis we assumed that $k \leq \frac{|\Sigma_P|}{2}$. This assumption was made to bound the number of text locations that may pass the filtering stage. However, bounding k by $\frac{\sqrt{m}}{2\sqrt{\log m}}$, gives the same running time, because the filtering stage is done only when $|\Sigma_P| \geq \sqrt{\frac{m}{\log m}}$, which bounds the number of text locations that passed the filtering by $O\left(\frac{n}{|\Sigma_P|}\right)$. Hence, the constraint on k is that it should be less than $O\left(\max\{\frac{|\Sigma_P|}{2}, \sqrt{\frac{m}{\log m}}\}\right)$.

4 Algorithm for Two-Side

The algorithm showed above, for *one-side pair correlation distance*, can be extended to solve the problem of *two-side pair correlation distance*. We use a common technique in pattern matching, and divide the text into $\frac{n}{m}$ overlapping segments of size $2m$. In each segment there is a sub-segment that its alphabet is bounded by $O\left(|\Sigma_P| + k\right)$, otherwise there is no match.

The reason is that each text location that has more than $|\Sigma_P| + k$ different symbols is a mismatch. Hence, if a match exists there is a segment of size m with alphabet of size $|\Sigma_P| + k$. Because each segment is of size $2m$ there are no more than $O\left(|\Sigma_P| + k\right)$ different text symbols in each text segment. We assume that $k \leq \frac{|\Sigma_P|}{2}$. Using this assumption, the number of convolutions made is no more than $O\left(|\Sigma_P|^2\right)$.

As in *one-side pair correlation* we set a parameter x such that if $|\Sigma_P| < x$ we make $O\left(|\Sigma_P|^2\right)$ convolutions, otherwise we do the following.

We define a frequent symbol as in definition 5, namely, a symbol that appears more than $\frac{m}{x+1}$. In contrast to *one-side pair correlation* where we have to deal with pattern symbols only, here we have to handle also text symbols. There are four groups of symbols we have to check:

1. Frequent pattern symbols with frequent text symbols
2. Frequent pattern symbols with rare text symbols
3. Rare pattern symbols with frequent text symbols
4. Rare pattern symbols with rare text symbols

For the first group we use convolutions. There are no more than x frequent pattern symbols, and no more than $2x$ frequent text symbols, hence total number of convolutions is $O(x^2)$. To check all other groups we use the filtering and verification method. We apply algorithm **I** to eliminate text locations that got less than $|\Sigma_P| - k$ scores. After filtering stage there are no more than $O\left(\frac{n}{|\Sigma_P| - k}\right)$ locations to check. In the verification stage we check each rare pattern symbol naively. There are no more than $|\Sigma_P|$ rare symbols, and each one of them has no more than x occurrences, hence running time for each text location to count rare pattern symbols is $O\left(\frac{|\Sigma_P|m}{x+1}\right)$. The number of locations we have to check is no more than $O\left(\frac{n}{|\Sigma_P| - k}\right)$ (due to the filtering stage), hence total running time for rare pattern symbols is $O\left(\frac{nm}{x}\right)$. So far we checked frequent pattern symbols with frequent text symbols, and rare pattern symbols with frequent and rare text symbols. All we have to check is frequent pattern symbols with rare text symbols. This is done exactly as we checked rare pattern symbols - we check naively each text symbol. However, now we handle rare pattern symbols as don't care to avoid counting them twice. The running time for frequent pattern symbols with rare text symbols is also $O\left(\frac{nm}{x}\right)$.

The total time for all symbols, rare and frequent is $O\left(x^2 n \log m + \frac{nm}{x}\right)$. Optimizing over x values we get that the minimum is where $x = \sqrt[3]{\frac{m}{\log m}}$, yielding a running time of $O\left(n\left(m \log m\right)^{\frac{2}{3}}\right)$.

Conclusion: Total running time for *pair correlation distance* is $O\left(\min\{|\Sigma_P|^2 n \log m, n\left(m \log m\right)^{\frac{2}{3}}\}\right)$.

As in *one-side pair correlation* we assumed that $k \leq \frac{|\Sigma_P|}{2}$. Here, in *two-side pair correlation* this assumption has two reasons: to bound the number of different text symbols in each text segment when using convolutions (in case $|\Sigma_P| \leq \sqrt[3]{\frac{m}{\log m}}$), and to bound the number of text locations that may pass the filtering stage (otherwise).

Bounding k by $\frac{\sqrt[3]{m}}{2\sqrt[3]{\log m}}$, gives the same running time. In case $|\Sigma_P| \leq \sqrt[3]{\frac{m}{\log m}}$, the number of different text symbols is no more than $O\left(|\Sigma_P| + \frac{\sqrt[3]{m}}{2\sqrt[3]{\log m}}\right)$, hence the number of convolutions made is bounded by $O\left(\left(\frac{m}{\log m}\right)^{\frac{2}{3}}\right)$. In the other case, when $|\Sigma_P| > \sqrt[3]{\frac{m}{\log m}}$, the number of text locations that pass the filtering is bounded by $O\left(\frac{n}{|\Sigma_P|}\right)$ resulting in running time of no more than $O\left(n(m\log m)^{\frac{2}{3}}\right)$. Hence, the constraint on k is that it should be less than $O\left(\max\{\frac{|\Sigma_P|}{2}, \sqrt[3]{\frac{m}{\log m}}\}\right)$.

5 Summary

In this paper we presented and defined the problems of *one-side pair correlation* and *two-side pair correlation*. For *one-side pair correlation* we presented an algorithm that runs in $O\left(\min\{|\Sigma_P|\,n\log m, n\sqrt{m\log m}\}\right)$, for any k that is bounded by $O\left(\max\{\frac{|\Sigma_P|}{2}, \sqrt{\frac{m}{\log m}}\}\right)$. This algorithm was extended to solve the problem of *two-side pair correlation* with running time of $O\left(\min\{|\Sigma_P|^2\,n\log m, n(m\log m)^{\frac{2}{3}}\}\right)$, for any k that is bounded by $O\left(\max\{\frac{|\Sigma_P|}{2}, \sqrt[3]{\frac{m}{\log m}}\}\right)$.

References

[AAC+03] Amir, A., Aumann, Y., Cole, R., Lewenstein, M., Porat, E.: Function matching: Algorithms, applications, and a lower bound. In: Baeten, J.C.M., Lenstra, J.K., Parrow, J., Woeginger, G.J. (eds.) ICALP 2003. LNCS, vol. 2719, pp. 929–942. Springer, Heidelberg (2003)

[AFM94] Amir, A., Farach, M., Muthukrishnan, S.: Alphabet dependence in parameterized matching. Information Processing Letters 49, 111–115 (1994)

[ALLL98] Amir, A., Landau, G.M., Lewenstein, M., Lewenstein, N.: Efficient special cases of pattern matching with swaps. Information Processing Letters 68(3), 125–132 (1998)

[ALP00] Amir, A., Lewenstein, M., Porat, E.: Faster algorithms for string matching with k mismatches. In: Proc. 11th ACM-SIAM Symp. on Discrete Algorithms (SODA), pp. 794–803 (2000)

[Bak93] Baker, B.S.: A theory of parameterized pattern matching: algorithms and applications. In: Proc. 25th Annual ACM Symposium on the Theory of Computation, pp. 71–80 (1993)

[Bak96] Baker, B.S.: Parameterized pattern matching: Algorithms and applications. J. Comput. Syst. Sci. 52(1), 28–42 (1996)

[Bak97] Baker, B.S.: Parameterized duplication in strings: Algorithms and an application to software maintenance. SIAM J. Comput. 26(5), 1343–1362 (1997)

[HLS04] Hazay, C., Lewenstein, M., Sokol, D.: Approximate parameterized matching. In: Albers, S., Radzik, T. (eds.) ESA 2004, vol. 3221, pp. 414–425. Springer, Heidelberg (2004)

[Lev66] Levenstein, V.I.: Binary codes capable of correcting, deletions, insertions and reversals. Soviet Phys. Dokl. 10, 707–710 (1966)

[LV86]   Landau, G.M., Vishkin, U.: Introducing efficient parallelism into approximate string matching. In: Proc. 18th ACM Symposium on Theory of Computing, pp. 220–230 (1986)

[LW75]   Lowrance, R., Wagner, R.A.: An extension of the string-to-string correction problem. Journal of the ACM 22(2), 177–183 (1975)

[Nos04]   Nosovskij, G.V.: Mathematical analysis of stock market movement. In: 3rd International Conference on Cyberworlds (CW), pp. 320–321 (2004)

[SWW03]   Schleimer, S., Wilkerson, D.S., Winnowing, A.A.: Local algorithms for document fingerprinting. In: Proceedings of the International Conference on Management of Data (SIGMOD), pp. 76–85 (2003)

[SYHC⁺99]   Shmulevich, I., Yli-Harja, O., Coyle, E., Povel, D., Lemstrom, K.: Perceptual issues in music pattern recognition - complexity of rhythm and key finding. In: Proc. of AISB Symposium on Musical Creativity, pp. 64–69 (1999)

[Wag75]   Wagner, R.A.: On the complexity of the extended string-to-string correction problem. In: Proc. of the 7th ACM Symposium on the Theory of Computing (STOC), pp. 218–223 (1975)

Some Approximations for Shortest Common Nonsubsequences and Supersequences

Vadim G. Timkovsky

Faculty of Economics and Business
The University of Sydney, NSW 2006, Australia
v.timkovsky@econ.usyd.edu.au

Abstract. This paper is devoted to polynomial-time approximations for the problems of finding a shortest common nonsubsequence and a shortest common supersequence of given strings. The main attention is paid to the special case of the latter problem where all given strings are of length two. We show strong connections of this case to the feedback vertex set problem, the maximal network flow problem and the maximal multi-commodity network flow problem.

1 Introduction

Let L be a finite set of strings on an alphabet Σ. A *Shortest Common NonSubsequence* (SCNS) or a *Shortest Common Supersequence* (SCS) is a shortest string on Σ that is not a subsequence or, respectively, is a supersequence of every string in L. The problems of finding SCNS and SCS we call simply the SCNS *problem* and the SCS *problem*. Without loss of generality, we assume that no string in L is a subsequence of another string in L. We also assume that Σ^* and $|\sigma|$ stand for the set of all strings in Σ and the length of a string $\sigma \in \Sigma^*$, respectively, and that $|S|$ will denote the number of elements in a set S. For the rest of the paper, we set $n = |\Sigma|$, $k = |L|$, $l = \max\{|\sigma| : \sigma \in L\}$.

As well as the SCS problem, the SCNS problem has applications in manufacturing, bioinformatics and data processing but has a definitely shorter history and less studied than the SCS problem [8, 9, 12, 16, 17, 18, 19, 20].

The SCNS problem is NP-hard [16, 17] even if $n = 2$ [10] but can be solved in polynomial time if k is fixed [16, 17]. The problem originates from the study of the flexibility of a group technology presented by the set of technologies L with the set of technological operations Σ. If we are given a technology $\sigma \in \Sigma^*$ and asked whether σ can be fulfilled by L, we should check whether σ is a subsequence of one of the technologies in L. It is not always necessary to do it by a multiple sequence comparison operations if we know the length of an SCNS for L. Indeed, if $|\sigma| < |\text{SCNS}|$, then we can obviously avoid comparisons of σ with technologies in L and be sure that σ can be fulfilled by L. At our knowledge, approximations for the SCNS problem have not yet been studied.

The SCS problem is also NP-hard [9] even if $n = 2$ [14] but can also be solved in polynomial time if k is fixed [18, 19, 20]. For fixed l and r, let Ll, $SCSl$ and

A. Amir, A. Turpin, and A. Moffat (Eds.): SPIRE 2008, LNCS 5280, pp. 257–268, 2008.
© Springer-Verlag Berlin Heidelberg 2008

SCS$l(r)$ denote the set L with only strings of length l, an SCS of Ll, and an SCSl in the case where every symbol of Σ appears r times totally in all strings of Ll, respectively. Then the SCS2(2) problem is solvable in polynomial time, while the SCS2(3) and SCS3(2) problems are NP-hard [19].

Efficient polynomial-time approximations for the SCS problem are inherently impossible: a linear approximation implies P=NP, and an approximation with ratio $O(\log^\delta k)$ for a fixed $\delta > 0$ implies NP $\subseteq$ DTIME($2^{\text{polylog} k}$) [8]. There are known only trivial approximations with ratios n and l [8], tournament and greedy approximations with ratios $\frac{3k+2}{8}$ and $\frac{k+3}{4}$, respectively [6]. Another greedy approximation has length $|SCS| + O(|SCS|^{0.707})$ on the average [8]. The version of the SCS problem with the *SP-score* criterion is more tractable and has approximations with ratios $2 - \frac{2}{k}$ [7], $2 - \frac{3}{k}$ [12] and $2 - \frac{l}{k}$ for any fixed $l \le k$ [1]. Recent papers devoted to the SCS problem consider heuristics and computational experiments with them [2, 3, 4, 11, 13].

In this paper we show that, a polynomial algorithm for the *Shortest NonSubsequence* (SNS) problem, i.e., the SCNS problem with $k = 1$ [16, 17], leads to SCNS approximations with ratios n and l, i.e., the same approximations as for the SCS problem. For any fixed $m \le n - 2$, we propose an $O(n^{m+2})$ algorithm finding an SCS2 approximation with ratio $2 - \frac{3m}{n+m}$. Using a *maximal network flow model* [5], we show that this ratio can be reduced to $2 - \frac{4}{n+1}$ if $m = 1$.

Developing the network flow approach towards the SCS2 problem, we show how a nontrivial lower bound for $|SCS2|$ can be obtained by a *maximal multicommodity network flow model* [5].

We also propose a polynomial-time SCS approximation with ratios $n - \frac{1}{2}$ and $n - \frac{1}{4}$ for even and odd l, respectively, in the case where every string in L does not have symbol squares, as well as a polynomial-time SCS approximation with ratio $n(1 - \frac{m}{l+m})$ for any fixed $m \le |SCS| - l$. Possible ways of improving the results established here are discussed as well.

The acronyms used in this paper are collected in the following list:

NS = NonSubsequence	CS = Common Supersequence
SNS = Shortest NS	SCS = Shortest CS
CNS = Common NS	CS2 = CS of $L2$
SCNS = Shortest CNS	SCS2 = Shortest CS2

2   CS-CNS Relations and SCNS Approximations

Let $L = \{\omega_1, \ldots, \omega_k\}$, and let $\eta_1, \ldots, \eta_k$ be SNSs of $\omega_1, \ldots, \omega_k$, respectively. Note that an SNS of a string can be found in polynomial time [16, 17]. Let m be the length of a longest string among the SNSs.

Lemma 1. $m \le |SCNS|$.

Otherwise there exists i with $|\eta_i| = m$ such that an SCNS is a shorter NS of ω_i than η_i. Lemma 2 shows a simple way of constructing CNSs.

Lemma 2. Let ν be a CS of $\{\nu_1, \ldots, \nu_k\}$, where $\nu_1, \ldots, \nu_k$ are NSs of $\omega_1, \ldots, \omega_k$, respectively. Then ν is a CNS of $\{\omega_1, \ldots, \omega_k\}$.

Otherwise there exists i such that ν is a subsequence of ω_i, hence, ν_i is a subsequence of ω_i as well.

Let σ be an *alphabet string*, i.e., a string of length n that contains all symbols of Σ. Since each symbol of Σ appears in σ only once, and σ^m is a CS of $\{\eta_1, \ldots, \eta_k\}$, Lemmas 1 and 2 imply the inequality

$$\frac{|\sigma^m|}{|\text{SCNS}|} = \frac{m|\sigma|}{|\text{SCNS}|} \leq \frac{m|\sigma|}{m} = n$$

which proves the following theorem:

Theorem 1. σ^m *is an SCNS approximation with ratio* n.

We can also obtain an approximation with ratio l as follows. Let A_i be the set of symbols located in ith positions of strings $\eta_1, \ldots, \eta_k$, $i = 1, \ldots, m$. If $|\eta_j| < i$ for some j, then η_j contributes nothing to A_i. Let σ_i be a string in which the symbols of A_i are written in an arbitrary order. The string $\sigma_1 \ldots \sigma_m$ is a CS of $\{\eta_1, \ldots, \eta_k\}$. But $|\sigma_i| \leq l$ for $i = 1, \ldots, m$, so by Lemma 1,

$$\frac{|\sigma_1 \ldots \sigma_m|}{|\text{SCNS}|} = \frac{|\sigma_1| + \ldots + |\sigma_m|}{|\text{SCNS}|} \leq \frac{ml}{m} = l$$

Therefore, from Lemma 2 we have the following theorem:

Theorem 2. $\sigma_1 \ldots \sigma_m$ *is an SCNS approximation with ratio* l.

Lemma 3 establishes a relationship between SCNSs and SCSs and can be considered as dual to Lemma 2.

Lemma 3. *Let* ω *be an SCNS of* $\{\omega_1, \ldots, \omega_k\}$. *Then there exist NSs* $\nu_1, \ldots, \nu_k$ *of* $\omega_1, \ldots, \omega_k$, *respectively, such that* ω *is an SCS of* $\{\nu_1, \ldots, \nu_k\}$.

To prove it let us try to find the leftmost embedding of ω into ω_i for each $i = 1, \ldots, k$. Then ν_i can be defined as the shortest prefix of ω that is an NS of ω_i. Thus, ω is a CS of $\{\nu_1, \ldots, \nu_k\}$. Lemma 2 implies that ω is an SCS of $\{\nu_1, \ldots, \nu_k\}$ indeed. Note that ν_i is not necessarily an SNS of ω_i.

3   SCS2 and Feedback Vertex Set

Let us consider only CS2s where every symbol appears either once or twice. It is clear that all SCS2s are among them. Besides, if a symbol appears twice in a CS2 then it has its *left* position and its *right* position in it. There always exists an embedding of an L2 into its CS2 such that all the left/right positions (of symbols that appear twice) absorb only beginnings/ends of strings of L2. This embedding is obviously unique, and we call it a *regular embedding* of L2 into its CS2. Taking into account these observations, let us show a polynomial equivalence of the SCS2 problem and the following well-known graph problem.

Feedback Vertex Set (FVS). Given a directed graph G with the vertex set V and the arc set A, find a smallest subset $F \subseteq V$ covering all cycles in G.

If we identify vertices with symbols, arcs with strings of length two, and set $V = \Sigma$, $A = L2$, then the relationship between F and an SCS2 can be described as follows. Let $U \subseteq V$. Let us replace every vertex $u \in U$ by the pair of vertices u^{in} and u^{out} and switch all arcs going in/out of u to $u^{\text{in}}/u^{\text{out}}$. This operation converts G into a graph we call a *bifurcation* or, to be more specific, the U *bifurcation* of G. Vertices of U we call *bifurcated*, see Fig. 1.

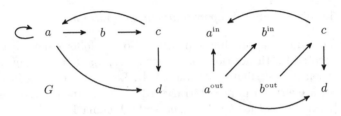

Fig. 1. The graph G for the string set $L = \{aa, ab, bc, ca, cd, ad\}$ on the alphabet $\Sigma = \{a, b, c, d\}$ and its $\{a, b\}$ bifurcation

The following lemma is a straightforward corollary from the definition of a bifurcation of a directed graph.

Lemma 4. U *covers all cycles in G if and only if the U bifurcation of G is acyclic.*

We call a sequence of all vertices of an acyclic directed graph a *chain* if it is compatible with its arcs. Note that a chain can be found by a *topological sorting* algorithm in polynomial time [15]. Chains naturally define strings on Σ.

Let C be a chain of an acyclic U bifurcation of G. Since C is compatible with arcs of G, every string in $L2$ is a subsequence of C, i.e., C defines a CS2, and the bifurcated vertices in C define symbols that appear twice in this CS2. On the other hand, since symbols appearing twice in CS2 define bifurcated vertices, a regular embedding of an $L2$ into its CS2 defines an acyclic bifurcation of G. Thus, we have

Theorem 3. CS2s *and only they define chains of acyclic U bifurcations of G. Hence* $|\text{CS2}| = n + |U|$.

Since Lemma 4 holds, Theorem 3 has the following corollaries that prove the required equivalence and a lower bound for $|\text{SCS2}|$.

Corollary 1. *An* SCS2 *defines a chain of an F bifurcation of G and vice versa. Hence* $|\text{SCS2}| = n + |F|$.

Corollary 2. *No m vertices cover all cycles in G if and only if* $|\text{SCS2}| > n + m$.

4 Binomial SCS2 Approximations

Now we describe an $O(n^{m+2})$ algorithm finding an SCS2 approximation with ratio

$$2 - \frac{3m}{n+m}$$

for any $m \le \frac{n}{2}$. As will will see, the algorithm produces an SCS2 if $m = \lfloor \frac{n}{2} \rfloor$. For each $k = 0, 1, 2, \ldots, m$ let us check whether there exist the vertex sets P_k and Q_k in G with k and $n - k$ vertices, respectively, such that each covers all cycles in G. We can do this by enumerating all $\binom{n}{k}$ subsets with k vertices and their compliments. If $P_0, P_1, \ldots, P_{k-1}$ do not exist but P_k does for some k, then $F = P_k$ and, by Corollary 1, a chain of the P_k bifurcation is an SCS2. Otherwise $P_0, P_1, \ldots, P_m$ do not exist, hence by Corollary 2, $|SCS2| > n + m$, and whether

(i) $Q_0, Q_1, \ldots, Q_k$ exist but Q_{k+1} does not for some $k < m$ or
(ii) $Q_0, Q_1, \ldots, Q_m$ exist.

In the case (i), we can obtain an SCS2 as a chain of the Q_k bifurcation. In the case (ii), we can take a chain of the Q_m bifurcation and, using Theorem 3, obtain an SCS2 approximation with length $2n - m$ and ratio

$$\frac{2n - m}{n + m} = 2 - \frac{3m}{n + m}.$$

In the worst case, we enumerate $2[\binom{n}{0} + \binom{n}{1} + \cdots + \binom{n}{m}] = O(n^m)$ subsets of V. For each subset, we need to check for cycles in the correspondent bifurcation in time $O(n^2)$ [15]. The total time requirement is $O(n^{m+2})$.

Thus, if $m = 1$, the ratio $2 - \frac{3}{n+1}$ can be reached in time $O(n^3)$. However, as we show in the following three sections, using a reduction to the maximal network flow problem, the ratio $2 - \frac{4}{n+1}$ can be reached in the same time.

5 Two Trivial SCS2 Approximations

Let, as before, σ be an alphabet string, and let γ_0 denote a chain of the graph G without cycles. Then we can define the string

$$\alpha_0 = \begin{cases} \gamma_0 & \text{if } G \text{ is acyclic} \\ \sigma\sigma & \text{otherwise} \end{cases}$$

Since $|\sigma\sigma| = 2n$ and an SCS2 repeats at least one symbol if $n > 1$, we have

Theorem 4. *If $n = 1$ then α_0 is an SCS2. If $n > 1$ then α_0 is an SCS2 approximation with ratio $2 - \frac{2}{n+1}$.*

Note that, if $L2$ contains strings aa for all $a \in \Sigma$, i.e., all vertices in G have loops, then $\alpha_0 = \sigma\sigma$ is an SCS2. Now we consider more precise approximations. The first one follows from the following evident fact.

Lemma 5. *If a directed graph has cycles and a vertex without a loop, the other vertices will cover all cycles in it.*

Let v be a vertex in G without a loop. Then Lemmas 4 and 5 imply that the $V-\{v\}$ bifurcation of G is acyclic. Let γ_1 denote a chain of such a graph G. Observing that γ_1 is an SCS2 if $n = 2$, we can define the string

$$\alpha_1 = \begin{cases} \gamma_0 & \text{if } G \text{ is acyclic} \\ \gamma_1 & \text{if } G \text{ has a vertex without a loop} \\ \sigma\sigma & \text{if all vertices in } G \text{ have loops} \end{cases}$$

Since $|\gamma_1| = 2n - 1$, we have

Theorem 5. *If $n \leq 2$ then α_1 is an SCS2. If $n > 2$ then α_1 is an SCS2 approximation with ratio $2 - \frac{3}{n+1}$.*

6 Rip Up, Sew Up, Cut Up

Without loss of generality, we assume that the graph G does not have loops (otherwise every vertex with a loop must be included in F) and is *strong*, i.e., *strongly connected* (otherwise, since each cycle in G is in a strong component of G, the FVS problem for G reduces to a number of separate FVS problems for strong components of G). Let us define the following derivatives from G:

$G - U$: the *reduction* of G on $U \subseteq V$, derived from G by deleting all vertices of U and adjacent arcs;

$G|G$: the V bifurcation of G we call further the *rip* of G; note that the rip has its vertices in pairs $\{v^{\text{in}}, v^{\text{out}}\}$, $v \in V$, and no cycles;

GG: the *sewn rip* of G, derived from $G|G$ by adding the *sewing arcs* $v^{\text{in}} \rightsquigarrow v^{\text{out}}$ for all $v \in V$;

$xGGx$: the *network* of G on $x \in V$, derived from GG by deleting the sewing arc $x^{\text{in}} \rightsquigarrow x^{\text{out}}$; we assume that x^{out} is the *source* and x^{in} is the *sink* in this network and that the *capacities* of sewing arcs and original arcs are one and unbounded, respectively, see Fig. 2.

The following lemma directly follows from the above definitions.

Lemma 6. *Any two sewing arcs in GG or $xGGx$ do not have common vertices. Besides, there are natural one-to-one correspondences between*

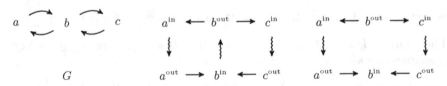

Fig. 2. The graph G for the string set $L = \{ab, ba, bc, cb\}$ on the alphabet $\Sigma = \{a, b, c\}$, the sewn rip GG and the network $bGGb$, where $F_b = \{b\}$, $C = \{a^{\text{in}} \rightsquigarrow a^{\text{out}}, c^{\text{in}} \rightsquigarrow c^{\text{out}}\}$

- vertices in G and sewing arcs in GG,
- cycles in G and cycles in GG,
- cycles through x in G and paths from x^{out} to x^{in} in $xGGx$.

Let $V(S)$ denote a vertex subset in G that corresponds to a sewing arc subset S in GG or $xGGx$, and let us consider the following auxiliary problems.

Feedback Cut (FC). Given a directed graph G with the vertex set V and a vertex $x \in V$, find a smallest set $F_x \subseteq V$ covering all cycles in G through x.

Network Cut (NC). Given a network N with the arc set E, integer capacities, a source s and a sink t, find a set $C \subseteq E$ of minimum total capacity covering all paths from s to t.

Note that the NC problem is dual to the *maximal network flow problem* with integer capacities that can be solved in polynomial time [5].

The equations $N = xGGx$, $s = x^{out}$, $t = x^{in}$ provide a polynomial reduction of the FC problem to the NC problem. Herein, C contains only sewing arcs, see Fig. 2. Thus, Lemma 6 implies $F_x = V(C)$ and

Lemma 7. *The FC problem can be solved in polynomial time.*

7   Cut-by-Cut Towards $2 - \frac{4}{n+1}$

The following algorithm finds a vertex set B as an approximation of F. Initially B is empty.

Cut-by-Cut Algorithm. Using a polynomial algorithm for the FC problem, choose a vertex x in G with the smallest *feedback cut* F_x, add F_x into B, and repeat this procedure for every nontrivial strong component of $G - F_x$.

Obviously, there are no cycles in $G - B$. The cut-by-cut algorithm grows a root tree that we call a *cut-by-cut tree* with the following properties:

- the *root* corresponds to the original strong graph G;
- the *leafs* correspond to nonbifurcated vertices of G;
- the *forks* correspond to nontrivial strong subgraphs of G, the vertices x chosen in them and feedback cuts F_x deleted from them; thus, it is convenient to denote a fork as x.

Moreover, every fork x has at least one leaf, it is x, and all *terminal* forks correspond to pairwise nonintersecting nontrivial strong subgraphs of G. The set of these subgraphs provides us a set of pairwise nonintersecting cycles in G. Besides, B does not contain leafs. So we have the following lemma.

Lemma 8. *Let f and g be the numbers of terminal forks and leafs in a cut-by-cut tree, respectively. Then $|F| \geq f$ and $|B| \leq n - g$.*

Since every feedback cut F_x is smallest, only one branch grows from a fork x if and only if G is a complete graph, and $G - F_x$ is a trivial graph with only one vertex x. Consequently, if G is not a complete graph then at least two branches grow from the root. This proves the following lemma.

Lemma 9. *If G is not a complete graph then $f + 1 \leq g$.*

Let r be the root of a cut-by-cut tree of height one. Since F_r is smallest, $|F_r| \leq |F|$, and since the height of the tree is one, the graph $G - F_r$ does not have cycles, i.e. F_r covers all cycles in G. This means that $B = F_r = F$.

Lemma 10. *If a cut-by-cut tree is of height one then $B = F$, i.e. the cut-by-cut algorithm finds an exact solution in this case.*

Lemma 11. (a) *If G is a complete graph then $B = F$ and $|B| = n - 1$.*
(b) *If G is not a complete graph and $n \leq 3$ then $B = F$ and $|B| = 1$.*
(c) *If G is not a complete graph and $n > 3$ then $|B| \leq n - 2$.*

The claims (a) and (b) immediately follow from Lemma 10: if G is a complete graph or $n \leq 3$, the cut-by-cut tree is of height one; if G is not a complete graph, then we get $|B| \leq n - 2$ from Lemmas 8 and 9 because $f \geq 1$.

Let, as before, G be a directed graph with the vertex set V, and let V_0 be the subset of vertices with loops. Let p and q denote the numbers of strong components of $G - V_0$ that are complete or have at most three vertices and, respectively, that are not complete and have more than three vertices.

Running the cut-by-cut algorithm for these $p+q$ strong graphs, we can produce the related vertex sets $B_1, \ldots, B_p$ and $C_1, \ldots, C_q$ for them, respectively, set $V_1 = B_1 + \ldots + B_p$, $V_2 = C_1 + \ldots + C_q$, and define the string

$$\alpha_2 = \begin{cases} \text{a chain of the } (V_0 + V_1) \text{ bifurcation} & \text{if } q = 0, \\ \text{a chain of the } (V_0 + V_1 + V_2) \text{ bifurcation} & \text{if } q > 0. \end{cases}$$

From Theorem 3, Corollary 2, and Lemmas 7 and 11 we obtain the ratio $\frac{2n-2}{n+1}$ and the following theorem.

Theorem 6. *If $q = 0$ then α_2 is an SCS2. If $q > 0$ then $n > 3$ and α_2 is an SCS2 approximation with ratio $2 - \frac{4}{n+1}$.*

8   Cut-by-Cut Umbrella

Let us imagine a root tree of height two that is a bunch of paths of length one and one more path of length one growing from the root. Holding the shortest path as a handle, we can really see an umbrella in hand. What is good in it?

The thing is the *umbrella* is the only version of a cut-by-cut tree of height at least two for which the inequality $f + 1 < g$ (i.e., the number of leafs exceeds the number of terminal forks by at least two) is false, see Lemmas 8 and 9. Therefore, if we improved the cut-by-cut algorithm so that the umbrella can be grown with the output $B = F$, we would increase Lemma 11, replacing $n-2$ by $n-3$, and get the ratio $2 - \frac{5}{n+1}$. The cut-by-cut algorithm grows an umbrella if, after deleting the first smallest feedback cut F_x from G, we obtain a graph with complete strong components. Herewith, there are at least one trivial component, x, and at least one nontrivial component among them. Perhaps, it is worth studying such graphs to attain better ratios.

9   SCS2 and Maximal Multi-commodity Flow

Let us consider again the sewn rip GG, see Section 6, and construct the network GVG placing intermediate vertices on all sewing arcs. In other words, we exchange the arcs $v^{in} \rightsquigarrow v^{out}$ in GG by pairs of the arcs $v^{in} \rightarrow v^{on}$, $v^{on} \rightarrow v^{out}$ for all $v \in V$, where v^{on} are new n vertices. Let v^{out} and v^{on} be the *sources* and the *sinks*, respectively, in GVG. Let the capacity of the arcs $v^{in} \rightarrow v^{on}$ be one, and let the capacity of the other arcs be unbounded.

It is not hard to verify a polynomial reduction from the FVS problem to the problem of finding an arc set F of minimal total capacity[1] that shuts off all n different flows in the network GVG from the sources v^{out} to the sinks v^{on}. The latter problem, however, can be formulated as the following program.

GVG Cut. Given a directed graph G with n vertices and the arc set A, find
nonnegative x_i^k, y_{ij}^k, z_i, where $i, j, k = 1, \ldots, n$, that minimize $z_1 + \ldots + z_n$
under the following conditions for all arcs $i \rightarrow j \in A$ and k:

$$\begin{cases} x_j^i - x_i^i - y_{ij}^i + z_i = 1 \\ x_j^k - x_i^k - y_{ij}^k + z_i = 0, \ i \neq k \\ z_i \in \{0,1\} \end{cases}$$

Note that $F = \{v^{in} \rightarrow v^{on} : z_v = 1\}$. If we treat arcs in G as tasks of an irregular (since G has cycles) network model for n projects, x_i^k, x_j^k, and y_{ij}^k as *start, completion,* and *processing times*, respectively, of the task $i \rightarrow j$ in project k, then the GVG cut problem is finding a smallest bifurcation of G (as a regular network model) in which all n projects can be finished. The tasks can be performed with delays determined by the variables z_i.

If the constraints $z_i \in \{0,1\}$ are replaced by the constraints $0 \leq z_i \leq 1$ for all $i = 1, \ldots, n$, the GVG cut problem turns into a linear program that is dual to the *maximal multi-commodity network flow problem* [5] for GVG.

Since the dual problem is a relaxation of the GVG cut problem, the quantity of a maximal multi-commodity network flow in GVG can be used as a lower bound for $|SCS2|$. This can lead to better SCS2 approximations.

10   SCS of Strings without Squares

Let us show how SCS2 approximations can be used for SCS approximations. Scanning strings of L from left to right let us cut them into segments of length two (of course, the last, i.e. the most right, segment may be of length one) and construct the string set L_1 from all the first, i.e., the most left, k segments, and the set L_2 from all the second segments, and so on. When constructing, the segments from shorter strings can be exhausted at certain point, and then the next string sets will have been formed at the expense of longer strings.

Thus, we obtain the sets $L_1, \ldots, L_t$, where $t = \frac{l}{2}$ for even l and $t = \frac{l-1}{2} + 1$ for odd l, where l is the length of a longest string in L. For even l, the sets consist

---

[1] Here we use the denotation F from the FVS problem to show its relevance.

of strings of length two. And only for odd l, the set L_t consists of one-symbol strings. Thus, an SCS for L_t, we denote it as τ_0, is trivial with $|\tau_0| \leq n$. If L consists of only strings without *squares*, i.e., aa, where $a \in \Sigma$, then strings in the sets L_i, $i = 1, \ldots, t$, have no squares as well. This means that the related graphs do not have loops (see Section 3).

Theorem 3 and Lemma 5 imply that there exist simple SCS2 approximations of length $2n - 1$ for all t sets like γ_1 (see Section 5). Let us find and concatenate them saving the numeration of the sets. And let us denote the resulted string as τ_1. Its length is at most $\frac{(2n-1)l}{2}$ for even l and at most $\frac{(2n-1)(l-1)}{2} + n$ for odd l. Dividing these two expressions by l, i.e., a lower bound for $|SCS|$, and setting

$$\tau = \begin{cases} \tau_0 & \text{if } l = 1 \\ \tau_1 & \text{if } l > 1 \end{cases}$$

we prove the following theorem:

Theorem 7. *Let L be a set of strings without squares on an alphabet with n symbols, and let l be the length of a longest string in L. Then the string τ is an SCS approximation for L with ratios $n - \frac{1}{2}$ and $n - \frac{1}{4}$ for even and odd l, respectively.*

Unfortunately, Lemma 11 does not work here.

11   Ratios Less Than Alphabet Size

Let, as before, σ be an alphabet string, and let l be the length of a longest string in L. Since l is a trivial lower bound for $|SCS|$ and $\frac{|\sigma^l|}{|SCS|} \leq \frac{nl}{l}$, the string σ^l is a trivial SCS approximation with ratio n. Decreasing the length of the SCS approximation and increasing the lower bound are two ways of reaching a better ratio. We will follow the latter, using the idea of Corollary 2.

Let $\circ$ be the empty symbol not belonging to Σ, and let Λ be the set of longest words in L. The denotations $\mathsf{Pref}_i(\lambda)$ and $\mathsf{Suff}_i(\lambda)$ will stand for the *prefix* and the *suffix* of length i of a string λ, respectively. Looking through the strings

$$\mu(a, \lambda, i) = \mathsf{Pref}_i(\lambda)\, a\, \mathsf{Suff}_{l-i}(\lambda)$$

for all triplets (a, λ, i), where $a \in \{\circ\} \cup \Sigma$ and $\lambda \in \Lambda$, where $i = 0, 1, \ldots, l$, we can check whether $\mu(a, \lambda, i)$ is an SCS of L. If such a string is found, we denote it as μ_1. Now let us consider the string

$$\mu_1' = \begin{cases} \mu_1 & \text{if the procedure founds } \mu_1 \\ \sigma^l & \text{otherwise} \end{cases}$$

If the procedure described above finds μ_1 then $\mu_1' = \mu_1$ is an SCS of L. Otherwise, $l + 1$ is a lower bound for $|SCS|$, and then μ_1' has ratio $\frac{nl}{l+1}$. This proves

Theorem 8. *Let L be a string set on an alphabet with n strings, and let l be the length of a longest string in L. Then the string μ_1' is an SCS approximation for L with ratio $n\left(1 - \frac{1}{l+1}\right)$.*

To improve the procedure let us replace strings $\mu(a, \lambda, i)$ by the strings

$$\lambda_{0i_1} a_1 \lambda_{i_1+1,i_2} a_2 \ \ldots \ a_{m-1} \lambda_{i_{m-1}+1,i_m} a_m \lambda_{i_m+1,l+1}$$

for all vectors $(a_1, \ldots, a_m \lambda, i_1, \ldots, i_m)$, where m is fixed, $a_1, \ldots, a_m \in \{\circ\} \cup \Sigma$, $\lambda \in \Lambda$, $0 \le i_1 < \ldots < i_m \le l + 1$, and λ_{uv} is a substring of $\circ\lambda\circ$ located in the positions $u, u + 1, \ldots, v - 1, v$. We assume that the positions 0 and $l + 1$ contain $\circ$. If one of these strings is an SCS, we denote it as μ_m. The procedure runs in polynomial time because m is fixed. Defining the string μ'_m in the same way as μ'_1 finishes the proof of the following theorem:

Theorem 9. *Let L be a string set on an alphabet with n symbols, and let l be the length of a longest string in L. Then for every fixed m, where $m \le |SCS| - l$, the string μ'_m is an SCS approximation for L with ratio $n\left(1 - \frac{m}{l+m}\right)$.*

12   Concluding Remarks

Section 2 shows that NSs prove to be connecting links between CNSs and CSs, and Lemma 3, as we suggest, can throw more light on the relationship between SCNS and SCS approximations. The cut-by-cut umbrella from Section 3 is an annoying hindrance in getting a better SCS2 approximation by the cut-by-cut algorithm. In fact, any procedure neutralizing the 'umbrella' effect in a combination with the cut-by-cut algorithm could give essentially more precise SCS2 approximations. The GVG cut problem from Section 9 can be useful not only for computing a lower bound for $|SCS2|$, but worthy of a special attention in studying the SCS problem by the multi-commodity network flow model. We also believe that an elaboration of the way of the decomposition of L into $L2$s can improve the results established in Sections 10 and 11.

Acknowledgements

The author would like to thank Anatoly Rubinov for his valuable contribution to the design of the multi-commodity network flow model used in this paper, and three anonymous referees for their critical comments.

References

1. Bafna, V., Lawler, E.L., Pevzner, P.A.: Approximation algorithms for multiple sequence alignment. Theoret. Comput. Sci. 182, 233–244 (1997)
2. Barone, P., Bonzzoni, P., Vedova, G.D., Mauri, G.: An approximation algorithm for the shortset common supersequence problem: an experimental analysis. In: Proc. of the 2001 ACM Symp. on Applied Comput., Las Vegas, March 11-14, pp. 56–60 (2001)
3. Cotta, C.: Memetic algorithms with partial lamarckism for the shortest common supersequence problem. In: Mira, J., Álvarez, J.R. (eds.) IWINAC 2005, vol. 3562, pp. 84–91. Springer, Heidelberg (2005)

4. Cotta, C.: A comparison of evolutionary approaches to the shortest common supersequence problem. In: Cabestany, J., Prieto, A.G., Sandoval, F. (eds.) IWANN 2005, vol. 3512, pp. 50–58. Springer, Heidelberg (2005)

5. Ford Jr., L.R., Fulkerson, D.R.: Flows in networks, Princeton, N.Y (1962)

6. Fraser, C.B., Irving, R.W.: Approximation alrorithms for the shortest common supersequence. Nordic J. Comput. 2(3), 303–325 (1995)

7. Gusfield, D.: Efficient methods for multiple sequence alignment with guaranteed error bounds. Bull. Math. Biol. 55(1), 141–154 (1993)

8. Jiang, T., Li, M.: On the approximation of shortest common supersequences and longest common subsequences. SIAM J. on Discrete Math. (1992)

9. Maier, D.: The complexity of some problems on subsequences and supersequences. J. ACM 25, 322–336 (1978)

10. Middendorf, M.: The shortest common non-subsequence problem is NP-complete. Theoret. Comput. Sci. 108, 365–369 (1993)

11. Ning, K., Leong, H.W.: Towards a better solution to the schortest common supersequence problem: the decomposition and reduction algorithm. BMC Bioinformatics 7(suppl. 4), S12 (2006)

12. Pevzner, P.A.: Multiple alignment, communication costs, and graph matching. SIAM J. on Appl. Math. 52, 1763–1779 (1992)

13. Pietrzak, K.: On the parameterized complexity of the fixed alphabet shortest common supersequence and longest common subsequence problems. J. Comput. System Sci. 67, 757–771 (2007)

14. Raiha, K.-J., Ukkonen, E.: The shortest common supersequence problem over binary alphabet is NP-complete. Theoret. Comput. Sci. 16, 187–198 (1981)

15. Reingold, E.M., Nievergelt, J., Deo, N.: Combinatorial algorithms. Prentice-Hall, Inc., Englewood Cliffs (1977)

16. Rubinov, A.R., Timkovsky, V.G.: Nonsimilarity combinatorial problems. BioSystems 30, 81–92 (1993); Pevzner, P.A., Gelfand, M.S. (eds.): Computer Genetics, pp. 81–92. Elsevier (1993)

17. Rubinov, A.R., Timkovsky, V.G.: String noninclusion optimization problems. SIAM J. Discrete Math. 11, 456–467 (1998)

18. Sankoff, D.: Minimum mutation tree of sequences. SIAM J. Appl. Math. 28, 35–42 (1975)

19. Timkovsky, V.G.: Complexity of common subsequnce and supersequence problems and related problems. Kibernetika 5, 1–13 (1989); Cybernetics 25, 565–580 (1990) (English translation)

20. Waterman, M.S., Smith, T.F., Beyer, W.A.: Some biological sequence metrics. Adv. Math. 20, 367–387 (1976)

On the Structure of Small Motif Recognition Instances

Christina Boucher, Daniel G. Brown, and Stephane Durocher

D. R. Cheriton School of Computer Science,
University of Waterloo, Waterloo, Ontario, Canada N2L 3G1
{cabouche,browndg,sdurocher}@cs.uwaterloo.ca

Abstract. Given a set of sequences, S, and degeneracy parameter, d, the CONSENSUS SEQUENCE problem asks whether there exists a sequence that has Hamming distance at most d from each sequence in S. A *valid motif set* is a set of sequences for which such a consensus sequence exists, while a *decoy set* is a set of sequences that does not have a consensus sequence but whose pairwise Hamming distances are all at most $2d$. At present, no efficient solution is known to the CONSENSUS SEQUENCE problem when the number of sequences is greater than three. For instances of CONSENSUS SEQUENCE with binary sequences and cardinality four, we present a combinatorial characterization of decoy sets and a linear-time exact algorithm, resolving an open problem posed by Gramm *et al.* [7].

1 Introduction

Understanding the structure and function of genomic data remains an important biological and computational challenge. *Motifs* are short sequences of genomic DNA responsible for controlling biological processes, such as gene expression. Motifs with the same function may not entirely match, due to random mutations or chemical properties. The *motif consensus* of the instances is a short sequence representing their shared pattern. Given a number of DNA sequences, *motif recognition* is the task of discovering motif instances in sequences without prior knowledge of the consensus or their placement within the sequence.

Closely related to the motif recognition problem is the CONSENSUS SEQUENCE problem that asks, given a parameter d and a set of sequences $S = \{s_1, \ldots, s_n\}$ each of length l, whether there exists a sequence s^*, which we call a consensus, that is of distance at most d from each sequence in S. Note that the consensus sequence need not be contained in S. In this context, the distance metric is the Hamming distance, $H(s_i, s_j)$, between two sequences s_i and s_j. CONSENSUS SEQUENCE is NP-complete, even for the case where each sequence is binary; therefore, no polynomial-time solution is possible unless $P = NP$ [5]. Clearly, a set for which the distance between any pair of sequences exceeds $2d$ cannot have a consensus. We say a set of sequences S is *pairwise bounded* if for all sequences $a, b \in S$, $H(a, b) \leq 2d$. Thus, the CONSENSUS SEQUENCE problem essentially reduces to discerning between pairwise bounded sets that have a consensus, and if so, finding one such sequence s^*, and those that do not. A set of sequences S

A. Amir, A. Turpin, and A. Moffat (Eds.): SPIRE 2008, LNCS 5280, pp. 269–281, 2008.
© Springer-Verlag Berlin Heidelberg 2008

is a *motif set* if there exists a consensus sequence, s^*. We say set S is a *decoy* set if S is pairwise bounded but does not have a consensus.

These problems – motif recognition and CONSENSUS SEQUENCE – have an extensive number of applications, due to the fact that many problems aim to determine if a set of sequences has a specific measure of similarity. For example, the CONSENSUS SEQUENCE problem arises in areas such as coding theory [3,5], data compression [6], and bioinformatics [7,8,10]. In the context of coding theory, a well-known problem related to CONSENSUS SEQUENCE asks if there exists a code that is not too far away from a given set of codes [3,5]. Given its applicability, the CONSENSUS SEQUENCE problem needs to be solved efficiently in practice. Li *et al.* [12] present a polynomial-time approximation scheme (PTAS) for CONSENSUS SEQUENCE. For a given value of r, all choices of r subsequences of length l are considered from the n sequences. The algorithm has $O(l(nm)^{r+1})$ run time, which is polynomial for any constant r. Many researchers have studied the algorithm due to Li *et al.* [12]; for a variant of CONSENSUS SEQUENCE there are known "weak" instances for which the approximation ratio is $1 + \Theta(1/\sqrt{r})$ [1], and "strong" instances for which the PTAS will be guaranteed to determine the correct answer in efficient time [2].

Another approach is to investigate the parameterized complexity of CONSENSUS SEQUENCE. A problem φ is said to be *fixed-parameter tractable* (FPT) with respect to parameter k if there exists an algorithm that solves φ in $f(k) \cdot n^{O(1)}$ time, where f is a function of k that is independent of n [8]. Gramm *et al.* [7] demonstrate that CONSENSUS SEQUENCE is FPT when the number of sequences remains fixed: the problem is polynomial-time solvable with a fixed number of sequences. This FPT result is based on an Integer Linear Programming (ILP) formulation with a constant number of variables (assuming n is fixed), and the application of the result of Lenstra [11], which states that ILP is polynomial-time solvable when the number of variables remains fixed. Unfortunately, such an ILP formulation is only of theoretical interest since the corresponding algorithms lead to very long running times even when the number of sequences is small (e.g., four sequences over a binary alphabet). Other parameterizations of the CONSENSUS SEQUENCE also exist; for example, when d is fixed, the problem can be solved in $O(nl + nd(d + 1)^d)$ time [8].

Gramm *et al.* [7] and Sze *et al.* [14] give direct (non-ILP based) combinatorial algorithms for solving CONSENSUS SEQUENCE exactly for three sequences. The algorithm of the former authors considers the possible combinations of alphabet symbols that can occur for three sequences, then specifies conditions for which a consensus sequence can be constructed [7]. Sze *et al.* [14] give a counting argument to demonstrate a condition for which a set of three sequences has a consensus and when it does not. In fact, a stronger property applies to binary sequences: any three pairwise-bounded binary sequences have a consensus.

Gramm *et al.* state that the problem of finding an efficient polynomial-time algorithm for solving CONSENSUS SEQUENCE on a set of four sequences remains open "due to the enormous combinatorial complexity [of the ILP-based solution]" [8, p. 13]. We resolve this open problem for binary sequences; specifically, we give an exact combinatorial algorithm for four binary sequences. This result is inspired by the combinatorial decomposition theorem for decoy sets that is also presented, which demonstrates that each decoy set can be characterized by containing two specific subsequences. Our aim is that these results might be extended to resolve the more general CONSENSUS SEQUENCE problem, in particular, for the four-symbol DNA alphabet, or for more than four sequences.

2 Preliminaries

We begin with some definitions concerning general sequence analysis. Let $l, d \leq l$, and n be positive integers and σ_i be a function that returns the ith symbol in a sequence. For any symbol $\beta \in \Gamma$ let β^l denote the l-length sequence of all β's. Given a set of sequences $S = \{s_1, \ldots, s_n\}$, each of which has length l, the ith *column* refers to the column vector $c_i = [\sigma_i(s_1), \ldots, \sigma_i(s_n)]^T$ in the $n \times l$ matrix representation of S. A sequence s^* is an *optimal sequence* for S if and only if there is no sequence s_2^* with $\max_{i=1,\ldots,n} H(s_2^*, s_i) < \max_{i=1,\ldots,n} H(s^*, s_i)$. Note that the optimal sequence for S is not unique; there may exist multiple. We formally define the CONSENSUS SEQUENCE problem as follows:

CONSENSUS SEQUENCE
INSTANCE: a set of n sequences, $S = \{s_1, s_2, \ldots, s_n\}$ over an alphabet Γ, each of length l, and a positive integer d.
FIND: a l-length sequence s^* over alphabet Γ where $H(s^*, s_i) \leq d$ for every s_i in S, or declare that no such s^* exists.

The difficulty of CONSENSUS SEQUENCE lies in distinguishing between decoy and valid motifs. In the context of coding theory, Frances and Litman show that CONSENSUS SEQUENCE remains NP-hard even when restricted to a binary alphabet; in this case, they refer to the corresponding problem as RADIUS DECISION [5]. We will be interested in the cardinality of a decoy set, that is, the number of sequences contained in the set. We say set $\hat{S} \subseteq S$ is a decoy of *minimal cardinality* if $\hat{S}$ is a decoy set such that for all $S' \subseteq S$, if $|S'| < |\hat{S}|$, then S' has a consensus.

Gramm *et al.* [7] refer to the process of permuting the columns of S such that these are grouped by column type as "normalization". A normalized instance can be derived from the input set of sequences by a simple linear-time algorithm. Given an optimal sequence for the normalized set of sequences, the inverse of this same permutation returns an optimal sequence for the original input [7].

Definitions Specific to Sets of Cardinality Four: Given a set $S = \{s_1, \ldots, s_4\}$ of binary sequences, the symbols in each column have either two, three, or four matching symbols. Sixteen types of columns are possible in general. We say a column belongs to *group i* if it has exactly i matching symbols. To reduce the number of possible types to eight, suppose without loss of generality that $s_4 = \beta^l$. Equivalently, create a new set S' by performing a logical exclusive-or of each

Table 1. The values $\lambda_{\alpha\beta\beta}$ through $\lambda_{\alpha\alpha\beta}$ denote the number of columns of each type in groups three and two. The symbol "-" implies that the value is undefined at these columns.

Group	Four	Three				Two		
# of columns.	$\lambda_{\beta\beta\beta}$	$\lambda_{\alpha\beta\beta}$	$\lambda_{\beta\alpha\beta}$	$\lambda_{\beta\beta\alpha}$	$\lambda_{\alpha\alpha\alpha}$	$\lambda_{\beta\alpha\alpha}$	$\lambda_{\alpha\beta\alpha}$	$\lambda_{\alpha\alpha\beta}$
s_1	β	α	β	β	α	β	α	α
s_2	β	β	α	β	α	α	β	α
s_3	β	β	β	α	α	α	α	β
s_4	β	β	β	β	β	β	β	β
maj_i	β	β	β	β	α	-	-	-

sequence in S with s_4 (say α corresponds to boolean true). A consensus sequence for S is found by performing another exclusive-or on a consensus sequence for S'. Let λ_{abc} denote the number of instances of column $(a, b, c, \beta)^T$, where $a, b, c \in \{\alpha, \beta\}$. See Table 1. Note that only columns of group three and two need to be considered, since any optimal sequence will have the majority vote at each column of group four. A pair of columns are considered to be *identical* if a pair of sequences in one column mismatch if and only if the same sequences mismatch in the second column. For example the column $[\alpha\alpha\beta\beta]^T$ is identical to $[\beta\beta\alpha\alpha]^T$, but neither is identical to $[\alpha\beta\beta\alpha]^T$.

Let maj_i denote the majority of the four symbols in column i. That is, $\text{maj}_i = \alpha$ if symbol α occurs three or more times in column i and $\text{maj}_i = \beta$ if symbol β occurs three or more times; maj_i is undefined if α and β each occur twice. Assuming that $s_4 = \beta^l$, only the columns associated with $\lambda_{\alpha\alpha\alpha}$ are such that $\text{maj}_i = \alpha$.

3   Ubiquitousness or Rareness of Bounded Decoy Sets

In this section, we consider the relative frequency, or infrequency, of decoy sets that do not have a proper subset that is also a decoy. Our empirical results demonstrate that the relative frequency of such decoy sets is minimal, and that the majority of decoy sets contain a decoy subset of cardinality four. Still, the results of Gramm et al. [8] imply that we cannot characterize all decoy sets of arbitrary size n as having a proper subset that is a decoy. We refer to a set Q of decoys, each of cardinality n, as having decoys of *bounded cardinality* if every decoy in Q has a proper subset that is a decoy.

Proposition 1. *Let Γ denote an alphabet of arbitrary fixed size. If $P \neq NP$, then for any n_0 there exist a decoy set S such that every subset of S of cardinality n_0 has a consensus.*

Proof. Suppose otherwise. That is, there exists an n_0 such that every decoy S of size $n \geq n_0$ has a subset of size n_0 that is a decoy. By Gramm et al. [8], for any fixed n_0, there exists an algorithm that decides whether a set of n_0 sequences is a decoy in $f(l, d)$ time, where $f(l, d)$ is polynomial in l and d. Consequently, for any set of $n \geq n_0$ sequences S, we can check each of the $\binom{n}{n_0}$ subsets of S of size n_0 to determine whether any is a decoy in time $O(n^{n_0} f(l, d)))$. That is, we can determine in polynomial time whether S is a decoy. Since CONSENSUS SEQUENCE is NP-complete, this is possible only if P = NP. □

It should be noted that this corollary does not preclude the fact that there may exist values of n where all decoys of cardinality n have the cardinality of the minimal decoy also as n. What the result implies is that there does not exist a threshold γ such that for all values of n greater than γ the minimal decoy sets have bounded cardinality below γ, if $P \neq NP$. Although Proposition 1 implies that no fixed n_0 exists, we conjecture that most decoys have a subset of size four that is a decoy. We provide evidence toward this property with an empirical study on random sets of binary sequences which we now describe. In turn, these results motivate the need for an efficient algorithm for determining whether a set of four sequences is a consensus; we describe such an algorithm in Section 4.2.

   We empirically investigate the rarity of the occurrence of decoy sets of cardinality n for which the cardinality of a minimal decoy set is large relative to n. We sampled without replacement 1000 times from the set of all possible pairwise-bounded sets of binary, l-length sequences; each set sampled has exactly n sequences taken from the binary alphabet. We varied the values for n, l and d. For each sample set, we determined whether the set is a decoy or a valid motif, with respect to the value of d, and determined the cardinality of the minimum decoy set. We repeated this experiment 10 times and calculated the mean values obtained. Table 2 outlines this data. One significant empirical trend demonstrates that as the number of sequences increased, the number of decoys that do not contain a minimal decoy of cardinality four became exponentially smaller; when

Table 2. Data obtained from calculating the average of 10 experiments that obtain a random sample, without replacement, of 1000 sequence sets and determine the size of the minimal decoy contained in each set decoy obtained in the sample. The first column is the cardinality of the minimum decoy set.

No. of sequences	$l = 8, d = 3$			$l = 10, d = 3$			$l = 15, d = 4$		
	$n = 6$	$n = 10$	$n = 12$	$n = 6$	$n = 10$	$n = 12$	$n = 6$	$n = 10$	$n = 12$
No. of valid motif	443.4	4.6	88.7	394.4	9.3	3.6	101.4	3.5	2.6
4	542.2	995.4	991.3	605.6	990.7	996.4	898.6	996.5	997.4
5	12.4	0	0	7	0	0	4.1	0	0
6	2	0	0	0.2	0	0	0.8	0	0
7	-	0	0	-	0	0	-	0	0
8	-	0	0	-	0	0	-	0	0
9	-	0	0	-	0	0	-	0	0
10	-	0	0	-	0	0	-	0	0
11	-	-	0	-	-	0	-	-	0
12	-	-	0	-	-	0	-	-	0

n was 10 and 12 the number of minimal decoys of size larger than four was 0. The only value of n for which decoys of size n were seen was 6. Further, the total number of decoys in 900,000 set of sequences sampled were approximately 1,500 in total. In summary, the empirical results appear to indicate that a large percentage of binary decoys can be characterized by containing a minimal decoy set of size four, the smallest size possible, further motivating the main results in this paper.

4   Investigating Binary Decoy Sets of Cardinality Four

Gramm *et al.* [7] suggest that a direct combinatorial approach to solve CONSENSUS SEQUENCE where n is fixed would be of practical and theoretical interest. Here we focus on partially resolving this open problem. We restrict interest to binary decoy sets of cardinality four, give a decomposition theorem and a linear-time, exact algorithm for these instances. We first prove that all binary sets of cardinality four can be decomposed into subsequences that have a specific characterization. The linear-time exact algorithm considers all possible combinations of symbols from the binary alphabet, and sequentially constructs a consensus or returns that no consensus exists.

4.1   A Decomposition Theorem

We will prove that each decoy of cardinality four can be decomposed into two subsequences that have a specific characterization. We begin by presenting the terminology and notation used to define these two subsequences. We define an $\alpha\beta$-set for an alphabet $\{\alpha, \beta\}$ as the set of all possible sequences of length two, that is, the set $\{\alpha\alpha, \alpha\beta, \beta\alpha, \beta\beta\}$. Given a set $S = \{s_1, s_2, s_3, s_4\}$ of cardinality four, we refer to S as containing an $\alpha\beta$-set if there are distinct indices i and j where the set of subsequences defined by columns i and j of S is an $\alpha\beta$-set. For example, the set of four sequences $\{\alpha\alpha\beta, \alpha\beta\beta, \beta\alpha\alpha, \beta\beta\beta\}$ contains an $\alpha\beta$-set at the first two columns. Next, we refer to a sequence s as *adequately far* if there exists a sequence, say $s_1 \in S$, such that $H(s, s_1) = d - 1$, and for all $s_i \in S$ the distance $H(s_i, s)$ is equal to either $d - 1$ or d. We refer to a sequence s^2 as *too close* if there exists some $s_i \in S$ with $H(s^2, s_i) \leq d - 2$ and $H(s^2, s_i) \leq d - 1$, for all $s_i \in S$. Putting these definitions together, we obtain the following property:

Definition 1. (Characterization of decoys of cardinality four) *A set of binary sequences S has property D if the following conditions hold:*

1. *S has an $\alpha\beta$-set realised at indices i and j, and*
2. *each optimal sequence for S', the set of sequences obtained from S by removing the columns i and j, is adequately far.*

We require Lemma 1 to prove our combinatorial decomposition theorem. The proof is omitted due to space constraints. We illustrate an example of property D in Figure 1: a decoy set where all sequences that have distance to the closest sequence equal to $d - 1$ and an $\alpha\beta$-set at the last two columns.

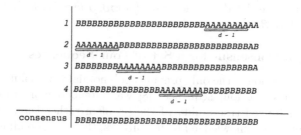

Fig. 1. Illustrates a decoy set such that optimal sequence to the set of sequences obtained from S by removing the last two columns that have the $\alpha\beta$-set, is adequately far, and more specifically each has distance equal to $d-1$ to the optimal sequence

Lemma 1. *Assume $d \geq 2$, $l \geq 2$ and $\Gamma = \{\alpha, \beta\}$. Every decoy set S of cardinality four contains an $\alpha\beta$-set.*

Theorem 1. (Decomposition theorem for cardinality four) *Assume $d \geq 2$, $l \geq 2$ and $\Gamma = \{\alpha, \beta\}$. A set S of cardinality four is a decoy if and only if there exists a set of subsequences contained in S such that property D holds.*

Lastly, we demonstrate the following result about the existence of an unique consensus for sets of arbitrary cardinality. It seems natural that as the cardinality of a valid motif set S increases relative to d, the number of consensuses for S decreases. As we observe in Proposition 2, a valid motif set of maximal cardinality has an unique consensus when $d < l$.

Proposition 2. *If a set S has a consensus but no superset of S has a consensus, then either:*

1. *$d < l$ and S has a unique consensus, or*
2. *$d \geq l$ and S is the set of all possible binary sequences of length l, each of which is a consensus for S.*

Proof. Case 1. Assume $d < l$. Assume S has two distinct consensuses, denoted by a and b. Since no superset of S has a consensus, all sequences c such that $H(a, c) \leq d$ must be in S. The same holds for all sequences e such that $H(b, e) \leq d$. Furthermore, sequences a and b must be in S. Consequently, $H(a, b) \leq d$. Let $\delta = H(a, b)$. Without loss of generality, assume that a and b differ in the first δ bits. Let f denote a binary sequence that agrees with a in the first δ bits, differs from a in the next d bits, and agrees with a in any remaining bits. Observe that $H(f, b) = H(f, a) + H(a, b) = d + \delta > d$. Consequently, f is not in S. Since $H(a, f) \leq d$, sequence a is a consensus of $S \cup \{f\}$. This derives a contradiction; our assumption must be false and the consensus of S must be unique.

  Case 2. Assume $d \geq l$. Any two binary sequences of length l differ in at most l bits. Since no superset of S has a consensus, $S = \Gamma^l$. $\square$

In some cases, a set of sequences S that has a consensus does not have a decoy as a superset whereas in other cases, every pairwise bounded superset of S is a decoy. For example, let $d = 1$, let $S = \{\beta\beta\beta\beta, \beta\beta\alpha\alpha, \beta\alpha\beta\alpha, \alpha\beta\beta\alpha\}$, and let $S' = \{\beta\beta\beta\beta, \beta\beta\beta\alpha, \beta\beta\alpha\alpha\}$.

4.2   Finding a Consensus for a Set of Four Sequences

As motivated in Section 1, the only previous polynomial-time solution for finding a consensus of a set of four sequences [8] was intended more to demonstrate the fixed-parameterized tractability of the problem rather than to provide an efficient solution. As acknowledged by its authors, the corresponding description (for which many details are omitted) results in an algorithm with extremely high (although theoretically linear) run time and, furthermore, does not lend itself well to simple or practical implementation. In this section, we present a simple linear-time algorithm for finding a consensus of a set of four binary sequences or determining that the set is a decoy. After describing the algorithm, we prove its correctness and show its worst-case run time is $O(l)$ for any arbitrary d.

Given a set of binary sequences $S = \{s_1, \ldots, s_4\}$, algorithm BINARYCONSEN-SUS4 identifies a consensus sequence s^* for S if one exists. Again, to simplify the algorithm's, description suppose $s_4 = \beta^l$. The algorithm greedily assigns symbols to s^*, one symbol at a time. Each column c_i is initially considered to be *free*; that is, no symbol has been assigned to $\sigma_i(s^*)$. Once it is assigned a symbol, we say column c_i is *fixed* and its value is not modified again. The algorithm has three phases in which columns of groups four, three, and two are fixed, respectively.

Phase One. Fix symbols of s^* in all columns of group four such that these agree with the symbol of the corresponding column.

Phase Two. The symbols of s^* in columns of group three are fixed sequentially. Say the first $i - 1$ columns of group three have been fixed and consider consider the ith such column. Let s_j denote the sequence of S that disagrees with the remaining three sequences in this column. Let s^+ denote the sequence given by the symbols of s^* in the fixed columns and the symbols of s_j in the free columns. If s^+ is a consensus for S, then let $s^* = s^+$ and return s^*. Otherwise, fix the current column of s^* to agree with the majority and continue to the next column of group three.

Phase Three. If phase three is reached, then only columns of group two remain, of which at most three types may be present. The free columns are fixed by selecting the number of columns of each type that will be assigned symbol α versus β. That is, a solution for columns of group two corresponds to a triple of integers (x, y, z), where $x \in [0, \lambda_{\beta\alpha\alpha}]$ denotes the number of columns of type $\lambda_{\beta\alpha\alpha}$ that will be assigned the symbol α and $\lambda_{\beta\alpha\alpha} - x$ represents the number that will be assigned the symbol β. The variables y and z are defined analogously. See Table 3. We denote the corresponding sequence by $s^*_{x,y,z}$. Therefore, the problem reduces to identifying an integer triple (x, y, z) selected from the region $R = [0, \lambda_{\beta\alpha\alpha}] \times [0, \lambda_{\alpha\beta\alpha}] \times [0, \lambda_{\alpha\alpha\beta}]$ that minimizes

$$f(x, y, z) = \max_{s_i \in S} H(s_i, s^*_{x,y,z}), \tag{1}$$

where

$$H(s_1, s^*) = \lambda_{\alpha\beta\beta} + x + \lambda_{\alpha\beta\alpha} - y + \lambda_{\alpha\alpha\beta} - z, \tag{2a}$$

$$H(s_2, s^*) = \lambda_{\beta\alpha\beta} + \lambda_{\beta\alpha\alpha} - x + y + \lambda_{\alpha\alpha\beta} - z, \tag{2b}$$

$$H(s_3, s^*) = \lambda_{\beta\beta\alpha} + \lambda_{\beta\alpha\alpha} - x + \lambda_{\alpha\beta\alpha} - y + z, \tag{2c}$$

$$H(s_4, s^*) = \lambda_{\alpha\alpha\alpha} + x + y + z. \tag{2d}$$

The sequence $s^*_{x,y,z}$ does not actually need to be constructed since the corresponding value of (1) is obtained in constant time upon fixing values for x, y, and z.

Table 3. The consensus s^* found by algorithm BINARYCONSENSUS4 (if one exists) is displayed in the last row. The values $\lambda_{\beta\beta\beta}$ through $\lambda_{\alpha\alpha\beta}$ denote the number of columns of each type and functions x, y, and z denote the number of occurrences of symbol α in the corresponding column as derived by the algorithm.

Column Group	Four	Three				Two		
Algorithm Phase	1	2				3		
Number of Columns	$\lambda_{\beta\beta\beta}$	$\lambda_{\alpha\beta\beta}$	$\lambda_{\beta\alpha\beta}$	$\lambda_{\beta\beta\alpha}$	$\lambda_{\alpha\alpha\alpha}$	$\lambda_{\beta\alpha\alpha}$	$\lambda_{\alpha\beta\alpha}$	$\lambda_{\alpha\alpha\beta}$
Set S s_1	β	α	β	β	α	β	α	α
s_2	β	β	α	β	α	α	β	α
s_3	β	β	β	α	α	α	α	β
s_4	β	β	β	β	β	β	β	β
Consensus s^*	β	β	β	β	α	x	y	z

Instead of evaluating all integer combinations for (x, y, z) (requiring $O(l^3)$ time), we identify a set $T \subseteq \mathbb{Q}^3 \cap R$ containing a constant number of triples such that the optimal (possibly non-integer) solution to (1) is a triple in T. Interpreted geometrically, (2a) through (2d) correspond to four respective hyperplanes in $\mathbb{R}^4$ whose maximum, $f(x, y, z)$,, defines a surface. Let

$$x_0 = \frac{1}{4}\left(-\lambda_{\alpha\beta\beta} + \lambda_{\beta\alpha\beta} + \lambda_{\beta\beta\alpha} - \lambda_{\alpha\alpha\alpha} + 2\lambda_{\beta\alpha\alpha}\right),$$

$$y_0 = \frac{1}{4}\left(\lambda_{\alpha\beta\beta} - \lambda_{\beta\alpha\beta} + \lambda_{\beta\beta\alpha} - \lambda_{\alpha\alpha\alpha} + 2\lambda_{\alpha\beta\alpha}\right),$$

$$z_0 = \frac{1}{4}\left(\lambda_{\alpha\beta\beta} + \lambda_{\beta\alpha\beta} - \lambda_{\beta\beta\alpha} - \lambda_{\alpha\alpha\alpha} + 2\lambda_{\alpha\alpha\beta}\right). \tag{3}$$

If $(x_0, y_0, z_0) \in R$, then let $T = \{(x_0, y_0, z_0)\}$. Otherwise, let T denote the set of triples that correspond to x-, y-, and z-coordinates of vertices of the intersection of the surface defined by (1) with the boundary of R. If this intersection is empty, then it follows that no consensus exists.

For each triple $(x, y, z) \in T$, evaluate the integer triples within unit ℓ_∞ distance of (x, y, z) in region R. That is, for every $(x, y, z) \in T$, consider the integer triples in $[\max(0, x - 1), \min(x + 1, \lambda_{\beta\alpha\alpha})] \times [\max(0, y - 1), \min(y + 1, \lambda_{\alpha\beta\alpha})] \times [\max(0, z - 1), \min(z + 1, \lambda_{\alpha\alpha\beta})]$, of which there are at most eight. Compute

(1) for each such integer triple (x, y, z) and store the corresponding minimizing sequence $s^*_{x,y,z}$. Let $s^* = s^*_{x,y,z}$.

Termination. Consider the maximum distance between s^* and a sequence in S, i.e., the minimum (integer) value of (1). If this value is at most d, then s^* is returned as a consensus sequence for S. Otherwise, S is a decoy set and no consensus sequence exists.

We now demonstrate that algorithm BINARYCONSENSUS4 correctly returns a consensus s^* for every set S that is a valid motif set. Furthermore, this is achieved in $O(l)$ time, independently of d. The proof of Theorem 2 refers to Lemmas 2 and 3 which follow.

Theorem 2. *Given any $d \in \mathbb{Z}^+$, any $l \in \mathbb{Z}^+$, and any set S of four binary sequences of length l, algorithm BINARYCONSENSUS4 returns a consensus for S with degeneracy parameter d if one exists or returns that S is a decoy in $O(l)$ time.*

Proof. The correctness of Phase 1 is straightforward. The correctness of Phase 2 follows by induction on i using Lemma 2. Consequently, if S has a consensus, then either a consensus has been found by the end of Phase 2 (i.e., $s^* = s^+$), or there exists a consensus s^* such that $\sigma_x(s^*) = \text{maj}_x$ for all columns of groups three and four. The optimal solution for the remaining free columns is found in Phase 3. The correctness of Phase 3 follows by Lemma 3. Therefore, algorithm BINARYCONSENSUS4 returns a consensus s^* if one exists, and returns that no consensus exists otherwise.

Each phase requires a single pass through the columns of S. Phase 1 simply requires counting the number of columns of each type. Phase 2 also requires maintaining the twelve distances $H(s_i, s_j^+)$ for each $\{i, j\} \subseteq \{1, \dots, 4\}$, where s_j^+ denotes s^+ for which the free columns are defined according to s_j (as described in Phase 2 of algorithm BINARYCONSENSUS4). Every time a column of group three is fixed, each of these twelve values can be updated in constant time. Phase 3 simply requires counting the number of columns of each type. Since (1) is defined by the maximum of four hyperplanes and region R is bounded by three pairs of parallel planes, the number of triples in T is constant and, furthermore, the coordinates of these triples are straightforward to compute in constant time. Finally, since any point in $\mathbb{R}^3$ has at most eight integer points within unit ℓ_∞ distance from it, the set of integer triples evaluated is also computed in constant time and space. Therefore, algorithm BINARYCONSENSUS4 terminates in $O(l)$ time. $\qquad\square$

Definition 2 (Majority Rule Property). *We say property $P(i)$ holds for a set S of four binary sequences if and only if either*

1. *S is a decoy, or*
2. *there exists a consensus of S for which the first i columns of group three have value maj_i.*

Lemma 2. *Let $S = \{s_1, \ldots, s_4\}$ denote a set of four binary sequences that has m columns of group three. If $P(i)$ holds for some $i \in \{0, \ldots, m-1\}$ and $s_j \in S$ denotes the sequence that mismatches in the $(i+1)$st column of group three, then either*

1. *$P(i+1)$ holds for S, or*
2. *s^+ is a consensus for S,*

where $\sigma_x(s^+) = \text{maj}_x$ in the first i columns of group three and $\sigma_x(s^+) = \sigma_x(s_j)$ in the remaining columns.

Proof. If s^+ is a consensus for S then the claim holds. Similarly, if S is a decoy then $P(i+1)$ is true and the claim holds. Therefore, suppose S is not a decoy and s^+ is not a consensus for S. By $P(i)$, S has a consensus s^* for which the first i columns of group three have value maj_x. Let k denote the index of the $(i+1)$st column of group three.

   Case 1. Suppose $\sigma_k(s^*) = \text{maj}_k$. Therefore, $P(i+1)$ is true, and the claim holds.

   Case 2. Suppose $\sigma_k(s^*) \neq \text{maj}_k$. That is, $\sigma_k(s^*) = \sigma_k(s_j)$. Since s^+ is not a consensus, s^* and s^+ must differ in at least one column; let k' denote the index of such a column. Let s^{**} denote a sequence of length l such that $\sigma_x(s^{**}) \neq \sigma_x(s^*)$ for $x \in \{k, k'\}$ and $\sigma_x(s^{**}) = \sigma_x(s^*)$ otherwise. That is, $\sigma_k(s^{**}) = \text{maj}_k$. Thus, $H(s_j, s^{**}) = H(s_j, s^*)$ and, furthermore,

$$\forall x \in \{1, \ldots, 4\}, \ H(s_x, s^{**}) \leq H(s_x, s^*).$$

Therefore, s^{**} is a consensus for S, $P(i+1)$ is true, and the claim holds. $\quad\square$

Lemma 3. *There exists an integer triple $(x, y, z) \in [0, \lambda_{\beta\alpha\alpha}] \times [0, \lambda_{\alpha\beta\alpha}] \times [0, \lambda_{\alpha\alpha\beta}]$ that minimizes (1) and is within unit ℓ_∞ distance from a triple in T, where set T contains either (3) or the set of triples that correspond to vertices of the intersection of the surface defined by (1) with the boundary of R.*

Proof. Since no two of the hyperplanes induced by (2a) through (2d) are parallel, $f(x, y, z)$ is a convex function whose surface includes a unique simplicial vertex located at the point of intersection of these four hyperplanes. Furthermore, this point minimizes $f(x, y, z)$ since f is increasing as it tends to infinity in any direction. Thus, $f(x, y, z)$ is minimized at a unique (possibly non-integer) point found by solving for x, y, and z in

$$H(s_1, s^*) = H(s_2, s^*) = H(s_3, s^*) = H(s_4, s^*). \tag{4}$$

The constraints of (4) corresponds to system of three linear equations with the unique solution (3).

   Since the coefficients of x in (2a) through (2d) are all ± 1, function $f(x, y, z)$ has slope ± 1 along the x-axis for any fixed y and z. The same holds for any fixed x and y or any fixed x and z. Consequently, since $f(x, y, z)$ is convex, a minimum integer solution to (1) lies within unit ℓ_∞ distance of its non-integer solution. The set T contains either the unique minimum (3) (if it lies within region R) or the set of triples that correspond to vertices of the intersection of the surface defined by $f(x, y, z)$ with the boundary of R, one of which must minimize $f(x, y, z)$ over R. Therefore, the claim holds.   □

5 Conclusion

Motif recognition, in which the objective is to identify meaningful patterns in biological data, is a fundamental problem of computational biology. We have obtained a combinatorial characterization of the consensus problem for instances of four binary sequences, and a linear-time algorithm for obtaining a consensus for this restricted set of instances. Our results generalize previous work and answer some open problems concerning CONSENSUS SEQUENCE [7]. We aim to generalize our current results to identify a combinatorial characterization of decoy sets over larger alphabets. Such a generalization would invite many open problems in motif recognition to be revisited, as their tractability might be determined more concretely, opening the possibility for more efficient algorithmic solutions.

Acknowledgements

We gratefully acknowledge research support of the National Sciences and Engineering Research Council of Canada.

References

1. Brejová, B., Brown, D., Harrower, I., López–Ortiz, A., Vinař, T.: Sharper upper and lower bounds for an approximation scheme for Consensus–Pattern. In: Apostolico, A., Crochemore, M., Park, K. (eds.) CPM 2005, vol. 3537, pp. 1–10. Springer, Heidelberg (2005)
2. Brejová, B., Brown, D., Harrower, I., Vinař, T.: New bounds for motif-finding in strong instances. In: Lewenstein, M., Valiente, G. (eds.) CPM 2006. LNCS, vol. 4009, pp. 94–105. Springer, Heidelberg (2006)
3. Cohen, G., Honkala, I., Litsyn, S., Sole, P.: Long packing and covering codes. IEEE Trans. Inf. Theory 43(5), 1617–1619 (1997)
4. Fellows, M., Gramm, J., Niedermeier, R.: On the parameterized intractability of motif search problems. Combinatorica 26(2), 141–167 (2006)
5. Frances, M., Litman, A.: On covering problems of codes. Th. Comp. Sys. 30, 113–119 (1997)
6. Graham, R.L., Sloane, N.J.A.: On the covering radius of codes. Trans. Inf. Theory 31, 385–401 (1985)
7. Gramm, J., Niedermeier, R., Rossmanith, P.: Exact solutions for Closest String and related problems. In: Eades, P., Takaoka, T. (eds.) ISAAC 2001, vol. 2223, pp. 441–453. Springer, Heidelberg (2001)
8. Gramm, J., Niedermeier, R., Rossmanith, P.: Fixed-parameter algorithms for closest string and related problems. Algorithmica 37(1), 25–42 (2003)

9. Gramm, J., Guo, J., Niedermeier, R.: Parameterized intractability of distinguishing substring selection. Th. Comp. Sys. 39(4), 545–560 (2006)
10. Lanctot, J.K., Li, M., Ma, B., Wang, S., Zhang, L.: Distinguishing string selection problems. In: Proc. SODA 1999, pp. 633–642 (1999)
11. Lenstra, W.H.: Integer programming with a fixed number of variables. Math. of OR 8, 538–548 (1983)
12. Li, M., Ma, B., Wang, L.: Finding similar regions in many strings. J. Comp. and Sys. Sci. 65(1), 73–96 (2002)
13. Pevzner, P., Sze, S.: Combinatorial approaches to finding subtle signals in DNA sequences. In: Proc. ISMB 2000, pp. 344–354 (2000)
14. Sze, S., Lu, S., Chen, J.: Integrating sample-driven and patter-driven approaches in motif finding. In: Jonassen, I., Kim, J. (eds.) WABI 2004. LNCS (LNBI), vol. 3240, pp. 438–449. Springer, Heidelberg (2004)

Exact Distribution of a Spaced Seed Statistic for DNA Homology Detection

Gary Benson[1,*] and Denise Y.F. Mak[2]

[1] Departments of Computer Science, Biology, Program in Bioinformatics,
Boston University, Boston, MA 02215
gbenson@bu.edu
[2] Graduate Program in Bioinformatics, Boston University, Boston, MA 02215
dyfmak@bu.edu

Abstract. Let a *seed*, S, be a string from the alphabet $\{1, *\}$, of arbitrary length k, which starts and ends with a 1. For example, $S = 11 * 1$. *S occurs* in a binary string T at position h if the length k substring of T ending at position h contains a 1 in every position where there is a 1 in S. We say that the 1s at the corresponding positions in T are *covered*. We are interested in calculating the probability distribution for the *number* of 1s covered by a seed S in an iid Bernoulli string of length n with probability of 1 equal to p. We refer to this new probability distribution as C_{nSp}, for *covered*, with S being the seed. We present an efficient method to calculate this distribution *exactly*. Covered 1s represent matching positions detected in DNA sequences when using multiple hits of a spaced seed. Knowledge of the distribution provides a statistical threshold for distinguishing true homologies from randomly matching sequences.

1 Introduction

Let a *seed*, S, be a string from the alphabet $\{1, *\}$, of arbitrary length k, which starts and ends with a 1. For example, $S = 11 * 1$ with length, $k = 4$. *S occurs* in a binary string T at position h if the length l substring of T ending at position h contains a 1 in every position where there is a 1 in S. We say that the 1s at the corresponding positions in T are *covered*.

More formally, let $S = s_1, \ldots, s_k$, and $T = t_1, \ldots, t_n$. Then S occurs at position h in T if for every $i \in [1..k]$ if $s_i = 1$ then $t_{h-k+i} = 1$. A 1 at t_j is *covered* by S if there exists some h and i such that S occurs at position h in T, $s_i = 1$, and $j = h - k + i$. For example, if $T = 101101111$ and $S = 11 * 1$, then S occurs in T at positions 6 and 9. Five 1s in T are covered, those at positions 3, 4, 6, 7, and 9:

```
1 2 3 4 5 6 7 8 9        1 2 3 4 5 6 7 8 9
1 0 1 1 0 1 1 1 1        1 0 1 1 0 1 1 1 1
    1 1 * 1                        1 1 * 1
```

* This research was supported in part by NSF grant IIS-0612153 and NIH grant 1 R01 GM072084.

A. Amir, A. Turpin, and A. Moffat (Eds.): SPIRE 2008, LNCS 5280, pp. 282–293, 2008.
© Springer-Verlag Berlin Heidelberg 2008

We are interested in calculating the probability distribution for the *number* of 1s covered by a seed S in binary strings of length n randomly generated by an iid (independent and identically distributed) Bernoulli process with p equal to the probability of 1 (success). For a *contiguous* seed (consisting entirely of 1s, for example $S = 111$), this reduces to the R_{nkp} probability distribution for the number of 1s which occur in *runs* of length k or longer (k is the number of 1s in S) in an iid Bernoulli string of length n [4,8]. For an example of these distributions, see figure 3. We will refer to the new probability distribution as C_{nSp}, for *covered*, with S being the seed.

The R_{nkp} distribution has proven useful as a criterion for detecting tandem and inverted repeats in DNA sequences [2,11]. In those applications, the binary string (here called the *representative string*) represents a gapless alignment between two homologous sequences (homologous means that the sequences have a common evolutionary ancestor) which have undergone random point mutations. As in the example below, a 1 represents a position where the two sequences remain the same and a 0 represents a position where they differ.

```
Sequence 1:             A C G T G C G T A A T T T C G
Sequence 2:             A C C A G C T T T A T T C C G
Representative string:  1 1 0 0 1 1 0 1 0 1 1 1 0 1 1
```

The parameter p models the expected probability that any two aligned positions (*i.e.* in a column) remain the same. A typical assumption might be that 80% of the positions are expected to remain the same (*i.e.* $p = 0.8$).

Algorithms that seek to detect mutated homologous sequences typically start by finding small matching words. A match is called a *hit* and each hit is tested to see if it is part of an extended region of homology. The *seed*, S, describes the shape of the small matching words. If $S = 111$, then in the example sequences above, the nucleotide triplet ATT is considered a hit because it appears in both sequences and shows up as 111 in the representative string. Hit testing for extended homology uses dynamic programming alignment or some approximation to alignment and is usually the most costly step in homology detection programs.

While searching with seeds is an effective method, it suffers from a serious drawback. A short seed (in the example, three matching letters) will occur frequently at random, especially in long genomic sequences, and as a result, too much time is spent pointlessly extending hits. One solution to this problem involves using a longer seed (say 11 or 12 letters), which significantly reduces the number of random hits, but also reduces the sensitivity of the detection program because many homologous sequences do not contain long unmutated stretches of nucleotides.

Another solution uses smaller seeds, but requires multiple seed hits clustered closely together. This is where the R_{nkp} distribution comes in. It provides an estimate for the expected number of matching aligned positions that participate in those hits. This permits the setting of a reasonable threshhold criterion to distinguish between clustered hits occurring randomly in unrelated sequences and clustered hits occurring in homologous sequences. Fortuitously, the R_{nkp} distribution for homologous sequences of distinct sizes can be obtained merely by varying the parameter n.

Spaced seeds. A few years ago, interest turned to spaced seeds [9,5,7,10] which increase single hit sensitivity without simultaneously increasing the number of random hits. While a contiguous seed, such as $S = 111$, is a short word, a spaced seed consists of a number of explicit positions which must match separated by "don't care" positions which may or may not match. The seed $11 * 1$ at the beginning of this section is a spaced seed. The 1s indicate matching positions and the $*$ indicates a wildcard position which may align with either a 1 or 0 (match or mismatch) in the representative string. In the aligned sequences above, the string pairs (GCGT, GCTT) and (TTTC,TTCC) are hits for the seed $11 * 1$ because in each pair, the first, second, and fourth letters match.

Spaced seeds have been studied primarily in terms of a single hit paradigm, in part, because there was no straightforward way to compute the distribution C_{nSp}. In this paper, we present an efficient method to calculate this distribution *exactly*, which also yields an efficient method for the exact computation of the R_{nkp} distribution. (On the way to this result, we developed a less efficient method for calculating the distribution, presented in [3].)

Importantly, our method is *probability independent* in that the calculation yields a formula in terms of the probability parameter p, and does not require fixing the value of p in advance. Previous methods for computing R_{nkp} have all required that p be specified in advance, and in that case, if the model parameter p changes, then the computation must be run again. The probability independent method is based on *probability equivalence classes* which partition the set of representative strings. This method has previously been used in a different form by us to identify *all* optimal spaced seeds for single hit homology detection [10]. These seeds are optimal in the sense that they have the highest sensitivity of all seeds with equivalent random hit probability.

Our coverage concept is similar to coverage defined in [6], although that paper deals with *minimum* coverage using a *fixed number* of spaced seeds, rather than, as here, a probability distribution on coverage for an unconstrained number of seeds. A dynamic programming algorithm is presented in [6] (which is similar in some ways to [3]) for computing minimum number of spaced *seeds* which hit alignments in which the number of mismatches is fixed. This minimum number of seeds is then used to compute a minimum *coverage* with a branch and bound algorithm.

The remainder of the paper is organized as follows. In section 2 we show how to efficiently calculate C_{nSp}, first in the probability independent manner, and then in the probability dependent manner. We give time and space complexities for the calculation on spaced and contiguous seeds. In section 3 we show how to modify the calculation to restrict the distribution to binary strings that represent confirmable alignments in the hit extension phase of a homology detection program. Finally in section 4 we present several graphs illustrating the distribution.

2   Methods

Probability equivalence classes. Binary strings (representative strings) which have the same probability, when the success probability parameter p is specified, belong to the same probability equivalence class. For strings of length n, there are $n + 1$ classes, corresponding to the possible number of 1s, *i.e.* $0, 1, 2, \ldots, n$. For example, when $n=100$, each string with 70 1s, *no matter how arranged*, has probability $p^{70}(1 - p)^{30}$.

For the purposes of this work, we further divide the equivalence classes. Let

$$X[n, i, j, b]$$

be the *number* of binary strings with i ones, j of which are covered by a given seed S when considering all strings of length n such that the strings all begin with the same prefix, represented by X, and all have the same initial covered positions, as indicated by b. For example if X represents 111111, and $b = 10101$, then *all* the strings counted in $X[n, i, j, b]$ start with 11111 and all have positions 1, 3, and 5 covered and positions 2 and 4 uncovered.

Modified Aho-Corasick tree data structure. Our data structure for counting strings in the equivalence classes is a simple modification of the Aho-Corasick tree (AC tree [1]) built on the patterns of the seed S. A *pattern* of a seed is any string of 1's and 0's which is obtained by replacing each wildcard in the seed by either a 1 or 0. For a seed with r stars, there are 2^r patterns.

The AC tree is modified as follows. Refer to figure 1. In a normal AC tree, one unlabeled directed fail edge exits every node. We dispense with these edges and for each node which now has less than two out edges, referred to as a *fail node*, we add one or two new directed *fail edges* to bring the total number of out edges to two. Each new fail edge is labeled with either a 0 or a 1 such that the two out edges at each node have distinct labels. Each fail edge terminates on a *fail-to node* determined as follows. Let the *node string* for node X be the concatenation of edge labels from the root to X. The node string has length depth(X), where the root has depth zero. If X is a fail node and the new fail edge out of X has label σ, then the fail-to node, Y, is the node whose node string is the longest proper suffix of the node string for X concatenated with σ.

Fail nodes and fail-to nodes are here called *marked nodes*. The X in the probability equivalence class notation above is a marked node in the AC tree. The string represented by X is the node string for X. b is a binary string of length $\min(\text{depth}(X) - 1, 0)$ which indicates a possible pattern of initial covered positions in a string that starts with the node string for X. For example, in figure 1, the node string for node X is 111111, and one possible pattern of covered positions is $b = 10101$.

The root, R, of the AC tree is a special node. It's probability equivalence class holds all the information necessary to determine C_{nSp}. If we fix n, p, S, and j (and realize that b for the root is the empty string b_ϵ), then the combined probability of all strings which have j covered 1s is

$$\sum_{i=j}^{n} (R[n, i, j, b_\epsilon]) \, p^i (1 - p)^{n-i}.$$

Allowing j to vary and summing the terms as above gives us C_{nSp} directly.

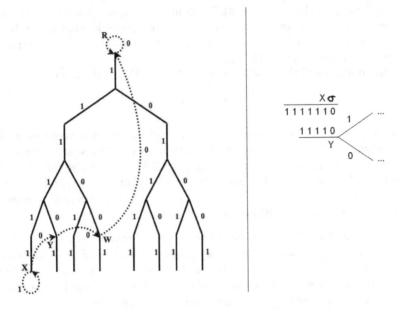

Fig. 1. Left: Modified Aho-Corasick tree for seed 1*1**1. Fail edges are shown for nodes X, Y, W, and R (labeled with --→.) Right: Idea for recursion. Node string to X followed by label σ of edge from X to Y replaces node string to Y in one set of strings through Y.

2.1 Recurrence Formula

Recurrence idea. The calculation of $X[n, i, j, b]$ for any marked node X is accomplished using a dynamic programming recurrence. We assume that the counts in the equivalence classes for each marked node for string lengths from 1 to $n-1$ have been computed. Each equivalence class saves some additional information about the extreme left ends of its strings (the prefix string represented by X and the coverage pattern by b). Then, inductively, computing an equivalence class for strings of length n consists of following out edges from node X in the AC tree, and accumulating the counts in the equivalence classes down those edges. The effect is to substitute the node string for X in place of an existing prefix, possibly resulting in some additional covered bits (figure 1).

Computing the equivalence class counts. Assume that the recurrence is being computed for marked node X. It has two out edges and depending on X, each out edge is either a fail edge or an edge included in the original AC tree as follows:

- two fail edges (X is a leaf)
- one fail edge and one edge included in the AC tree (X is an internal node preceding a 1 in the seed)
- two edges included in the AC tree (X is an internal node preceding a * in the seed)

The recurrence formula for X depends on its node type. For brevity, we show the formula for the second category above. The fail edge yields a summation using information stored at the fail-to node Y. The AC tree edge yields a summation using information stored at one or more marked descendant nodes $Z_1, Z_2, \ldots$ (the first marked descendant nodes down any path below X in the AC tree edge). An explanation for the notation is given after the formula. Note that the base cases are omitted due to space limitations.

$$\forall i \in [0, n], j \in [0, i], b_x \in B_x$$

$$X[n, i, j, b_x] = \sum_{b_y \in BF_y(b_x)} Y[n - dd_{xy}, i - od_{xy}, j - cv(b_y), b_y]$$

$$+$$

$$\sum_{Z_l,\; l \in \{1,2,\ldots\}} \left(\sum_{b_{z_l} \in BD_{z_l}(b_x)} Z_l[n, i, j, b_{z_l}] \right). \tag{1}$$

Depth difference: The difference in recursion level between the nodes as determined by the AC tree.

$$dd_{xy} = \text{depth}(X) + 1 - \text{depth(Y)}.$$

For example, in figure 1 $dd_{xy} = 2$. The 1 in the formula accounts for the fail edge from X to Y.

Ones difference: The increase in 1s in the strings of the equivalence class when substituting the node string for X followed by the fail edge label ($\text{depth}(X) + 1$ bits) for the node string to Y (the leftmost $\text{depth}(Y)$ bits).

$$od_{xy} = \text{ones}(X) + I - \text{ones}(Y) \quad \text{where } I = \begin{cases} 1 & \text{if edge } \boldsymbol{XY} \text{ is labeled 1} \\ 0 & \text{if edge } \boldsymbol{XY} \text{ is labeled 0} \end{cases}$$

where $\text{ones}(X)$ is the number of 1s in X.

Covered difference: The increase in covered positions when replacing the leftmost bits as above. Note that there is only an increase if X is a leaf, otherwise there are no new covered positions. An example is shown on the left side of figure 2.

$$cv(b_y) = \begin{cases} \text{number of 1 bits in } ([b_s \text{ OR } b_y] \text{ XOR } b_y) & \text{if } X \text{ is a leaf} \\ 0 & \text{if } X \text{ is not a leaf} \end{cases}$$

where

- b_s is the bit string obtained from the seed S by replacing every star (*) by 0. These are the positions covered by an occurrence of the seed. For example, if $S = 1 * 1 * *1$, then $b_s = 101001$.
- in the OR and XOR operations, b_y is left filled with zeros so that it contains as many bits as b_s,

```
        101001  b_s                    101110  b_x right filled with zeros
     OR 001010  b_y                 OR* 101001  b_s
        ---------                       ---------
        101011                          *0*11*
    XOR 001010  b_y                      *11*   rightmost 4 positions
        ---------                               (length of strings in B_y)
        100001
number of 1 bits = 2                 {*11*} = {0110, 0111, 1110, 1111}
                                     and intersection with B_y yields {1111}
```

Fig. 2. Left: Example of $cv(b_y)$ calculation. $S = 1 * 1 * *1$, $b_y = 1010$, and X is a leaf. Then $b_s = 101001$ and b_y left filled with zeros is 001010. The number of 1 bits in 100001 is 2 so $cv(1010) = 2$, which means replacing the node string to Y with the node string to X followed by the label of the fail edge from X to Y yields two additional covered 1's. Right: Example of the calculation of the compatible subset $BF_y(b_x)$ when X is a leaf. $S = 1 * 1 * *1$, $b_x = 10111$, $B_y = \{0000, 0101, 1010, 1011, 1111\}$. *11* produces four strings, of which one string intersects with the set B_y.

Compatible subsets of bit strings: In equation 1, B_x is the set of bit strings which indicate initial covered positions compatible with starting a string with the node string for X. The bit strings in B_x have length $\text{depth}(X) - 1$. If this value is negative, B_x only contains the empty string. For example, in figure 1 the compatible bit strings for node X are 10100, 10101, 10110, 10111, 11110 and 11111. For every marked node X, B_x can be determined by a recursive formula based on the modified AC tree. Details are omitted.

$BF_y(b_x)$ is the set of bit strings in B_y that can lead to bit string b_x when Y follows X along a fail edge (the F stands for fail edge). The method for constructing the set depends on whether or not X is a leaf. Also, in some cases, not all b_x can be achieved down a fail edge to a particular Y. An example for the case where X is a leaf is given on the right side of figure 2.

$$BF_y(b_x) = \phi \text{ if truncate-right}(\text{depth}(Y) - 1, (b_x \text{ XOR } b_s)) \text{ contains 1s}$$
$$\text{and } X \text{ is a leaf.}$$

otherwise

$$BF_y(b_x) = \begin{cases} B_y \cap \{\text{pats-right}(\text{depth}(Y) - 1, (b_x \text{ OR}^* b_s))\} & \text{if } X \text{ is a leaf} \\ B_y \cap \{\text{pats-right}(\text{depth}(Y) - 1, (b_x || *))\} & \text{if } X \text{ is not a leaf} \end{cases}$$

where

- ϕ is the empty set (not the same as the set containing the empty string),
- in the XOR and OR* operations, b_x is right filled with zeros so that it contains as many bits as b_s,
- in the OR* operation, if b_s is 1, the result is *, if b_s is 0, the result is the same as the bit in b_x,
- truncate-right(v, s) removes the rightmost v bits from s,

- pats-right(v, s) yields all patterns (bit strings) which can be generated from a string s over the alphabet $\{*, 0, 1\}$ by replacing each * by a 0 or a 1. Only the rightmost v bits are retained in each generated string.
- $\cap$ means intersection,
- $\|$ means concatenate,

$BD_z(b_x)$ is the set of bit strings in B_z that can lead to bit string b_x when Z follows X along edges in the AC tree (the D stands for descendant).

$$BD_z(b_x) = B_z \cap \{\text{pats-right}(\text{depth}(Y) - 1, (b_x \|*^{\text{depth}(Z)-\ \text{depth}(X)}))\}$$

where

- $*^{\text{depth}(Z)-\ \text{depth}(X)}$ is a string of stars of length $\text{depth}(Z) - \text{depth}(X)$.

2.2   Time and Space Complexity

Theorem 1. *The C_{nSp} distribution for strings of length n and a seed S of length k with r stars can be computed in time $O(2^r(k - r + 1)n^3 B_{max})$, and space $O(2^r(k - r + 1)(k + 2)n^2 \overline{B} \cdot B_{max})$ where $\overline{B}$ is the average size of the set B_x for marked nodes X and B_{max} is the size of the largest such set.*

Although $\overline{B}$ and B_{max} are potentially as large as 2^{k-1}, (the b_x coverage bit strings are always 1 bit shorter than the depth of node X), in our calculations, we have found that they are typically small integers.

Corollary 1. *The R_{nkp} distribution for strings of length n and a seed S consisting of k 1s can be computed in time $O(kn^3 B_{max})$ and space $O(2kn^2 \overline{B} \cdot B_{max})$.*

2.3   Probability Dependent Recursion

The recursion can be made probability *dependent* which requires picking p first. The formulation is similar to that shown in equation 1, except the variable i (the number of 1s in the strings), is omitted since equivalence classes now only depend on the number of covered positions j. Also, the array elements hold probabilities rather than sequence counts. Formulas are omitted due to space limitations. The time and space complexities are reduced by a factor of n.

3   Restricting C_{nSp} to Confirmable Alignments

The calculations described above assume that alignments of homologous sequences can be represented by all possible bit strings. Yet, representative strings that contain too many 0s reflect homologies that *can not be confirmed* in the hit extension step of homology detection programs because there are too many mismatches. For example, "homologous sequences" in which only 30% of the positions match could never be confirmed because the alignment score would be too low. Although the probabilities of representative strings that contain many

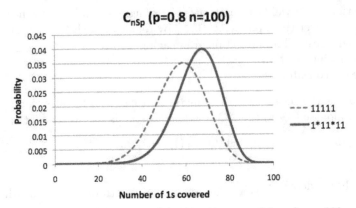

(a) C_{nSp} for seeds 11111 and $1 * 11 * 11$ at $p = 0.8$ and $n = 100$.

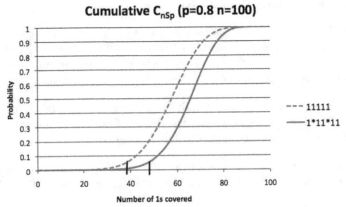

(b) Cumulative C_{nSp} for seeds 11111 and $1 * 11 * 11$ at $p = 0.8$ and $n = 100$. The small vertical lines mark the 5% threshold.

Fig. 3.

0s are relatively small when p is set high, for example, $p = 0.8$, these probabilities nonetheless have a significant impact on the C_{nSp} distribution as we now show.

We recalculated the C_{nSp} distribution with a restriction that eliminates the contribution from strings with many 0s. Different approaches are possible. We chose to use the following. Define a representative string of length n, as *confirmable* if for every suffix of the bit string, of length $\geq n_{min}$, the fraction of 1s is $\geq f$. In the recurson arrays, this corresponds to the elements of $X[n, i, j, b_x]$ in which $i/n \geq f$. Therefore, to restrict our results to confirmable alignments, any element of an array with $n \geq n_{min}$ that has a ratio $i/n < f$ is not used at higher recursion levels (and is therefore not computed). Note that the distribution C_{nSp} assumes that the total probability across all j is 1. Here we must normalize the

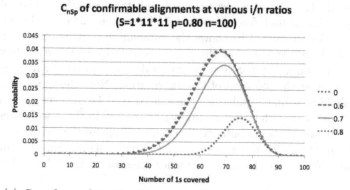

(a) C_{nSp} for seed $1 * 11 * 11$ at ratio $i/n >= 0.0$, 0.6, 0.7, and 0.8 with $p = 0.8$ and $n = 100$ and $n_{min} = 14$.

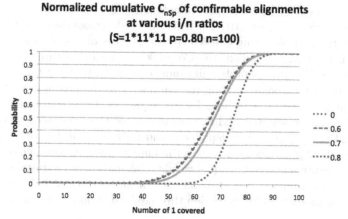

(b) Normalized cumulative C_{nSp} for seed $1 * 11 * 11$ at ratio $i/n >= 0.0$, 0.6, 0.7, and 0.8 with $p = 0.8$ and $n = 100$ and $n_{min} = 14$ (twice the seed length).

Fig. 4.

distribution so that the total probability is the sum of the probabilities of the confirmable alignments.

4 Results

Figure 3(a) shows the exact probability distribution C_{nSp} for two seeds, 11111, and 1*11*11 for strings of length 100 and $p = 0.8$. These seeds have *equal weight*, that is, they have equal expected probability of random hits (determined by the number of 1s in the seed). As expected, the spaced seed which is more sensitive in the single hit paradigm shifts the probability distribution to the right. Note that for seed 11111, C_{nSp} is identical to the distribution R_{nkp}. The calculation for the spaced seed took 5 seconds on a 64-bit system, dual, dual core 2.8GHZ with 4GB RAM. Figure 3(b) shows the cumulative probabilities for the same seeds shown in figure 3(a). For the purposes of DNA homology detection, the threshold for number of matching positions detected to trigger a hit extension is typically set at the 5% level of the distribution. Note the significant shift in this threshold to a higher value for the spaced seed (47.5) over the contiguous seed (38). Recall that the average number of 1s is 80 in these sequences.

Figure 4(a) shows the probability distribution C_{nSp} for the seed 1*11*11 under several assumptions for the ratio $i/n \geq f$. Figure 4(b) shows the cumulative probabilites. Note again the shift to a higher threshold as the value of f increases.

5   Conclusion

This paper presents a dynamic programming method for calculating the probability distribution C_{nSp} exactly. The distribution describes the expected number of 1s in a random bit string covered by occurrences of a spaced seed. The recurrence is built on a modified Aho-Corasick tree data structure. Two methods are presented. The first is probability independent, in that the time consuming steps are performed before setting the parameter p, the probability of a match between aligned positions in homologous sequences. Once completed, different values of p can be selected and the calculation of C_{nSp} proceeds quickly. The second method is probability dependent and faster overall, but only applies to a single value of p. An example shows the improvement in coverage of a spaced seed over a contiguous seed of equal weight. Calculation of the distribution over bit strings which represent confirmable alignments is discussed. An example illustrates the increase in hit extension threshold value under this assumption.

Acknowledgments

The authors would like to thank Brandon Mensing, Harshith Chennamaneni, and Won-Beom Kim for their assistance in validating our calculations.

References

1. Aho, A.V., Corasick, M.J.: Efficient string matching: An aid to bibliographic search. Comm. ACM 18, 333–340 (1975)
2. Benson, G.: Tandem repeats finder: a program to analyze DNA sequences. Nucleic Acids Research 27, 573–580 (1999)
3. Benson, G., Mak, D.Y.F.: Exact distribution of a spaced seed statistic for applications in DNA repeat detection. In: Proceedings of the 2008 International Workshop on Applied Probability (IWAP 2008) (2008)
4. Benson, G., Su, X.: On the distribution of k-tuple matches for sequence homology: a constant time exact calculation of the variance. J. Computational Biology 5, 87–100 (1998)

5. Buhler, J., Keich, U., Sun, Y.: Designing seeds for similarity search in genomic DNA. Journal of Computing and System Sciences 70, 342–363 (2005)
6. Burkhardt, S., Kärkkäinen, J.: Better filtering with gapped q-grams. Fundam. Inform. 56(1-2), 51–70 (2003)
7. Keich, U., Li, M., Ma, B., Tromp, J.: On spaced seeds for similarity search. Discrete Applied Mathematics 138(3), 253–263 (2004)
8. Lou, W.Y.W.: The exact distribution of the k-tuple statistic for sequence homology. Statistics and Probability Letters 61(1), 51–59 (2003)
9. Ma, B., Tromp, J., Li, M.: Patternhunter: faster and more sensitive homology search. Bioinformatics 18, 440–445 (2002)
10. Mak, D.Y.F., Benson, G.: All hits all the time: Parameter free calculation of seed sensitivity. In: Proceedings of the 5th Asia-Pacific Bioinformatics Conference, pp. 327–340. Imperial College Press (2007)
11. Warburton, P., Giordano, J., Cheung, F., Gelfand, Y., Benson, G.: Inverted repeat structure of the human genome: the X chromosome contains a preponderance of large highly homologous inverted repeats which contain testes genes. Genome Res., 1861–1869 (2004)

Author Index